Disinfection By-products in Drinking Water

Disinfection By-products in Drinking Water

Edited by

Clive Thompson
ALcontrol UK Ltd., Rotherham, UK
clive.thompson@alcontrol.co.uk

Simon Gillespie
Scottish Water, Edinburgh, UK
simon.gillespie@scottishwater.co.uk

Emma Goslan
Cranfield University, Cranfield, UK
E.H.Goslan@cranfield.ac.uk

ROYAL SOCIETY OF CHEMISTRY

THE QUEEN'S AWARDS FOR ENTERPRISE: INTERNATIONAL TRADE 2013

Based on proceedings of the meeting DBP 2014: Disinfection By-products in drinking water held on 27th-29th October 2014 at Municipal Hall, Mülheim an der Ruhr, Germany.

Special Publication No. 352

Print ISBN: 978-1-78262-088-4
PDF eISBN: 978-1-78262-271-0

A catalogue record for this book is available from the British Library

Published by The Royal Society of Chemistry,
Thomas Graham House, Science Park, Milton Road,
Cambridge CB4 0WF, UK

Registered Charity Number 207890

Visit our website at www.rsc.org/books

Printed in the United Kingdom by CPI Group (UK) Ltd, Croydon, CR0 4YY, UK

Preface

The international conference, Disinfection By-products in Drinking Water (DBP2014 was held between October 27th and 29th, 2014 in the Municipal Hall, Mülheim an der Ruhr, alongside the banks of the River Ruhr. It brought together international experts on disinfection by-products (DBPs) to review and debate recent developments and current thinking in the field. It was attended by delegates from around 20 different countries from Europe, Asia, North America and Australasia.

The conference focussed on the challenges faced by water utilities and regulatory authorities around the world in balancing the risk of microbial contamination against the potential health risks associated with DBPs. It brought delegates up to date on the latest technology for monitoring DBPs, on treatment processes for minimising their formation, on research into the health and toxicological concerns of DBPs, and on the potential direction of future regulation.

Chlorination was first used to disinfect public drinking water supplies in the UK in Maidstone as early as 1897 and in the USA in 1908, with ozone and other disinfectants introduced in Europe around the same time. However it wasn't until the early 1970s that Johannes Rook in Rotterdam and Tom Bellar in Ohio independently discovered higher levels of chloroform in treated drinking water than in the source water, and the whole field of disinfection by-product research was born.

Whether it is in the production of safe, clean drinking water, or the management of swimming pools in communal recreational facilities, the disinfection of water to protect against microbial contamination carries the risk of producing harmful DBPs. Chlorination of the naturally occurring organic matter in water results in a wide range of toxic halogenated organic compounds. Carcinogenic nitrosamines are a potential by-product of chloramination, and ozonation oxidises bromide to bromate as well as producing a plethora of small organic molecules. There are at least 600 known disinfection by-products, and the count continues to rise.

Worldwide, regulation governing the control of DBPs has become more stringent. Most countries in the developed world have set regulatory standards for trihalomethanes and bromate. In the USA, the EPA additionally sets standards for haloacetic acids and chlorite. In Europe the disinfection by-product rule of the 1998 EU Drinking Water Directive requires water utilities to "design, operate and maintain the disinfection process so as to keep disinfection by-products as low as possible without compromising the effectiveness of the disinfection; and to verify the effectiveness of the disinfection process." Water treatment works designed in Europe since January 2010 should have taken the disinfection by-product rule into account and companies should have included this requirement in their regulatory risk assessments. As our understanding of DBPs and their associated health effects increases, so to do the challenges of ensuring that they are effectively regulated, monitored and minimised.

The DBP2014 conference was a cooperative venture organised by IWW Water Centre (DE), RSC (UK) and SCI (UK). This book represents the proceedings of that conference, and we, the editors, hope that you enjoy reading it.

Simon Gillespie
K Clive Thompson
Emma Goslan

Contents

Disinfection By-products Formation

Disinfection By-products Prevention and Treatment

Disinfection By-products Formation

THE NEXT GENERATION OF DRINKING WATER DISINFECTION BY-PRODUCTS: OCCURRENCE, FORMATION, TOXICITY, AND NEW LINKS WITH HUMAN EPIDEMIOLOGY

Susan D. Richardson[1]* and Cristina Postigo[2]

[1] Department of Chemistry and Biochemistry, University of South Carolina, JM Palms Center, 631 Sumter St., Columbia, SC 29208
[2] Department of Environmental Chemistry, Institute for Environmental Assessment and Water Research, (IDAEA-CSIC), Carrer Jordi Girona 18-26, 08034 Barcelona, Spain.

1 INTRODUCTION

Drinking water disinfection by-products (DBPs) are an unintended consequence of using disinfectants to kill harmful pathogens in drinking water. They are formed primarily by the reaction of disinfectants with natural organic matter (NOM) and bromide or iodide [1], but can also be formed from pollutants, such as pesticides, pharmaceuticals, antibacterial agents, estrogens, textile dyes, bisphenol A, parabens, surfactants, and algal toxins [2-4]. Popular disinfectants for drinking water include chlorine, chloramines, ozone, chlorine dioxide, and UV.

Eleven DBPs are regulated in the United States [5], but more than 600 have been identified [1]. And, despite this large number of DBPs identified, more than 50% of the halogenated material formed in chlorinated drinking water is still unknown, as well as the toxicological risk that it poses to human health [2, 6].

Adverse health concerns include bladder cancer, miscarriage, and birth defects [1, 7-12]. While bladder cancer is the primary cancer observed in humans, none of the 11 regulated DBPs cause bladder cancer in animals; consequently, many researchers believe that the regulations are not adequately controlling for these effects in humans. As a result, there is intensive research in emerging, unregulated DBPs. These include iodo-trihalomethanes, iodo-acids, haloamides, halonitromethanes, halofuranones, haloacetonitriles, haloacetaldehydes, nitrosamines, and halobenzoquinones. Many of these unregulated DBPs are more genotoxic or cytotoxic than those currently regulated [1, 13-15]. For example, iodoacetic acid is the most genotoxic DBP identified to-date, and is 2x more genotoxic than bromoacetic acid [1]. Iodoacetic acid was also recently shown to be tumorigenic in mice [16].

Nitrogen-containing DBPs (N-DBPs) have become a major focus because they are generally more toxic than DBPs that do not contain nitrogen. For example, many nitrosamines are known to be carcinogens [1]. Nitrosamines were on the U.S.

Environmental Protection Agency's (EPA's) Unregulated Contaminant Monitoring Rule [17], and are currently being considered for regulation in the United States.

DBPs can also form in disinfected swimming pool water [18-22]. Some swimming pool DBPs are the same as those found in drinking water, but some are different, due to the additional human precursors that can be present in pools (e.g., urine, sweat, hair, sunscreens, lotions, personal care products etc.) [22, 23]. For example, trichloramine is a common DBP found in chlorinated swimming pools, and it is formed by the reaction of urea (from urine or sweat) with chlorine [24]. While trichloramine is produced in the water, it is quickly transported to the air phase above, due to its high Henry's Law constant. Trichloramine is suspected as the causal agent in the increased asthma in epidemiologic studies of elite swimmers [25, 26]. One study has also shown increased incidence of bladder cancer with heavy exposure from swimming pools [27].

Newer potential health concerns include severe skin rashes, and respiratory and digestive issues resulting from exposure to chloraminated drinking water. However, there has not yet been a controlled scientific study to examine these issues. Chloramination has become a popular disinfectant in the U.S., due to tightened DBP regulations [5], and is also used in other countries, including the UK and Australia. The use of chloramines can result in ~90% reduction in the levels of regulated trihalomethanes (THMs) and haloacetic acids (HAAs) compared to chlorination, and it also allows longer residual disinfection in distribution systems.

DBP formation depends on the type of disinfectant, dose, and the type of organic matter or other precursors present in the water [1, 28-31]. Formation mechanisms for several DBPs and DBP classes have been investigated, including iodo-DBPs, halonitromethanes, nitrosamines, haloamides, halopyrroles, and halobenzoquinones. A review of their formation follows.

2 IODO-DBPs

Iodo-DBPs identified to-date include iodo-THMs (dichloroiodomethane, bromochloroiodomethane, dibromoiodomethane, chlorodiiodomethane, bromodiiodomethane, and iodoform); iodo-acids (iodoacetic acid, bromoiodoacetic acid, chloroiodoacetic acid, diiodoacetic acid, (*Z*)-3-bromo-3-iodopropenoic acid, (*E*)-3-bromo-3-iodopropenoic acid, and (*E*)-2-iodo-3-methylbutenedioic acid) [6, 32-35]; iodo-amides (bromoiodoacetamide and chloroiodoacetamide) [1, 14, 36]; and the recently reported iodoacetaldehyde [37, 38].

Iodo-THMs were the first iodo-DBPs to be discovered, back in the mid-1970s, [39] and have been measured in drinking waters treated with chlorination or chloramination [6, 31, 40-44]. Highest levels are consistently observed in chloraminated water (up to 15 μg/L), and total iodo-THM levels can be as much as 81% of the regulated THMs [6]. Point-of-use treatment with iodine [45], and chlorination or chloramination of hydraulic fracturing (HF) wastewater can also produce iodo-THMs [46].

Iodo-acids, which are the most genotoxic of the iodo-DBPs [35, 47], were first identified in a U.S. Nationwide Occurrence Study [6, 41, 47], and levels up to 1.7 μg/L were reported in a 23 city survey of chloraminated and chlorinated drinking water from the U.S. and

Canada [35]. Chlorine dioxide was also reported to form iodoacetic acid when reacted with source waters [48], and iodo-acids were also tentatively identified in simulated drinking waters treated with chlorine, monochloramine, and chlorine-chloramine [32]. Finally, iodoacetic acid and chloroacetic acid can form when chlorinated tap water is allowed to react with iodized table salt (containing potassium iodide) or with potassium iodide itself [32]. The rank order for genotoxicity is iodoacetic acid >> diiodoacetic acid > bromoiodoacetic acid > (*E*)-2-iodo-3-methylbutenedioic acid > (*E*)-3-bromo-3-iodopropenoic acid > (*E*)-3-bromo-2-iodopropenoic acid. Iodoacetic acid is also teratogenic, producing developmental effects (neural tube closures) in mouse embryos, at low nM levels similar to levels that induce DNA damage in mammalian cells [49, 50]. Iodoacetic acid is also tumorigenic in mice [16].

Iodoacetamides—bromoiodoacetamide and chloroiodoacetamide—have been identified in drinking water treated with chloramines or chlorine. Bromoiodoacetamide was originally found in chloraminated drinking water from several cities in the U.S. [14]; later, both bromoiodoacetamide and chloroiodoacetamide were found in chloraminated and chlorinated drinking water from three provinces in China [36]. Haloacetamides can form by hydrolysis of the corresponding haloacetonitriles [51, 52], or by an independent pathway [53]; they are preferentially formed with chloramination vs. chlorination [53]. Both of these iodoacetamides are highly cytotoxic and genotoxic in mammalian cells [14]. As a class, haloamides are the most cytotoxic of all DBP classes measured to-date, and they are the second-most genotoxic DBP class, close behind the halonitriles.

Chloramination increases the formation of all of these iodo-DBPs. In practice, drinking water plants can add preformed monochloramine (NH_2Cl, formed by the reaction of chlorine and ammonia), but generally chlorine is added first and allowed to react for a certain amount of time (free chlorine contact time) before the ammonia is added, to enable a higher level of microbial inactivation. Most research shows that a lower free chlorine contact time (increased NH_2Cl contact time) increases the formation of iodo-DBPs [6, 29, 32, 35, 54, 55], consistent with a mechanism proposed by Bichsel and von Gunten, which involves competing mechanisms to form iodate and organic iodo-DBPs [56, 57]. Reaction of aqueous chlorine (HOCl) with iodide initially forms hypoiodous acid (HOI), which then reacts quickly with HOCl to form iodite and iodate. The corresponding reactions to form organic iodo-DBPs (e.g., iodo-THMs and iodo-acids) are much slower, favoring the formation of iodate instead of organic iodo-DBPs. On the other hand, reactions of NH_2Cl with HOI to form iodite and iodate are much slower, such that NH_2Cl favors the formation of organic iodo-DBPs over iodate. New research indicates that ozone pretreatment at lower pH might be used to minimize iodo-DBP (and bromate) formation by selectively oxidizing iodide to iodate [58].

Natural iodide is believed to be the major source of iodine in the formation of iodo-DBPs, but, new research has revealed that compounds used for medical imaging (i.e., iodinated X-ray contrast media (ICM)) can also be a source of iodine [59-61]. ICM are excreted within ~24 h after medical imaging, are stable during wastewater treatment, and can be present up to 100 μg/L in rivers and creeks [62] and up to 2.7 μg/L in drinking water reservoirs [59]. These ICM structures have 3 iodines attached to a benzene ring that also contains 3 amide side chains, and can react with chlorine or chloramine to form iodo-THMs and iodo-acids. NOM and pH can significantly impact their formation, and OCl^- is hypothesized as the reacting disinfectant species. Moreover, new controlled laboratory studies indicate that iodo-THMs are favored at low chlorine doses [61].

Iopamidol appears to be more reactive that other ICM investigated (e.g., iopromide, iohexol, iomeprol, diatrizoate, hiztodenz, and iodixanol) [59-61]. New mechanistic research using liquid chromatography (LC)-high resolution MS/MS and nuclear magnetic resonance (NMR) spectroscopy has revealed the initial points of reaction on the iopamidol structure, along with the initial high molecular weight DBPs formed [60]. Proposed reaction mechanisms involve cleavage of one of side chains, substitution of chlorine for iodine on the benzene ring, amide hydrolysis, cleavage of the other side chains, and oxidation of NH_2 to NO_2. Structures for 19 high molecular weight DBPs were proposed.

3 NITROSAMINES

N-Nitrosodimethylamine (NDMA) was discovered to be a DBP in 2002 [63, 64], and created significant interest due to its potent carcinogenicity [1]. NDMA was initially discovered in chlorinated drinking waters from Ontario, Canada [65], and was later found in many other locations [63, 64, 66]. Other nitrosamines have also been found as DBPs, including *N*-nitrosopiperidine, *N*-nitrosodiphenylamine, *N*-nitrosopyrrolidine, and *N*-nitrosomorpholine [67, 68]. A new total nitrosamine (TONO) assay indicates that the nitrosamines identified so far only represent 5-10% of the total nitrosamines formed in drinking water and recreational waters [69, 70]. Tobacco-specific nitrosamines—4-(methylnitrosamino)-1-(3-pyridyl)-1-butanone and 4-(methylnitrosamino)-1-(3-pyridyl)-1-butanol—were also recently discovered as chloramination DBPs [71].

California regulates NDMA at 10 ng/L [72] and Ontario, Canada at 9 ng/L [73]. A Canadian national drinking water guideline was also recently established, limiting NDMA to 40 ng/L in drinking water [74], and the U.S. EPA is considering its regulation. NDMA, *N*-nitrosodiethylamine, *N*-nitrosodibutylamine, *N*-nitrosodipropylamine, *N*-nitrosomethylethylamine, and *N*-nitrosopyrrolidine were measured in drinking waters across the U.S. due to inclusion on the U.S. EPA's second Unregulated Contaminants Monitoring Rule (UCMR-2). National occurrence data revealed a maximum level of 530 ng/L for NDMA in chloraminated drinking water [17], which surpassed the previous maximum (180 ng/L) observed in Canadian chloraminated drinking water [67]. NDMA and 4 other nitrosamines are also on the U.S. EPA's Contaminant Candidate List (CCL-3), a priority list of drinking water contaminants [75].

NDMA is generally maximized with chloramination. Early research indicated that the nitrogen in monochloramine (NH_2Cl) was incorporated into the structure of NDMA [63], and subsequent research revealed that it was actually dichloramine ($NHCl_2$), which always coexists with NH_2Cl, that was the primary reactant [76]. Nitrite can also be a nitrosamine precursor in reactions with chlorine [77, 78]. NDMA can also sometimes form in chlorination reactions when nitrogen precursors are present (e.g., natural ammonia in source waters or when nitrogen-containing coagulants or ion-exchange resins are used in water treatment) [79-82]. Consumer products, including shampoos, laundry detergents, dish washing liquids, and fabric softeners, can also be precursors, with quaternary amine polymers surprisingly more reactive than the monomers [83]. Similarly, diallyldimethylammonium chloride (DADMAC) polymers, which are used as coagulants in drinking water treatment are more reactive than the monomers [82, 84, 85]. Amino acids, hydrophilic/low molecular weight dissolved organic nitrogen, amine-based pharmaceuticals, and diphenylamine can also serve as NDMA precursors [86]. The Lifestraw®, a point-of-use device that uses I_3^- complexed resins with an activated carbon filter, can also produce NDMA, but levels rapidly decline to low ng/L levels after the first few flushes of water [45]. An excellent review was recently published on the formation, precursors, control, and occurrence of nitrosamines in drinking water [87].

NDMA was recently identified in ozonated drinking water from Germany, which was a surprise because it had not been found previously in drinking water treated with ozone [88]. A fungicide, tolylfluanide, containing a dimethylamine group in its structure, was found to be the precursor [91], and trace levels of bromide can catalyze its formation [89]. In addition, dithiocarbamate pesticides can react with NH_2Cl, O_3, Cl_2, or ClO_2 to form NDMA [90]. Finally for the tobacco-specific nitrosamines just identified in 2014, tobacco alkaloids, including nicotine, nornicotine, and anabasine were determined to be precursors to their formation in chloraminated drinking water [71].

4 HALOAMIDES

Haloamides can form in reactions with either chlorine or chloramine, [6, 14, 41, 91, 92], but preferentially form with monochloramine [53]. The mechanism can involve hydrolysis of the corresponding haloacetonitriles [51, 52], or reaction of organic nitrogen precursors with NH_2Cl [53]. ^{15}N-labeled NH_2Cl studies showed an initial rapid formation of both dichloroacetamide and dichloroacetonitrile, with the nitrogen originating from organic nitrogen precursors. Reactions of asparagine as a model precursor also suggested that dichloroacetamide can be formed without a dichloroacetonitrile intermediate, and humic substances were more potent precursors for dichloroacetamide formation, while wastewater effluents and algal substances were more potent precursors for dichloroacetonitrile formation [53]. Therefore, independent mechanisms are also involved in the formation of haloacetamides, in addition to the hydrolysis of the haloacetonitriles. Amino acids, including aspartic acid, histidine, tyrosine, tryptophan, glutamine, asparagine, and phenylalanine, can also react with chlorine to form dichloroacetamide [93].

5 HALONITROMETHANES

Chloropicrin (trichloronitromethane) is the most commonly measured halonitromethanes, but a total of 9 have been identified in drinking water. Brominated nitromethanes show significant toxicity [34], with a cytotoxicity and genotoxicity greater than most DBPs currently regulated, and they have been found at levels up to 3 µg/L [6, 34, 41, 94, 95]. Halonitromethanes are also mutagenic in *Salmonella* [96], with potencies greater than that of the regulated THMs [97]. The greater cytotoxic and mutagenic activities of the halonitromethanes is likely due to the greater intrinsic reactivity conferred by the nitro group [97]. Pre-ozonation used prior to chlorination or chloramination vastly increases the formation of bromonitromethanes [6, 41]. Nitrite can also play a role in the formation of the nitro group in these DBPs [98].

Trihalonitromethanes (including tribromo, bromodichloro-, and chlorodibromonitromethane) require particular analytical conditions for their analysis because they can thermally decompose with commonly used injection port temperatures during gas chromatography (GC)-electron capture detection (ECD) or GC/mass spectrometry (MS) analysis [91].

6 HALOPYRROLES

2,3,5-Tribromopyrrole was first reported as a DBP in 2003, and was found in finished drinking water treated with pre-chlorination followed by treatment with combined ClO_2-Cl_2 or ClO_2-NH_2Cl [99]. This identification resulted from the first study of chlorine dioxide DBPs formed under high bromide/iodide conditions. 2,3,5-Tribromopyrrole is 8x more cytotoxic than dibromoacetic acid (a regulated DBP) and has about the same genotoxic potency as MX [99], a carcinogen [100]. Tribromopyrrole forms primarily from humic acid (vs. fulvic acid), and it is interesting that a soil humic model proposed by Schulten and Schnitzer [101] includes a pyrrole group in its structure [101]. In none of the samplings from this research was tribromopyrrole found in waters treated with chlorine only, therefore, the combination of chlorine dioxide and chlorine (or chloramines) may be necessary for its formation, or chloramination may also be important in its formation. New research also shows that halopyrroles can form as DBPs in chlorinated saline wastewater effluents [102]. Tri- and tetra-halopyrroles were identified, including brominated, chlorinated, and iodinated analogues.

7 HALOBENZOQUINONES

Four halobenzoquinones (HBQs)—2,6-dichlorobenzoquinone, 2,6-dibromobenzoquinone, 2,6-dichloro-3-methylbenzoquinone, and 2,3,6-trichlorobenzoquinone—were recently identified as new DBPs in drinking water treated with chlorine, chloramines, chlorine-chloramines, ozone-chloramines, and chloramines-UV [103, 104]. Levels ranged up to 275 ng/L. 2,6-Dichlorobenzoquinone, 2,3,6-Trichloro-1,4-benzoquinone, 2,3-dibromo-5,6-dimethyl-1,4-benzoquinone, and 2,6-dibromo-1,4-benzoquinone were also recently identified in chlorinated swimming pools [105], and subsequent research revealed the formation of hydroxylated HBQs with UV treatment [106]. Haloquinones are predicted to be highly toxic, with the chronic lowest observed adverse effect levels (LOAELs) of haloquinones predicted at low μg/kg body weight per day, which is 1000x lower than most regulated DBPs, except bromate. Controlled laboratory studies using phenol as a precursor showed highest levels of 2,6-dichlorobenzoquinone with chlorination, with highest levels of 2,6-dibromobenzoquinone resulting from preozonation in the presence of bromide. In the formation of HBQs in swimming pools, UV filters and other aromatic compounds found in lotions and sunscreens were found to be important precursors [105].

8 NEW LINKS WITH HUMAN EPIDEMIOLOGY

In 2012, for the first time, chemistry and toxicology and epidemiology were combined for a multicountry epidemiology study of drinking water and adverse birth outcomes in Europe. In this study called "HiWATE" (Health Impacts of Long-Term Exposure to Disinfection Byproducts in Drinking Water), Jeong et al. published an integrated study of chemistry (target DBP quantification and comprehensive identification) with toxicology (in vitro mammalian cell toxicity of complex drinking water samples) and epidemiology findings [107]. In total, 11 drinking water samples were collected from five European countries, each corresponding to a separate epidemiology study for the HiWATE program. More than 90 DBPs were identified, and THMs and HAAs were quantified. The range and type of DBPs reflected the diverse collection sites, the different disinfection processes, and the different characteristics of the source waters. Results showed a correlation of the mammalian cell cytotoxicity with the number of DBPs identified and several DBP chemical classes. DBP occurrence, cytotoxicity, and genotoxicity also correlated with epidemiology results (low birth weight and small for gestational age) from two countries. Specifically, % small for gestational age (SGA) correlated with the number of DBPs and the genotoxicity potency index values for cities in Spain, and risk of low birth weight (LBW) correlatied with the number of DBPs, levels of DBPs, cytotoxic potency, and genotoxicity potency in two locations in Lithuania.

References

1. S. D. Richardson, M. J. Plewa, E. D. Wagner, R. Schoeny and D. M. DeMarini, *Mutat. Res.* 2007, **636**, 178.
2. S. D. Richardson and C. Postigo in *Emerging Organic Contaminants and Human Health - The Handbook of Environmental Chemistry*, ed. D. Barcelo, Springer-Verlag Berlin Heidelberg, Germany, 2012, Vol. 20, p 93.
3. C. Postigo and S. D. Richardson, *J. Hazard. Mater.* 2014, **279**, 461.
4. S. D. Richardson and C. Postigo in *In Occurrence, Formation, Health Effects, and Control of Disinfection By-Products*, ed. T. Karanfil, W. Mitch, P. Westerhoff and X. Yuefeng, ACS Symposium Series, ACS Publications, Washington, DC, 2014, in press.
5. U.S. Environmental Protection Agency (U.S. EPA). Drinking Water Contaminants. http://goo.gl/90OBMZ (accessed November, 2014).
6. S. W. Krasner, H. S. Weinberg, S. D. Richardson, S. J. Pastor, R. Chinn, M. J. Sclimenti, G. D. Onstad and A. D. Thruston Jr., *Environ. Sci. Technol.* 2006, **40**, 7175.
7. K. P. Cantor, C. M. Villanueva, D. T. Silverman, J. D. Figueroa, F. X. Real, M. Garcia-Closas, N. Malats, S. Chanock, M. Yeager, A. Tardon, R. Garcia-Closas, C. Serra, A. Carrato, G. Castaño-Vinyals, C. Samanic, N. Rothman and M. Kogevinas, *Environ. Health Perspect.* 2010, **118**, 1545.
8. C. M. Villanueva, K. P. Cantor, S. Cordier, J. J. K. Jaakkola, W. D. King, C. F. Lynch, S. Porru and M. Kogevinas, *Epidemiology* 2004, **15**, 357.
9. K. Waller, S. H. Swan, G. DeLorenze and B. Hopkins, *Epidemiology* 1998, **9**, 134.
10. D. A. Savitz, P. C. Singer, K. E. Hartmann, A. H. Herring, H. S. Weinberg, C. Makarushka, C. Hoffman, R. Chan and R. Maclehose, *Drinking Water Disinfection By-Products and Pregnancy Outcome*; AWWA Research Foundation, Denver, CO., 2005; p 212.
11. M. J. Nieuwenhuijsen, M. B. Toledano, N. E. Eaton, J. Fawell and P. Elliott, *Occup. Environ. Med.* 2000, **57**, 73.
12. F. Bove, Y. Shim and P. Zeitz, *Environ. Health Perspect.* 2002, **110**, 61.
13. M. J. Plewa and E. D. Wagner, *Mammalian Cell Cytotoxicity and Genotoxicity of Disinfection By-Products*. Water Research Foundation: Denver, CO, 2009; p 134, 2009.

14. M. J. Plewa, M. G. Muellner, S. D. Richardson, F. Fasano, K. M. Buettner, Y. T. Woo, A. B. McKague and E. D. Wagner, *Environ. Sci. Technol.* 2008, **42**, 955.
15. M. J. Plewa, J. E. Simmons, S. D. Richardson and E. D. Wagner, *Environ. Mol. Mutagen.* 2010, **51**, 871.
16. X. Wei, S. Wang, W. Zheng, X. Wang, X. Liu, S. Jiang, J. Pi, Y. Zheng, G. He and W. Qu, *Environ. Sci. Technol.* 2013, **47**, 5913.
17. U.S. Environmental Protection Agency (U.S. EPA) Unregulated Contaminant Monitoring Rule 2 - UCMR 2. http://goo.gl/EQD8bI (accessed November, 2014).
18. J. Li and E. R. Blatchley III, *Environ. Sci. Technol.* 2007, **41**, 6732.
19. C. Zwiener, S. D. Richardson, D. M. DeMarini, T. Grummt, T. Glauner and F. H. Frimmel, *Environ. Sci. Technol.* 2007, **41**, 363.
20. S. D. Richardson, D. M. DeMarini, M. Kogevinas, P. Fernandez, E. Marco, C. Lourencetti, C. Ballesté, D. Heederik, K. Meliefste, A. B. McKague, R. Marcos, L. Font-Ribera, J. O. Grimalt and C. M. Villanueva, *Environ. Health Perspect.* 2010, **118**, 1523.
21. W. A. Weaver, J. Li, Y. Wen, J. Johnston, M. R. Blatchley and E. R. Blatchley III, *Water Res.* 2009, **43**, 3308.
22. S. Chowdhury, K. Al-hooshani and T. Karanfil, *Water Res.* 2014, **53**, 68.
23. H. Kim, J. Shim and S. Lee, *Chemosphere* 2002, **46**, 123.
24. C. Schmalz, F. H. Frimmel and C. Zwiener, *Water Res.* 2011, **45**, 2681.
25. V. Bougault and L. P. Boulet, *Brit. J. Sport. Med.* 2012, **46**, 402.
26. A. Florentin, A. Hautemaniere and P. Hartemann, *Int. J. Hyg. Environ. Health.* 2011, **214**, 461.
27. M. Kogevinas, C. M. Villanueva, L. Font-Ribera, D. Liviac, M. Bustamante, F. Espinoza, M. J. Nieuwenhuijsen, A. Espinosa, P. Fernandez, D. M. DeMarini, J. O. Grimalt, T. Grummt and R. Marcos, *Environ. Health Perspect.* 2010, **118**, 1531.
28. L. Liang and P. C. Singer, *Environ. Sci. Technol.* 2003, **37**, 2920.
29. G. Hua and D. A. Reckhow, *Water Res.* 2007, **41**, 1667.
30. A. D. Shah and W. A. Mitch, *Environ. Sci. Technol.* 2012, **46**, 119.
31. Z. Zhang, S. Echigo, R. A. Minear and M. J. Plewa in *Natural Organic Matter and Disinfection By-Products: Characterization and Control in Drinking Water*, ed. S. E. Barrett, S. W. Krasner and G. L. Amy, American Chemical Society, Washington, D.C., 2000, p 299.
32. G. Ding and X. Zhang, *Environ. Sci. Technol.* 2009, **43**, 9287.
33. A. Becalski, B. P. Y. Lau, T. J. Schrader, S. W. Seaman and W. F. Sun, *Food Addit. Contam.* 2006, **23**, 957.
34. M. J. Plewa, E. D. Wagner, P. Jazwierska, S. D. Richardson, P. H. Chen and A. B. McKague, *Environ. Sci. Technol.* 2004, **38**, 62.
35. S. D. Richardson, F. Fasano, J. J. Ellington, F. G. Crumley, K. M. Buettner, J. J. Evans, B. C. Blount, L. K. Silva, T. J. Waite, G. W. Luther, A. B. McKague, R. J. Miltner, E. D. Wagner and M. J. Plewa, *Environ. Sci. Technol.* 2008, **42**, 8330.
36. W. Chu, N. Gao, D. Yin, S. W. Krasner and M. R. Templeton, *J. Chromatogr. A* 2012, **1235**, 178.
37. C. H. Jeong, C. Postigo, S. D. Richardson, J. E. Simmons, S. Y. Kimura, B. J. Marinas, D. Barcelo, E. D. Wagner and M. Plewa, *Environ. Sci. Technol.* 2014, submitted.
38. C. Postigo, C. H. Jeong, S. D. Richardson, E. D. Wagner, M. Plewa, J. E. Simmons and D. Barcelo in *Occurrence, Formation, Health Effects, and Control of Disinfection By-Products*, ed. T. Karanfil, W. Mitch, P. Westerhoff and X.

Yuefeng, ACS Symposium Series, ACS Publications, Washington, DC, 2014, submitted.
39. H. J. Brass, M. A. Feige, T. Halloran, J. W. Mello, D. Munch and T. R.F. in *Drinking Water Quality Enhancement Through Source Protection,* ed. R. B. Pojasek, Ann Arbor Science, Ann Arbor, MI, 1977, p 393.
40. S. W. Krasner, M. J. McGuire, J. G. Jacangelo, N. L. Patania, K. M. Reagan and E. M. Aieta, *J. Am. Water Works Assoc* 1989, **81**, 41.
41. H. S. Weinberg, S. W. Krasner, S. D. Richardson and J. A. D. Thruston, *The Occurrence of Disinfection By-Products (DBPs) of Health Concern in Drinking Water: Results of a Nationwide DBP Occurrence Study*; National Exposure Research Laboratory, Office of Research and Development, U.S. Environmental Protection Agency, Athens, GA. : 2002.
42. B. Cancho, F. Ventura, M. Galceran, A. Diaz and S. Ricart, *Water Res.* 2000, **34**, 3380.
43. X. Wei, X. Chen, X. Wang, W. Zheng, D. Zhang, D. Tian, S. Jiang, C. N. Ong, G. He and W. Qu, *PLoS ONE* 2013, **8**.
44. J. Criquet, S. Allard, E. Salhi, C. A. Joll, A. Heitz and U. von Gunten, *Environ. Sci. Technol.* 2012, **46**, 7350.
45. E. M. Smith, M. J. Plewa, C. L. Lindell, S. D. Richardson and W. A. Mitch, *Environ. Sci. Technol.* 2010, **44**, 8446.
46. K. M. Parker, T. Zeng, J. Harkness, A. Vengosh and W. A. Mitch, *Environ. Sci. Technol.* 2014, **48**, 11161.
47. M. J. Plewa, E. D. Wagner, S. D. Richardson, A. D. Thruston Jr., Y. T. Woo and A. B. McKague, *Environ. Sci. Technol.* 2004, **38**, 4713.
48. T. Ye, B. Xu, Y. L. Lin, C. Y. Hu, L. Lin, T. Y. Zhang and N. Y. Gao, *Water Res.* 2013, **47**, 3006.
49. I. E. S. Hunter, E. H. Rogers, J. E. Schmid and A. Richard, *Teratology* 1996, **54**, 57.
50. E. S. I. Hunter and J. A. Tugman, *Teratology* 1995, **52**, 317.
51. V. Glezer, B. Harris, N. Tal, B. Iosefzon and O. Lev, *Water Res.* 1999, **33**, 1938.
52. D. A. Reckhow, A. L. MacNeill, T. L. Platt, A. L. MacNeill and J. N. McClellan, *J. Water Supp. - Res. Technol. - AQUA* 2001, **50**, 1.
53. H. Huang, Q. Y. Wu, H. Y. Hu and W. A. Mitch, *Environ. Sci. Technol.* 2012, **46**, 10624.
54. D. B. Jones, A. Saglam, A. Triger, H. Song and T. Karanfil, *Environ. Sci. Technol.* 2011, **45**, 10429.
55. D. B. Jones, H. Song and T. Karanfil, *Water Res.* 2012, **46**, 5491.
56. Y. Bichsel and U. von Gunten, *Environ. Sci. Technol.* 1999, **33**, 4040.
57. Y. Bichsel and U. von Gunten, *Environ. Sci. Technol.* 2000, **34**, 2784.
58. S. Allard, C. E. Nottle, A. Chan, C. Joll and U. von Gunten, *Water Res.* 2013, **47**, 1953.
59. S. E. Duirk, C. Lindell, C. C. Cornelison, J. Kormos, T. A. Ternes, M. Attene-Ramos, J. Osiol, E. D. Wagner, M. J. Plewa and S. D. Richardson, *Environ. Sci. Technol.* 2011, **45**, 6845.
60. F. M. Wendel, C. Lütke Eversloh, E. J. Machek, S. E. Duirk, M. J. Plewa, S. D. Richardson and T. A. Ternes, *Environ. Sci. Technol.* 2014, **48**, 12689.
61. T. Ye, B. Xu, Z. Wang, T. Y. Zhang, C. Y. Hu, L. Lin, S. J. Xia and N. Y. Gao, *Water Res.* 2014, **66**, 390.
62. T. A. Ternes and R. Hirsch, *Environ. Sci. Technol.* 2000, **34**, 2741.
63. J. Choi and R. L. Valentine, *Water Res.* 2002, **36**, 817.

64. W. A. Mitch and D. L. Sedlak, *Environ. Sci. Technol.* 2002, **36**, 588.
65. D. B. Jobb, R. Hunsinger, O. Meresz and V. Y. Taguchi in *Proceedings of the Fifth National Conference on Drinking Water*, Winnipeg, Manitoba, Canada, 1992.
66. J. M. Boyd, S. E. Hrudey, X. F. Li and S. D. Richardson, *TrAC - Trends Anal. Chem.* 2011, **30**, 1410.
67. J. W. A. Charrois, M. W. Arend, K. L. Froese and S. E. Hrudey, *Environ. Sci. Technol.* 2004, **38**, 4835.
68. Y. Y. Zhao, J. Boyd, S. E. Hrudey and X. F. Li, *Environ. Sci. Technol.* 2006, **40**, 7636.
69. P. Kulshrestha, K. C. McKinstry, B. O. Fernandez, M. Feelisch and W. A. Mitch, *Environ. Sci. Technol.* 2010, **44**, 3369.
70. N. Dai and W. A. Mitch, *Environ. Sci. Technol.* 2013, **47**, 3648.
71. M. Wu, Y. Qian, J. M. Boyd, S. Leavey, S. E. Hrudey, S. W. Krasner and X. F. Li, *Environ. Sci. Technol.* 2014, **48**, 1828.
72. California Department of Health - Drinking Water Program. Drinking Water Notification Levels and Response Levels: An Overview. http://goo.gl/XvsT74 (accesed November, 2014).
73. Ontario Regulation 169/03. Ontario Drinking Water Quality Standards. Safe Drinking Water Act, 2002. http://goo.gl/6OTlkl (accessed November 2014).
74. Health Canada. Guidelines for Canadian Drinking Water Quality: Guideline Technical Document - N-Nitrosodimethylamine (NDMA). Water, Air and climate Change Bureau, Healthy Environments and Consumer Safety Branch, Health Canada, Ottawa, Ontario (Catalogue No H128-1/11-662E) 2011 http://goo.gl/yAUstU (accessed November, 2014).
75. U.S. Environmental Protection Agency (U.S. EPA) Contaminant Candidate List 3 - CCL 3. http://goo.gl/6M90rO (accesed November, 2014).
76. I. M. Schreiber and W. A. Mitch, *Environ. Sci. Technol.* 2006, **40**, 6007.
77. J. Choi and R. L. Valentine, *Environ. Sci. Technol.* 2003, **37**, 4871.
78. I. M. Schreiber and W. A. Mitch, *Environ. Sci. Technol.* 2007, **41**, 7039.
79. W. A. Mitch, A. C. Gerecke and D. L. Sedlak, *Water Res.* 2003, **37**, 3733.
80. W. A. Mitch, J. O. Sharp, R. R. Trussell, R. L. Valentine, L. Alvarez-Cohen and D. L. Sedlak, *Environ. Eng. Sci.* 2003, **20**, 389.
81. J. M. Kemper, P. Westerhoff, A. Dotson and W. A. Mitch, *Environ. Sci. Technol.* 2009, **43**, 466.
82. A. Wilczak, A. Assadi-Rad, H. H. Lai, L. L. Hoover, J. F. Smith, R. Berger, F. Rodigari, J. W. Beland, L. J. Lazzelle, E. G. Kincannon, H. Baker and C. T. Heaney, *J. Am. Water Works Assoc* 2003, **95**, 94.
83. J. M. Kemper, S. S. Walse and W. A. Mitch, *Environ. Sci. Technol.* 2010, **44**, 1224.
84. S. H. Park, P. Piyachaturawat, A. E. Taylor and C. H. Huang, *Water Sci. Technol.* 2009, **9**, 279.
85. S. H. Park, S. Wei, B. Mizaikoff, A. E. Taylor, C. Favero and C. H. Huang, *Environ. Sci. Technol.* 2009, **43**, 1360.
86. E. Pehlivanoglu-Mantas and D. L. Sedlak, *Water Res.* 2008, **42**, 3890.
87. S. W. Krasner, W. A. Mitch, D. L. McCurry, D. Hanigan and P. Westerhoff, *Water Res.* 2013, **47**, 4433.
88. C. K. Schmidt and H. J. Brauch, *Environ. Sci. Technol.* 2008, **42**, 6340.
89. U. V. Gunten, E. Salhi, C. K. Schmidt and W. A. Arnold, *Environ. Sci. Technol.* 2010, **44**, 5762.

90. L. P. Padhye, J. H. Kim and C. H. Huang, *Water Res.* 2013, **47**, 725.
91. P. H. Chen, S. D. Richardson, S. W. Krasner, G. Majetich and G. L. Glish, *Environ. Sci. Technol.* 2002, **36**, 3362.
92. W. Chu, N. Gao, D. Yin and S. W. Krasner, *J. Hazard. Mater.* 2013, **260**, 806.
93. W. H. Chu, N. Y. Gao, Y. Deng and S. W. Krasner, *Environ. Sci. Technol.* 2010, **44**, 3908.
94. S. D. Richardson, A. D. Thruston Jr., T. V. Caughran, P. H. Chen, T. W. Collette, T. L. Floyd, K. M. Schenck, B. W. Lykins, G.-R. Sun and G. Majetich, *Environ. Sci. Technol.* 1999, **33**, 3368.
95. S. W. Krasner, R. Chinn, C. J. Hwang and S. Barrett in *Proceedings of the 1990 American Water Works Association Water Quality Technology Conference*, American Water Works Association, Denver, CO, 1991.
96. B. Kundu, S. D. Richardson, P. D. Swartz, P. P. Matthews, A. M. Richard and D. M. DeMarini, *Mutat. Res.* 2004, **562**, 39.
97. B. Kundu, S. D. Richardson, C. A. Granville, D. T. Shaughnessy, N. M. Hanley, P. D. Swartz, A. M. Richard and D. M. DeMarini, *Mutat. Res.* 2004, **554**, 335.
98. J. Choi and S. D. Richardson in *Proceedings of the International Workshop on Optimizing the Design and Interpretation of Epidemiologic Studies to Consider Alternative Disinfectants of Drinking Water*, Raleigh, NC, 2005.
99. S. D. Richardson, A. D. Thruston Jr., C. Rav-Acha, L. Groisman, I. Popilevsky, O. Juraev, V. Glezer, A. B. McKague, M. J. Plewa and E. D. Wagner, *Environ. Sci. Technol.* 2003, **37**, 3782.
100. T. A. McDonald and H. Komulainen, *J. Environ. Sci. Health - C* 2005, **23**, 163.
101. H. R. Schulten and M. Schnitzer, *Naturwissenschaften* 1993, **80**, 29.
102. M. Yang and X. Zhang, *Environ. Sci. Technol.* 2014, **48**, 11846.
103. F. Qin, Y. Y. Zhao, Y. Zhao, J. M. Boyd, W. Zhou and X. F. Li, *Angew. Chem. Int. Ed.* 2010, **49**, 790.
104. Y. Zhao, J. Anichina, X. Lu, R. J. Bull, S. W. Krasner, S. E. Hrudey and X. F. Li, *Water Res.* 2012, **46**, 4351.
105. W. Wang, Y. Qian, J. M. Boyd, M. Wu, S. E. Hrudey and X. F. Li, *Environ. Sci. Technol.* 2013, **47**, 3275.
106. Y. Qian, W. Wang, J. M. Boyd, M. Wu, S. E. Hrudey and X. F. Li, *Environ. Sci. Technol.* 2013, **47**, 4426.
107. C. H. Jeong, E. D. Wagner, V. R. Siebert, S. Anduri, S. D. Richardson, E. J. Daiber, A. B. McKague, M. Kogevinas, C. M. Villanueva, E. H. Goslan, W. T. Luo, L. M. Isabelle, J. F. Pankow, R. Grazuleviciene, S. Cordier, S. C. Edwards, E. Righi, M. J. Nieuwenhuijsen and M. J. Plewa, *Environ. Sci. Technol.* 2012, **46**, 12120.

N-NITROSODIMETHYLAMINE PRECURSORS IN WASTEWATER: REMOVAL VIA ACTIVATED SLUDGE TREATMENT

K.L. Linge[1], D. Liew[1], E.J. Marti[2,3], E. Dickenson[2], A. Heitz[4], and C.A. Joll[1]

[1] Curtin Water Quality Research Centre, Department of Chemistry, Curtin University, GPO Box U1987, Perth, Western Australia, 6845, Australia

[2] Southern Nevada Water Authority, Applied Research & Development Center, PO Box 99954, Las Vegas, NV 89193, USA

[3] University of Nevada, Las Vegas, Civil & Environmental Engineering and Construction, Las Vegas, NV, USA

[4] School of Civil and Mechanical Engineering, Department of Civil Engineering, Curtin University, GPO Box U1987, Perth, Western Australia, 6845, Australia

1. INTRODUCTION

Disinfection by-products (DBPs) are recognized as a potential health concern in treated drinking water, but have not been as extensively studied in recycled water. However, our analysis of chemicals in secondary wastewater after treatment with microfiltration (MF) and reverse osmosis (RO) has demonstrated that DBPs are often the most frequently detected chemicals in recycled water, and that DBPs may be more frequently detected in RO-treated recycled water than in secondary wastewater.[1-3] As part of the MF/RO treatment process, it is standard practice to chloraminate to minimize RO membrane fouling.[4] Dosed chlorine reacts with ammonia present in, or added to, the wastewater to form monochloramine, and this process leads to formation of DBPs.[1,3,5] Despite increased detection, the concentrations of most DBPs in recycled water are well below health values.[1,3] However, this is not the case for *N*-nitrosodimethylamine (NDMA), which has been identified as a potential health concern for both drinking water[6,7] and recycled water.[2,5,8] NDMA has primarily been reported by water utilities using chloramination,[9,10] though it is now recognised that NDMA can be formed during both chlorine and chloramine treatment, depending on the precursors present and the process conditions.[9,11] More recently, ozonation has also been identified to result in a significant NDMA formation from some precursors.[12-14]

The importance of *N*-nitrosamines as disinfection by-products of health concern is illustrated by the publication of a number of recent literature reviews describing their occurrence and control,[15-17] and increasing recognition of their cancer potencies and toxicities.[18,19] While *N*-nitrosamine concentrations in recycled water can be minimised by additional treatment with UV irradiation and advanced oxidation processes,[20,21] it is also possible that concentrations could be minimised by reduction of precursors present in the secondary wastewater.[22] While NDMA precursors are often modeled using dimethylamine,, the majority of NDMA precursors in wastewater are unidentified organic nitrogen species that have not been fully characterized.[23] In this study, we investigated the factors that influence the removal of NDMA precursors during activated sludge treatment, and the potential impact of different chloramination or ozonation strategies on NDMA formation in recycled water.

2. METHODOLOGY

2.1. Sampling Sites

Samples were collected from Beenyup Wastewater Treatment Plant (WWTP) and Beenyup Advanced Water Recycling Plant (AWRP) in Perth, Western Australia during 2010 and 2011. Beenyup WWTP is a conventional activated sludge treatment plant with continuous flow that treats predominantly domestic wastewater, with a small proportion of industrial wastewater (~2%).[1] The activated sludge system has a pre-anoxic zone for denitrification, combined with an aerated zone to promote nitrification (Figure 1). Recycling pumps located at the end of each tank pump aerated mixed liquors, which are rich in nitrate from the end of the tank back to the head without oxygen addition to increase denitrification. Additional denitrification occurs in the secondary sedimentation tanks, where liquid-solid separation occurs. The plant has a sludge handling system that uses polyacrylamide-based polymeric flocculants for the dewatering of the excess sludge using dissolved air flotation thickeners as well as for the dewatering of the anaerobic digested sludge by centrifugation. Effluent from the dissolved air flotation thickeners is returned to the treatment process just before primary sedimentation, while centrate from sludge handling is returned to the head of the WWTP (Figure 1). Beenyup WWTP has a capacity of 135 ML/day, and flow into the plant has two maxima in each 24 hour period, one around midday and another one between 9 and 11 pm. The retention time in the aeration tanks and the clarifiers varies between 6 to 8 hours depending on flow.[24] Samples from Beenyup WWTP were collected at the primary wastewater sampling point and at the combined secondary wastewater sampling point (Figure 1).

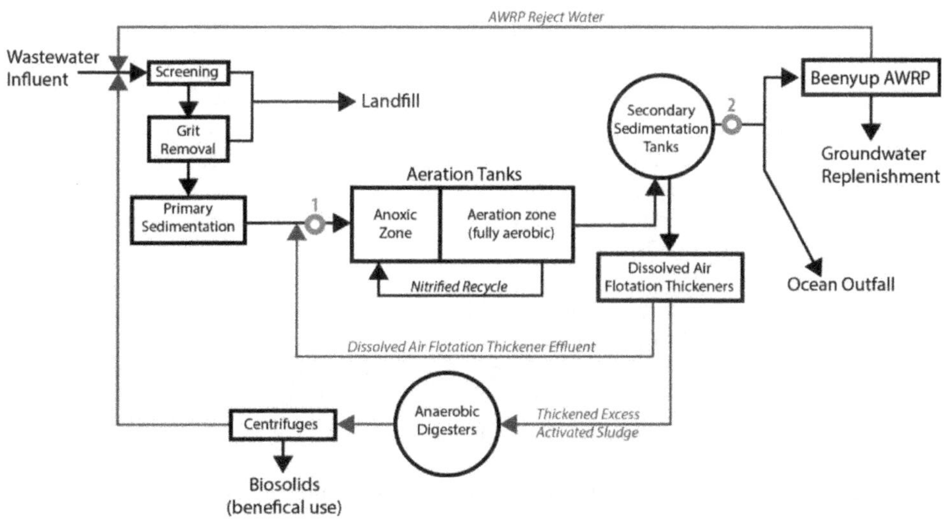

Figure 1 *Schematic of Beenyup Wastewater Treatment Plant, showing 1) primary and 2) secondary wastewater sampling points. Pathways by which polyacrylamide-based polymeric flocculants are introduced into the treatment plant are in grey.*

Primary wastewater has undergone solids removal by screening, grit removal and primary sedimentation. Therefore, the sample obtained had a lower solids content than the raw influent into the WWTP. However, previous research has shown close agreement between total NDMA precursor concentrations for wastewater influent and primary wastewater,[25] and therefore, precursor removal during primary treatment is expected to be minimal. Most secondary wastewater from Beenyup WWTP is discharged into the Indian Ocean. However, between March 2010 and September 2014, 7 ML/day of secondary wastewater from Beenyup WWTP fed into the Beenyup AWRP[26,27] as part of the Water Corporation of Western Australia's Groundwater Replenishment Trial. Beenyup AWRP was commissioned between March and November 2010, after which time reinjection into a deep underground aquifer commenced. Treatment at the AWRP consisted of chloramination to minimise biofouling on membranes, ultrafiltration (UF), RO, and UV disinfection. After UF/RO/UV treatment, about 4.5 ML/d of treated water was injected into the aquifer for replenishment, while the RO reject (about 2.5 ML/d) was sent to the head of the WWTP.

All samples were collected as grab samples. Samples for *N*-nitrosamine precursor or concentration analysis were collected in glass containers, filled to zero headspace, kept cool (in an ice box) and away from light until reaching the laboratory, where they were then stored at 4 °C until analysis. Samples for *N*-nitrosamine concentration contained ascorbic acid (20 mg/L) as a quenching agent. Trip and field blanks containing ultrapure water and ascorbic acid (20 mg/L), if appropriate, were also included during each sampling occasion to determine if there was any contamination through the sampling process, storage and transport. Trip blanks remained unopened until analysis, and field blanks were opened at each sampling location. Samples for total organic carbon (TOC) analysis were subsampled (40 mL) from glass containers used to collect samples for *N*-nitrosamine precursor analysis. Samples for nutrients were collected in high density polyethylene (HDPE) bottles.

N-nitrosamine precursor concentrations throughout both Beenyup WWTP and Beenyup AWRP were measured in July 2010 (during Beenyup AWRP commissioning) and February 2011 (see Figure 2 for sampling points). In addition, to investigate the effect of nitrification and denitrification on NDMA precursor removal in Beenyup WWTP, matched primary and secondary wastewater were collected in two separate sampling campaigns, as described in Table 1.

Sampling was undertaken during high, medium and low flow conditions, related to the demand on the WWTP at the time of sampling. The retention time in the aeration tanks and sedimentation tanks was taken into account when estimating the time lag between the sampling of the primary wastewater and the secondary wastewater, calculated to be 6, 7 and 8 hours for high, medium and low flow events, respectively.

2.2. *N*-Nitrosamine Precursor Analysis and Formation Studies

The total concentration of *N*-nitrosamine precursors was determined using a previously published NDMA precursor analysis protocol,[28] also known as the NDMA formation potential test. This protocol has been employed in many studies to quantify *N*-nitrosamine precursors in natural waters and wastewaters,[23,28,29] and applies a high dose of monochloramine (140 mg/L Cl_2) to a buffered sample with a contact time of 10 days. A modified version with 7 days contact time was used in this study, as preliminary tests indicated there was little *N*-nitrosamine formation after 7 days.

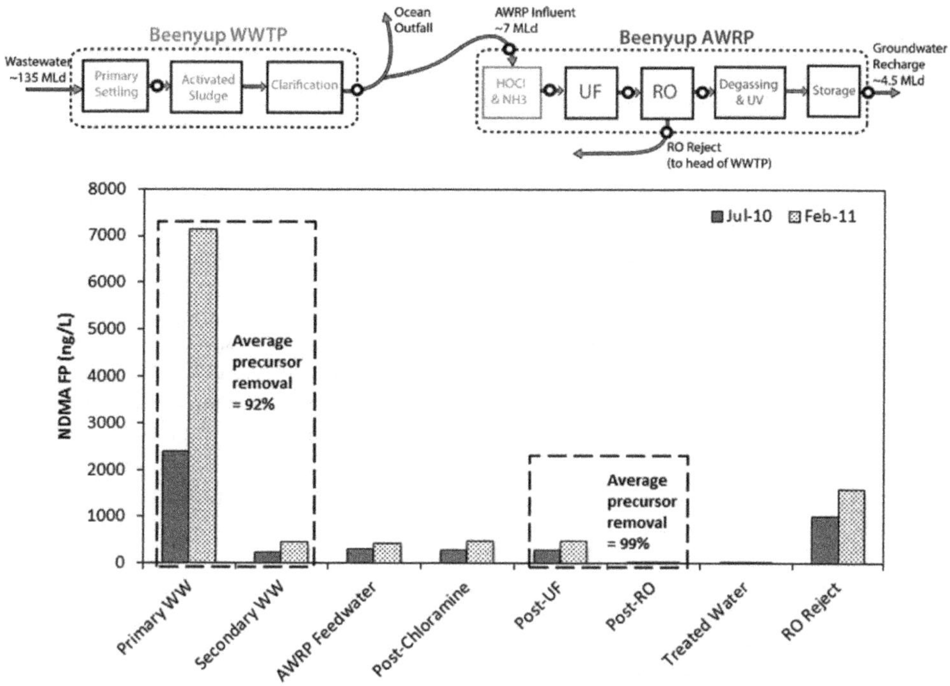

Figure 2 *NDMA precursors, as measured by formation potential, in Beenyup WWTP and AWRP. Sampling points for each plant are identified (bold circles) in the schematic.*

Table 1 *Overview of sampling dates in Beenyup WWTP including detention times*

	Target Flow rate	Detention time	Flow Rate ($L\ s^{-1}$)				
			Initial	Final	Ave.	Min.	Max.
			2010 Sampling Campaign				
Tue, 7 Sep 10	Low	8 hr	612	1597	1687	612	2126
Tue, 14 Sep 10	Low	8 hr	618	1688	1746	618	2197
Sat, 18 Sep 10	Medium	7 hr	2496	1924	2053	1813	2513
Sat, 11 Sep10	Medium	7 hr	2654	1990	2165	1903	2667
Mon, 20 Sep 10	High	6 hr	1867	1968	1563	1387	1968
Mon, 11 Oct 10	High	6 hr	1901	1918	1543	1346	1957
			2011 Sampling Campaign				
Mon, 13 Jun 11	Medium	7 hr	1843	1962	1605	1409	1962
Tue, 14 Jun 11	Low	8 hr	606	1815	1520	597	1901
Thu, 16 Jun 11	Low	8 hr	607	1528	1536	602	1913
Sat, 18 Jun 11	High	6 hr	2550	1981	2110	1889	2551
Mon, 20 Jun 11	Medium	7 hr	1805	1947	1617	1437	1953
Tue, 21 Jun 11	Low	8 hr	617	1655	1562	617	1928
Thu, 23 Jun 11	Low	8 hr	603	1521	1505	596	1907
Sat, 25 Jun 11	High	6 hr	2529	2004	2141	1903	2534

In addition to the *N*-nitrosamine precursor analysis, a series of experiments was conducted to measure *N*-nitrosamine formation in Beenyup secondary wastewater using lower chloramine concentrations (1, 3 and 10 mg/L Cl_2), closer to the typical concentrations used in water recycling plants (3 mg/L Cl_2 at Beenyup AWRP). These chloramination experiments used either preformed monochloramine, or simultaneously-dosed hypochlorite and ammonium chloride solutions, to form monochloramine inline. All reactions were carried out in amber glass bottles at pH 7 with phosphate buffer (10mM), where the disinfectant stock solution was added in one-tenth of the total sample volume, to make up the desired dose. Samples were collected after 24 hours, quenched with ascorbic acid to stop the reaction, and analysed for *N*-nitrosamines.

For ozonation experiments, ozone stock solutions were prepared using an oxygen-fed generator (American Ozone Systems, Inc., Chicago, Illnois) to diffuse ozone into ultrapure water. The concentration of the ozone stock solution was measured on an Agilent Technologies Cary 60 UV/Vis spectrophotometer at 600 nm using the indigo method.[30] An appropriate volume was added into samples to achieve the desired ozone to TOC ratios (0.1-0.5), taking dilution into account. Samples were sealed, mixed by inverting, and left at room temperature for one day to ensure ozone residual was zero before analysing for NDMA.

2.3. *N*-Nitrosamine Analysis

N-Nitrosamines were analysed after sample preconcentration on an Agilent Technologies 7890A gas chromatograph coupled with a 5975C mass spectrometer, operating in positive chemical ionization mode with ammonia as the reagent gas (flow = 0.5 mL/min). Sample preconcentration was undertaken either by liquid-liquid extraction (LLE) for NDMA precursor analysis samples, when higher concentrations of *N*-nitrosamines were expected, or by solid-phase extraction (SPE) for lower *N*-nitrosamine concentrations or experiments with lower disinfectant doses (1-10 mg/L Cl_2). For LLE, an aliquot of sample (50 mL) containing deuterated *N*-nitrosamines (80 ng/L) as the surrogate standards, was adjusted to pH 8 with sodium hydrogen carbonate or sodium carbonate. Sodium chloride (15 g) was added, and the sample vial shaken until all salt dissolved. The sample was then extracted with dichloromethane (5 mL) and the organic layer dried through anhydrous magnesium sulphate. The extract was concentrated to ~200 μL in a heating block (40 °C) under a gentle stream of nitrogen. The final extract was transferred to a GC microvial containing diphenylamine-d_{10} (50 μg/L) as the internal standard.

The SPE procedure was based on the method of Charrois et al.[19] with minor modifications. SPE cartridges were prepared in-house with LiChrolut® EN (0.35 g) and Carboxen™ 572 (0.5 g) and were conditioned with hexane (10 mL), dichloromethane (20 mL), methanol (20 mL) and ultrapure water. An aliquot of sample (1L) containing deuterated *N*-nitrosamines (14 ng/L) as the surrogate standards, was adjusted to pH 8 with sodium hydrogen carbonate. Each sample was passed through a SPE cartridge. After transfer, the cartridge was dried under vacuum for a few hours and eluted with dichloromethane (18 mL). The extract was concentrated to ~200 μL using the same procedure as for LLE. While the method used is able to measure 8 different *N*-nitrosamines, only NDMA was detected in significant concentrations. Limits of detection for NDMA were calculated based on the 95% confidence interval for triplicates of a low concentration standard according to the USEPA procedure [31] and were typically 1-3 ng/L for LLE preconcentration and ≤1 ng/L for SPE preconcentration.

2.4. Other Chemical Analyses

Conductivity, pH, and temperature were all measured in the field using portable probes (HQ40d, Hach Co., USA). Other water quality parameters measured by standard methods[32] included TOC (5310-TOC C), ammonia (4500-NH_3 G, LOR = 0.01 mg/L), nitrate and nitrite (4500-NO_3^- F, LOR = 0.01 mg/L), and total nitrogen (4500-N C, LOR = 0.01 mg/L). Dissolved organic nitrogen (DON) was determined as the difference between total nitrogen and the sum of nitrate, nitrite and ammonia.

2.5. Data Analysis

The normality of each chemical dataset was tested by calculation of the Kolmogorov-Smirnov Statistic and the Shapiro-Wilk Statistic, with most found to have a poor fit to the normal distribution curve. Therefore, non-parametric tests were used to determine whether the dataset showed significant differences with respect to either plant flow rate, or sampling event. The Kruskal–Wallis one-way analysis of variance by ranks was used to determine whether nitrogen removal was affected by low, medium or high flow rates, while the Mann–Whitney *U*, the non-parametric equivalent of the student's t-test, was used to determine whether nitrogen removal was different in the 2010 and 2011 sampling events. All statistical analysis was undertaken using SPSS Statistics v22 (IBM).

3. RESULTS AND DISCUSSION

3.1. Removal of NDMA Precursors

N-Nitrosamine precursor concentrations through both Beenyup WWTP and Beenyup AWRP were measured in July 2010 (during Beenyup AWRP commissioning) and February 2011 via the *N*-nitrosamine formation potential test (Figure 2). Formation of *N*-nitrosamines other than NDMA was always below 25 ng/L, except for *N*-piperidine (average 104 ng/L formation) and *N*-pyrrolidine (average 479 ng/L formation), both detected in primary wastewater but not secondary wastewater. Overall NDMA formed at least 90% of *N*-nitrosamine precursors on a molar basis. The concentration of NDMA precursors in primary wastewater was very variable (2400 ng/L in July 2010, and 7100 ng/L in February 2011). This variation might be due to seasonal changes, as other studies have found higher concentrations of NDMA precursors during summer sampling events than in winter sampling,[33,34] but more summer sampling in particular would be required to confirm this. Despite the difference in NDMA precursor concentrations in primary wastewater, there was consistent removal of NDMA precursors for both sampling events during the activated sludge treatment (average of 92% removal) and also during RO treatment (average 99% removal). The removal of NDMA precursors by RO treatment was comparable to that found in other studies,[23,35] and considerably better than the removal of NDMA itself, typically around 50%.[1,8,36] The rejection measured for NDMA precursors in this study was similar to measured rejection of dimethylamine and other alkylamines by both nanofiltration and RO membranes.[37] Rejection of these small molecules is high because they are positively charged at neutral pH.

In absolute terms, the largest removal of NDMA precursors was during activated sludge treatment in the Beenyup WWTP, comparable or higher to that found in other studies.[33,34,38] During activated sludge treatment, indigenous waterborne microorganisms utilise dissolved and suspended organic material, producing a secondary wastewater with lower C and N content. The two key processes for transformation and, ultimately, removal of N from wastewater are nitrification and denitrification. During nitrification, ammonia is oxidised into nitrite by nitrifying bacteria in the presence of oxygen, and the nitrite is then further oxidised to nitrate.[39] During denitrification, nitrate is reduced to nitrogen gas under anoxic conditions, thus completely removing N from the system. Biodegradable organic N, particularly urea, amino acids and proteins, can also be removed in this process, through conversion to ammonia by either hydrolysis or ammonification through heterotrophic bacteria,[40] although typically removal of organic N is less efficient than inorganic N.[41] Removal of NDMA precursors has also been previously linked to biological treatment, although variations in removal efficiency could not be linked to differences in treatment processes.[33,34] Formation of other DBPs from effluent organic matter has also previously been linked to the level of wastewater treatment.[22]

3.2. Nitrogen Removal During Activated Sludge Treatment

While activated sludge treatment involves both nitrification and denitrification steps, it is nitrification that has been found to strongly control concentrations of inorganic and organic nitrogen in secondary wastewater, and also correlate with biodegradable dissolved organic carbon (BDOC), dissolved organic carbon (DOC) and DBP formation potential.[42] Assessment of the nitrifying and denitrifying efficacy of Beenyup WWTP, using the quantitative classification scheme of Krasner et al.[22,42] (Figure 3), indicated that, in general, Beenyup WWTP has good nitrification (ammonia < 2 mg/L-N) and partial denitrification (nitrate > 5 mg/L-N). At higher flow rates, hydraulic overloading, which reduces the retention time, and higher organic matter loading has been found to reduce the efficacy of the nitrification and denitrification processes at Beenyup WWTP.[43]

To investigate the effect of nitrification and denitrification on NDMA precursor removal at Beenyup WWTP, matched primary and secondary wastewater were collected over a variety of different flow rates in two separate sampling campaigns, and a variety of N species, as well as NDMA precursors, were measured (Table 2). The total N in primary wastewater was dominated by ammonia and DON, while total N in secondary wastewater was dominated by nitrate, with some DON and ammonia. While NDMA precursors make up a component of DON, the calculated percentage by mass in Beenyup secondary wastewater was always less than 0.02%. Previous analysis by Pehlivanoglu-Mantas and Sedlak[44] suggests that DON in secondary wastewater is relatively low molecular weight (<10 kDA) and hydrophilic, which is consistent with its persistence in wastewater through both primary and secondary wastewater treatment. The same study suggested NDMA precursors are typically <1 kDa and may be more hydrophobic than bulk DON.

The Kruskal–Wallis one-way analysis of variance by ranks was used to determine whether nitrogen concentrations (i.e., ammonia, nitrate, nitrite, DON, total N, and NDMA precursors) in either primary or secondary wastewater were significantly different during low, medium or high flow rates. However, the null hypothesis (that there was no difference in distribution over different flow categories) was retained for all variables tested, except for NDMA precursors in secondary wastewater ($p = 0.044$). There was also a weak positive correlation between NDMA precursors and average event flow rate (Spearman's rho = 0.563, $p = 0.03$), which would indicate that there is less NDMA precursor removal at shorter retention times. However, overall there were much more significant differences between the two sampling campaigns than between different flow rates.

While N concentrations in primary wastewater were higher in the June 2011 sampling campaign, the concentration of ammonia and NDMA precursors in the secondary wastewater was lower, suggesting better removal in this second sampling campaign.

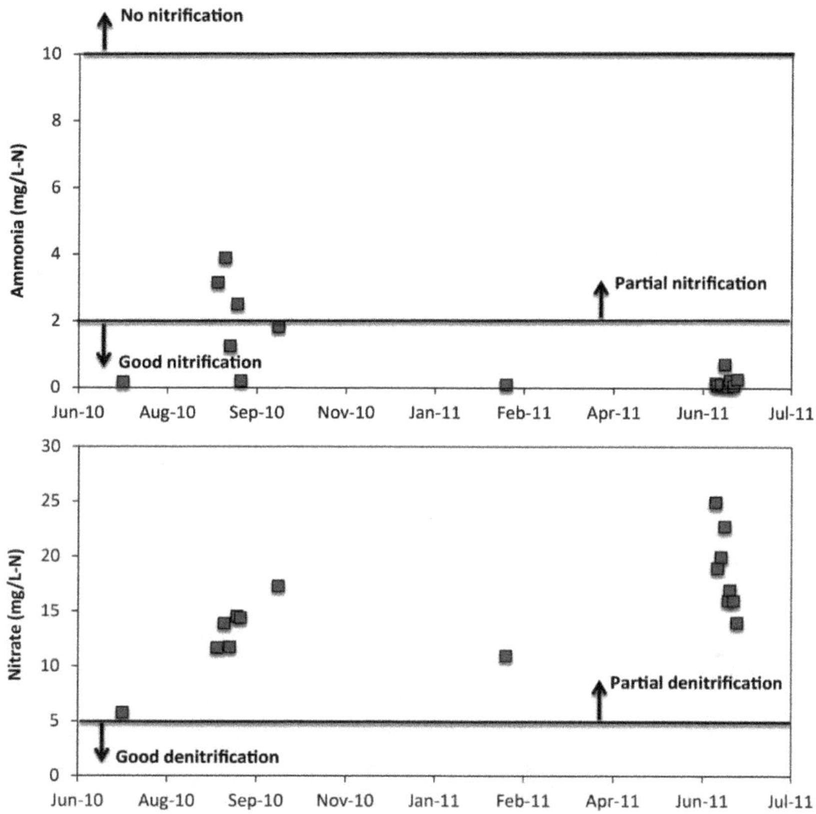

Figure 3 *Concentration of ammonia and nitrate in Beenyup WWTP secondary wastewater*

Table 2 *Average concentration of target N species in matched primary and secondary wastewater samples during the 2010 and 2011 sampling campaigns.*

	Avg Sept 2010 (n = 6)		Avg June 2011 (n = 8)	
	Primary WW	Secondary WW	Primary WW	Secondary WW
Ammonia (mg/L-N)	57 ± 7.8	2.0 ± 1.3	65 ± 16	0.2 ± 0.2
Nitrate (mg/L-N)	0.8 ± 0.47	14 ± 2.1	0.02 ± 0.007	19 ± 3.7
Nitrite (mg/L-N)	0.06 ± 0.08	0.18 ± 0.06	< 0.01	0.05 ± 0.08
DON (mg/L-N)	7 ± 1.7	1.6 ± 0.02	14 ± 13	4.8 ± 4.3
Total N (mg/L-N)	71 ± 7.9	19 ± 2.6	79 ± 14	24 ± 2.6
NDMA FP (ng/L)	4100 ± 397	420 ± 22	4900 ± 620	394 ± 38

In order to remove the influence of primary wastewater concentrations, percentage removals of ammonia, nitrate, DON, total N, and NDMA precursors were calculated for activated sludge treatment (Figure 4). The percentage removal of nitrate was estimated by assuming that all removed ammonia was transformed to nitrate. Comparison using the Mann Whitney *U* test confirmed that there was a significant difference ($p < 0.05$) in removal between the two sampling campaigns for all N species, except DON, which had much higher variability. This variability was probably related to both the relatively low DON concentrations, and also the method of estimation. The percentage removal of both ammonia and NDMA precursors was higher in the 2011 sampling campaign, when good nitrification was always achieved. In contrast, both total N and nitrate had poorer removal in the 2011 sampling campaign, reflecting poorer denitrification and a higher concentration of nitrate in the secondary wastewater.

In contrast to previous studies,[35] there was a correlation between NDMA precursors and ammonia concentrations in secondary wastewater (Spearman's rho = 0.648, p = 0.012), which provides additional support to the hypothesis that NDMA precursor reduction is related to the nitrification process. There was no correlation between NDMA precursors and DON in either primary (Spearman's rho = 0.196) or secondary (Spearman's rho = 0.033) wastewater, confirming that NDMA precursors are a minor and specific fraction of DON.

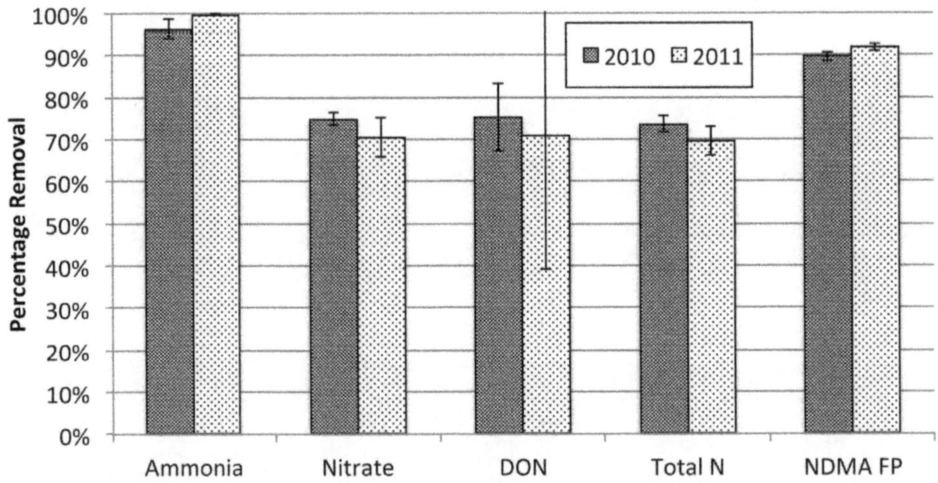

Figure 4 *Percentage removals of ammonia, nitrate, DON, TN, and NDMA precursors between primary and secondary wastewater, for the 2010 and 2011 sampling campaigns.*

3.3. NDMA Precursors in Primary Wastewater

As previously discussed, there was significant variability in the concentration of NDMA precursors measured in primary WW, ranging from 2400 to 7100 ng/L. Previous researchers have attributed variability in primary wastewater precursors to seasonal impacts,[33,34] or the presence of specific industries in the wastewater catchment, such as circuit board manufacture.[25] Another potential source of NDMA precursors within the treatment plant is amine-based cationic polyacrylamide polymers, often used for coagulation in WWTPs. Previous studies have shown formation of NDMA in the order of μg/L after chloramination of polyacrylamide compounds.[23] Increases in NDMA precursor concentrations between primary and secondary treatment have been attributed to the use of cationic polymers during clarification in other studies.[25] At Beenyup WWTP and AWRP, there are three polymers that may contribute to NDMA precursors within the primary wastewater: Flopam FO 4350 SH is used in the dissolved air flotation thickeners to thicken excess sludge before digestion, EMA 8845 MBL is used in Beenyup sludge handling, and NALCO® PC-1521T is an anti-scalant used in the Beenyup AWRP. Effluent from the dissolved air flotation thickeners is returned to the treatment process just before primary sedimentation, while centrate from sludge handling and reject water from Beenyup AWRP are returned to the head of the WWTP (Figure 1). Assessment of NDMA precursors in polymer solutions (10 mg/L) made with ultrapure water, using the 7 day precursor analysis protocol produced NDMA concentrations between 56 and 372 ng/L (Figure 5), much

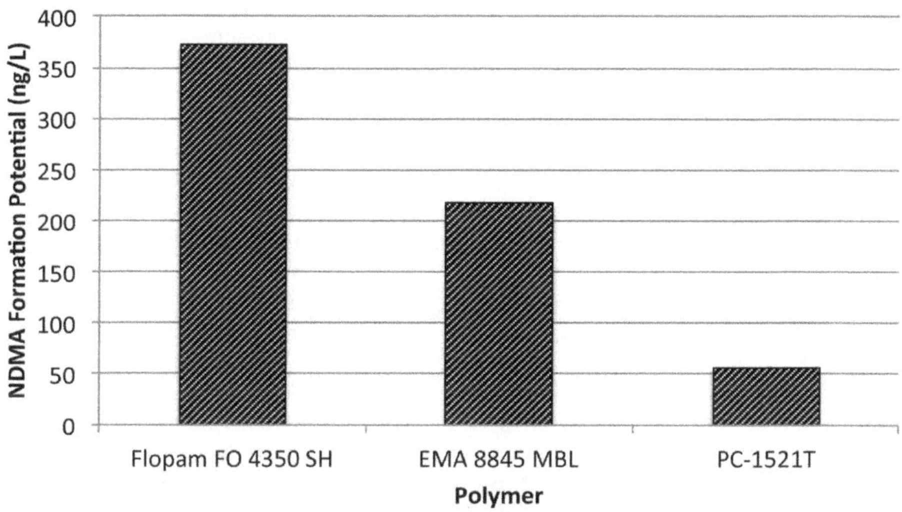

Figure 5 *Seven day NDMA precursor experiments for polymers (10 mg/L solutions) used in Beenyup WWTP and AWRP. Flopam FO 4350 SH is used in the dissolved air flotation thickeners to thicken excess sludge before digestion, EMA 8845 MBL is used in Beenyup sludge handling, and NALCO® PC-1521T is an anti-scalant used in Beenyup AWRP.*

lower than the concentrations of NDMA precursor measured in primary wastewater. Therefore, it appears that NDMA precursors in Beenyup primary wastewater are not significantly affected by the use of polymers. Previous research has attributed most NDMA precursors to domestic sources which are difficult to control[25] and therefore source control activities may have limited effect on reducing NDMA precursors in primary wastewater.

3.4. NDMA Formation from Ozonation and Chloramination of Secondary Wastewater

To investigate the potential formation of NDMA during water recycling treatment, laboratory experiments were undertaken using either ozonation or chloramination. Ozone is of benefit for potable water reuse applications as it reduces the concentrations of both pathogens and trace organic chemicals, and it can be implemented before and after traditional disinfection or membrane processes. While typical secondary amines, such as dimethylamine, and related compounds have been shown to be NDMA precursors upon ozonation, molar conversion yields are typically very low (i.e. 0.02%).[12,45] In contrast, ozonation of precursors in which dimethylamine is bonded directly to a nitrogen atom or separated with a good leaving group (e.g. CO_2, SO_2) leads to much higher molar conversion yields of NDMA (>50%), particularly when the reaction is catalysed by bromide.[12,13] Ozone is not currently part of the treatment strategy at Beenyup WWTP or AWRP, and therefore the effect of ozone on NDMA formation in treated wastewaters could not be tested in the full-scale plant. In laboratory experiments, relatively low concentrations of NDMA formed in Beenyup secondary wastewater after ozonation (Figure 6a), though the net NDMA formation did increase from 2 ng/L to 12 ng/L as the ozone to TOC ratio increased from 0.1 to 0.5. Previous research has found that ozonation can remove up to 95% of NDMA precursors, as measured using the NDMA formation potential test,[45,46] and so there may be opportunities to reduce the concentration of NDMA precursors in the AWRP feedwater before UF/RO treatment, without significant NDMA formation upon ozonation. Ozone dose optimisation is required to minimise production of ozone-DBPs,[47,48] and further investigations are required to measure the effects of the presence of ammonia, bromide and polyacrylamide sludge treatment polymers, which have recently been implicated in increased NDMA formation during ozonation of wastewater.[49]

Chloramination in water recycling plants is achieved by addition of preformed monochloramine, or by separate addition of ammonia and chlorine to the wastewater stream (inline-formation), which can lead to formation of both monochloramine and dichloramine.[50] Understanding chloramine speciation is important for minimising DBP formation, particularly formation of NDMA. For example, dichloramine enhances (between 3 to 30 times) NDMA formation compared to monochloramine during chloramination of secondary wastewaters, depending on the wastewater and contact time.[5,51] The effect of monochloramine dose was investigated by treating secondary wastewater with preformed or inline-formed monochloramine at 1, 3, and 10 mg/L Cl_2. Comparison of formation results after 24 hours (Figure 6b) showed similar concentrations (25-28 ng/L) of NDMA at 1 mg/L Cl_2 and 3 mg/L Cl_2 of both pre-formed and inline-formed monochloramine. While the concentrations of NDMA formed in these experiments are higher than the current Australian Guidelines for Water Recycling (10 ng/L),[52] additional laboratory tests using the conditions employed at Beenyup AWRP (3 mg/L Cl_2, pH 6-7 and contact time of 3 hours) demonstrated that NDMA formation was well below 5 ng/L for both pre-formed and inline-formed monochloramine, and therefore no further optimisation of the chloramination process is required for NDMA minimisation. These laboratory results are consistent with extensive monitoring of final product water from the Beenyup AWRP which showed that NDMA concentrations were almost always <10 ng/L.[53]

Increasing the dose of monochloramine from 3 to 10 mg/L Cl_2 resulted in 3 times (90 ng/L) and 7.6 times (190 ng/L) higher NDMA concentrations from preformed and inline-formed monochloramine, respectively. Choi et al.[10] has previously evaluated the effect of monochloramine dose on NDMA formation and observed a generally linear increase in NDMA concentration when preformed monochloramine was applied from 0.01-0.2 mM (0.7-14 mg/L Cl_2) to a fixed concentration of dimethylamine (0.1 mM). This is in contrast to the non-linear increase in NDMA formation reported in this study, and also by Farré et al.[5] during the application of preformed monochloramine from 4-15 mg/L Cl_2 to secondary wastewater. This non-linear increase in formation emphasises the presence of other organic matter (i.e. besides dimethylamine) can also act as NDMA precursors. At 10 mg/L Cl_2, inline-formed monochloramine produced twice as much NDMA as preformed monochloramine, which is also in agreement with previous research.[5] Higher concentrations of dichloramine and organic chloramines have been observed in Beenyup wastewater dosed with 10 mg/L Cl_2 inline-formed monochloramine, compared to 3 mg/L Cl_2 inline-formed monochloramine.[54] Overall, the results suggest that the monochloramine dose should be controlled carefully in full-scale plants that use inline-formed monochloramine in order to minimize NDMA formation.

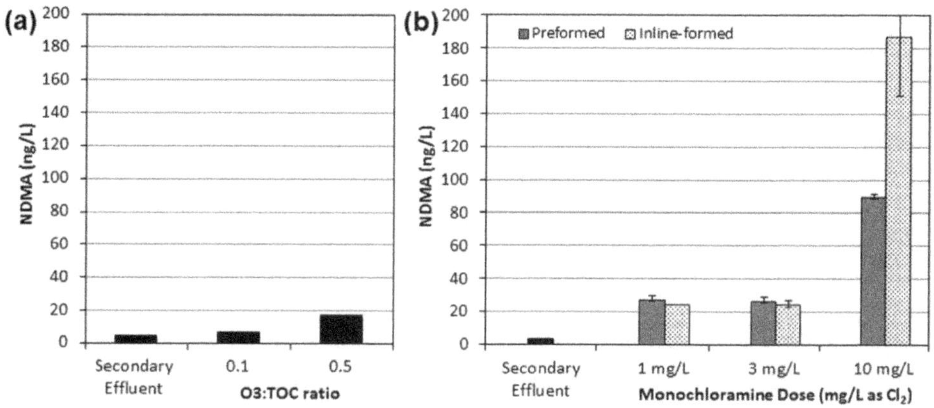

Figure 6 *NDMA formation from a) ozonation and b) preformed and inline-formed chloramination*

4. CONCLUSIONS

The two key barriers for *N*-nitrosamine precursor removal were found to be activated sludge treatment and RO. Reverse osmosis is much more effective in removing NDMA precursors than NDMA itself. Precursor removal during activated sludge treatment was related to the removal of ammonia, suggesting that nitrification is the most important removal mechanism for NDMA precursors. Hence, ensuring that the WWTP is operating well (in terms of nitrification) will promote maximum removal of NDMA precursors. Significant variability in the concentration of NDMA precursors in the primary wastewater led to variability in the concentration of NDMA precursors in secondary wastewater. Further investigations are required to understand the contribution of seasonal influences, influent flow rates or other sources of precursors to this variability. Analysis of treatment polymers used at the Beenyup WWTP and AWRP suggest that there would be little contribution from the polymers to NDMA formation.

Actual NDMA concentrations in secondary wastewater were much lower than NDMA precursor concentrations and did not correlate with precursor concentrations.

Although inline-formed monochloramine has been found to produce higher levels of dichloramine and organic chloramines than preformed monochloramine, the impact on *N*-nitrosamine formation was dose dependent. Increasing the chloramine dosage from 3 to 10 mg/L Cl_2, resulted in significantly more NDMA from inline-formed monochloramine than from preformed monochloramine. It is therefore important to control the dose at the treatment plant since a slight increase in dose could result in a significant increase in NDMA formation. Ozonation of Beenyup secondary wastewater did not lead to significant NDMA formation and therefore may have promise for NDMA precursor removal. However, ozone will produce its own suite of DBPs, and therefore it is important to test the formation of other DBP classes before any oxidation pre-treatment step is employed.

Acknowledgements

The authors thank E. Cosa, S. Higginson and M. Handyside (Water Corporation of Western Australia) and A. Chan and G. Chidlow (Curtin University) for their valuable assistance in sampling, analysis and method validation. Research reported in this paper was undertaken as part of ARC Linkage projects LP0989326 and LP130100602, and also supported by the National Science Foundation under Grant No. 1415103. The authors thank industry partners GHD, Water Research Australia, and Water Corporation of Western Australia for their collaboration.

References

1. P. Van Buynder, R. Lugg, C. Rodriguez, M. Bromley, J. Filmer, P. Blair, M. Handyside, S. Higginson, N. Turner, O. Lord, P. Taylor, K. Courtney, C. Newby, A. Heitz, K. Linge, J. Blythe, F. Busetti and S. Toze, *Premier's Collaborative Research Program (2005-2008): Characterising Treated Wastewater For Drinking Purposes Following Reverse Osmosis Treatment. Technical Report* ISBN: 978-0-9807477-0-6, Department of Health, Western Australia, 2009.
2. K. L. Linge, P. Blair, F. Busetti, C. Rodriguez and A. Heitz, *J. Water Supply Res. T.*, 2012, **61**, 494.
3. K. L. Linge, J. W. Blythe, F. Busetti, P. Blair, C. Rodriguez and A. Heitz, *Sep. Purif. Technol.*, 2013, **104**, 221.
4. C. R. Bartels, M. Wilf, K. Andes and J. Iong, *Water Sci. Technol.*, 2005, **51**, 473.
5. M. J. Farré, K. Döderer, L. Hearn, Y. Poussade, J. Keller and W. Gernjak, *J. Hazard. Mat.*, 2011, **185**, 1575.
6. J. M. Boyd, J. W. A. Charrois, R. Hofmann and S. E. Hrudey, in *Disinfection By-Products – Relevance to Human Health*, eds. S. Hrudey and J. W. A. Charrois, IWA Publishing, London, 2012.
7. J. W. A. Charrois, J. M. Boyd, K. L. Froese and S. E. Hrudey, *J. Environ. Eng. Sci.*, 2007, **6**, 103.
8. T. Fujioka, S. J. Khan, J. A. McDonald, A. Roux, Y. Poussade, J. E. Drewes and L. D. Nghiem, *Water Res.*, 2013, **47**, 6141.
9. S. D. Richardson, *Global NEST J.*, 2005, **7**, 43.
10. J. Choi and R. L. Valentine, *Water Res.*, 2002, **36**, 817.
11. Y. Y. Zhao, J. M. Boyd, M. Woodbeck, R. C. Andrews, F. Qin, S. E. Hrudey and X. F. Li, *Environ. Sci. Technol.*, 2008, **42**, 4857.
12. E. J. Marti, A. N. Pisarenko, J. R. Peller and E. R. V. Dickenson, *Water Res.*, in press, http://dx.doi.org/10.1016/j.watres.2014.08.047.

13. U. Von Gunten, E. Salhi, C. K. Schmidt and W. A. Arnold, *Environ. Sci. Technol.*, 2010, **44**, 5762.
14. K. Kosaka, M. Asami, K. Ohkubo, T. Iwamoto, Y. Tanaka, H. Koshino, S. Echigo and M. Akiba, *Environ. Sci. Technol.*, 2014, **48**, 11243.
15. T. Bond, J. Huang, M. R. Templeton and N. Graham, *Water Res.*, 2011, **45**, 4341.
16. J. Nawrocki and P. Andrzejewski, *J. Hazard. Mat.*, 2011, **189**, 1.
17. S. W. Krasner, W. A. Mitch, D. L. McCurry, D. Hanigan and P. Westerhoff, *Water Res.*, 2013, **47**, 4433.
18. *Monographs on the evaluation of carcinogenic risk of chemicals to humans: Some N-Nitroso compounds* Vol. 17, International Agency for Research on Cancer, World Health Organization, Lyon, France, 1978.
19. J. W. A. Charrois, M. W. Arend, K. L. Froese and S. E. Hrudey, *Environ. Sci. Technol.*, 2004, **38**, 4835.
20. C. M. Sharpless and K. G. Linden, *Environ. Sci. Technol.*, 2003, **37**, 1933.
21. B. Xu, Z. Chen, F. Qi, J. Ma and F. Wu, *J. Hazard. Mat.*, 2009, **168**, 108.
22. S. W. Krasner, P. Westerhoff, B. Y. Chen, G. Amy, S. N. Nam, Z. K. Chowdhury, S. Sinha and B. E. Rittmann, *Contribution of Wastewater to DBP Formation* ISBN 978-1-60573-005-9, AWWA Research Foundation, Denver, 2008.
23. W. A. Mitch and D. L. Sedlak, *Environ. Sci. Technol.*, 2004, **38**, 1445.
24. *Beenyup WWTP - Operation Manual*, Water Corporation of Western Australia, Perth, Western Australia, 2004.
25. D. L. Sedlak, R. A. Deeb, E. L. Hawley, W. A. Mitch, T. D. Durbin, S. Mowbray and S. Carr, *Water Environ. Res.*, 2005, **77**, 32.
26. J. Y. M. Tang, F. Busetti, J. W. A. Charrois and B. I. Escher, *Water Res.*, 2014, **60**, 289.
27. C. Loi, F. Busetti, K. L. Linge and C. Joll, *J. Chrom. A*, 2013, **1299**, 48.
28. W. A. Mitch, A. C. Gerecke and D. L. Sedlak, *Water Res.*, 2003, **37**, 3733.
29. A. C. Gerecke and D. L. Sedlak, *Environ. Sci. Technol.*, 2003, **37**, 1331.
30. K. L. Rakness, E. C. Wert, M. Elovitz and S. Mahoney, *Ozone-Science & Engineering*, 2010, **32**, 33.
31. USEPA, in *Title 40 Code of Federal Regulations, Part 136, Appendix B, Revision 1.11*, USEPA, 2005.
32. *Standard Methods For the Examination of Water & Wastewater*, 21st edn., APHA, AWWA, WEF, Maryland, United State of America, 2005.
33. S. Yoon, N. Nakada and H. Tanaka, *J. Hazard. Mat.*, 2011, **190**, 897.
34. S. Yoon, N. Nakada and H. Tanaka, *Water Sci. Technol.*, 2013, **68**, 2118.
35. M. J. Farre, J. Keller, N. Holling, Y. Poussade and W. Gernjak, *Water Sci. Technol.*, 2011, **63**, 605.
36. M. H. Plumlee, M. Lopez-Mesas, A. Heidlberger, K. P. Ishida and M. Reinhard, *Water Res.*, 2008, **42**, 347.
37. Y. Miyashita, S. H. Park, H. Hyung, C. H. Huang and J. H. Kim, *J. Environ. Eng.-ASCE*, 2009, **135**, 788.
38. M. Kavanaugh, M. Pirnie and D. Sedlak, *Removal and destruction of NDMA and NDMA precursors during wastewater treatment* ISBN: 0-97477586-8-X, WateReuse Foundation, Alexandria, VA, USA, 2006.
39. G. Bitton, *Wastewater Microbiology*, 3rd edn., John Wiley & Sons, Hoboken, New Jersey, 2005.
40. M. H. Gerardi, *Nitrification and denitrification in the activated sludge process*, Wiley-Interscience, New York, 2002.

41. E. Pehlivanoglu-Mantas and D. L. Sedlak, *Crit. Rev. Environ. Sci. Technol.*, 2006, **36**, 261.
42. S. W. Krasner, P. Westerhoff, B. Y. Chen, B. E. Rittmann, S. N. Nam and G. Amy, *Environ. Sci. Technol.*, 2009, **43**, 2911.
43. E. Cosa, Final Year Project Report, Curtin University, 2010.
44. E. Pehlivanoglu-Mantas and D. L. Sedlak, *Water Res.*, 2008, **42**, 3890.
45. P. Andrzejewski, B. Kasprzyk-Hordern and J. Nawrocki, *Water Res.*, 2008, **42**, 863.
46. C. Lee, C. Schmidt, J. Yoon and U. Von Gunten, *Environ. Sci. Technol.*, 2007, **41**, 2056.
47. U. Von Gunten, *Water Sci. Technol.*, 2007, **55**, 25.
48. U. Von Gunten, *Water Res.*, 2003, **37**, 1469.
49. M. Sgroi, P. Roccaro, G. L. Oelker and S. A. Snyder, *Environ. Sci. Technol.*, 2014, **48**, 10308.
50. W. A. Mitch, G. L. Oelker, E. L. Hawley, R. A. Deeb and D. L. Sedlak, *Environ. Eng. Sci.*, 2005, **22**, 882.
51. I. M. Schreiber and W. A. Mitch, *Environ. Sci. Technol.*, 2005, **39**, 3811.
52. *Australian Guidelines for Water Recycling: Managing Health and Environmental Risks. Augmentation of Drinking Water Supplies*, Environment Protection and Heritage Council, National Health and Medical Research Council, Natural Resource Management Ministerial Council, Canberra, 2008.
53. *Groundwater Replenishment Trial Final Report* ISBN 1 74043 829 9, Water Corporation of Western Australia, Perth, Western Australia, 2013.
54. K. L. Linge, D. Liew, C. Toulliou, S. Allard, A. Heitz and C. A. Joll, in preparation.

CHARACTERISATION OF ORGANIC NITROGEN IN ALGAL ORGANIC MATTER AND FORMATION OF N-DBPS AFTER CHLORINATION

Xiang Li [1], Cynthia Joll [2], Kathryn Linge [2], Stuart J. Khan [1] and Rita K. Henderson [3]

1. UNSW Water Research Centre, School of Civil and Environmental Engineering, The University of New South Wales, Sydney, NSW, Australia. E-mail: xiang.li@unsw.edu.au
2. Curtin Water Quality Research Centre, Curtin University, Perth, WA, Australia.
3. The bioMASS Lab, School of Chemical Engineering, The University of New South Wales, Sydney, NSW, Australia.

1 INTRODUCTION

Algae are ubiquitous in surface drinking water sources, including rivers, lakes and reservoirs. When an algal bloom occurs, both the algal cells and the associated algogenic organic matter (AOM) can significantly increase in concentration.[1] AOM is derived from algal metabolic excretions during growth, or release on cell lysis.[1-3] In natural waters, AOM is the major source of dissolved organic nitrogen (DON), which has been identified as an important N-DBP precursor during drinking water disinfection.[4-6]
Studies investigating the formation of DBPs from algae upon chlorination have typically focussed on carbonaceous (C-) DBPs, such as trihalomethanes and haloacetic acids.[2, 7] However, there has been a concern that emerging N-DBPs are more toxic than regulated C-DBPs,[8] prompting research focussing on N-DBP formation. To date, most studies have been limited to identifying only a few select N-DBPs, such as dichloroacetonitrile (DCAN) and trichloronitromethane (TCNM), from chlorination of cell suspensions. However, algal cells are normally removed prior to disinfection and the relationship between N-DBP formation and AOM concentration and character has not been widely studied.
While a wide variety of organic compounds have been identified as precursors for N-DBPs,[9, 10] most investigations have only focused on low molecular weight (LMW) substances,[3, 11] such as free amino acids and free amino sugars, due to analytical limitations.[12] Research focussing on N-DBP formation from HMW AOM is limited. However, in river water, 70% of nitrogen in natural organic matter (NOM) is found with high molecular weight (HMW, >1 kDa) substances, including proteins and polysaccharides.[13, 14] The LMW (70 to 1 kDa) substances only represent a small fraction. For example, the free amino acid concentration in river water was reported to be only 10% of that of dissolved combined amino acids.[15]
The aims of this study were therefore 1) to characterise organic nitrogen contained within AOM exuded from the green algae *Chlorella vulgaris* (*C. vulgaris*) and the cyanobacteria *Microcystis aeruginosa (M. aeruginosa) and Anabaena circinalis (A. circinalis)* using an advanced technique that facilitates the evaluation of the HMW fraction, and 2) to investigate the N-DBP formation potential of fractionated AOM after chlorination. Seven N-DBPs in

three different classes were analysed, including three haloacetonitriles (HANs), two halonitromethanes (HNMs) and two haloacetamides (HAAms).

2 MATERIALS AND METHODS

2.1 Cultivation of algae

Cultures of axenic algal species were obtained from the Australian National Algae Culture Collection, CSIRO (Hobart, Tasmania). Algae were sub-cultured in 250 mL flasks containing 100 mL sterilised Jaworski or MLA media.[1, 16] Flasks were illuminated with 600 ± 60 μmol $m^{-2}s^{-1}$, using an auto light/dark cycle of 16 h/8 h in an incubator at 21 °C (LABEC temperature cycling chamber, Australia). Daily mixing by hand was performed. All samples were grown in batches without replenishment of growth media. The evaporation rate during the growth period was recorded for correction of concentrations. Daily observations by microscope were undertaken to ensure the absence of contamination and to count cell population, which was determined using a light microscope (Leica ICC 50, Germany) and haemocytometer or Sedgewick Rafter cells. Cell populations were measured by counting at least 100 cells in triplicate.

2.2 Extraction and preparation of AOM

AOM was extracted from algae at the early stationary phase of growth, corresponding to day 20 under the laboratory conditions. Cell numbers were counted immediately before the extraction procedure. Algae suspensions (40 mL) were transferred into 50 mL centrifuge tubes and centrifuged (Allegra® X-12R, USA) at 10 000 g and 21°C for 15 min. The supernatant liquid was separated and filtered through a 1.2 μm glass microfiber filter (Whatman®). The deposited cell pellets were washed with modified growth media containing no nitrogen by adding the modified media to the 40 mL mark and then recentrifuging after mixing on a vortex mixer. The two supernatant liquids were then combined to provide the AOM sample, which was subsequently filtered through a 0.45 μm glass Whatman® microfiber filter. All measurements of AOM characterisation were undertaken within 24 hours of extraction. The AOM samples were diluted at pH 7 using phosphate buffer solution (pH 7, 13 mM) into suitable concentrations for further measurement and chlorination.

2.3 TOC/TN and DIN measurement

Dissolved concentrations of total organic carbon (TOC) and total nitrogen (TN) were determined using a TOC-VCSH & TNM-1 analyser (Shimadzu, Japan). TOC was determined via the non-purgeable organic carbon (NPOC) method. Concentrations of dissolved inorganic nitrogen (DIN) (nitrate, nitrite and ammonia) in AOM samples were measured via flow injection analysis (LACHAT® QuickChem 8500, USA). The difference between the TN and DIN concentrations was considered to be the concentration of DON.

2.4 LC-OCD-OND chromatography and ultrafiltration

A size exclusion liquid chromatography system with organic carbon and, uniquely, organic nitrogen detection (LC-OCD-OND, DOC Labor, Germany) was used to fractionate AOM samples into six major size fractions (biopolymers, humic-like, building blocks, low molecular weight organic acids and neutrals, hydrophobic organic carbon) using a Toyopearl TSK HW-50S column.[17]
Ultrafiltration was used to determine the size distribution of molecular weight in AOM extracted from different species.[1, 18] The AOM samples extracted from the three algal samples were fractionated into three portions of >50k, 1-50k and <1kDa by filtering the AOM solutions (<0.45 μm) through 50k, and then 1k, polyethersulfone ultrafiltration membranes (Sartorius stedim, Germany) in the ultrafiltration stirred cell using nitrogen gas at a constant pressure of 1 bar. The fractionated AOM samples were measured again using LC-OCD-OND. The associated chromatograms were used to determine the approximate retention time regarding the specific molecular size (50 k and 1 k Da, Figure 1). In total, five AOM fractions (bulk, 'dissolved', >50k, 1-50k, <1k Da) were prepared for the subsequent disinfection experiment.

2.5 Chlorination experiment

All the chemical solutions were prepared from analytical grade chemicals and ultrapure laboratory water. A free chlorine stock solution was prepared in an amber glass bottle from fresh sodium hypochlorite solution (Sigma-Aldrich, Australia). The free chlorine concentration was measured daily using a HACH Colorimeter (HACH). TOC concentrations of all five AOM fractions were measured and diluted to 2 mg L^{-1}. Disinfection was applied to each AOM fraction at a ratio of free chlorine and TOC to 8 mg L^{-1} : 2 mg L^{-1} alongside a phosphate buffer solution (pH 7, 13 mM), which ensured the pH was maintained during chlorination. The initial dilution of AOM samples was conducted in 1 L volumetric flasks in the dark. Immediately after adding chlorine, replicate samples were dispensed into 40 mL amber glass vials with PTFE lined/silicone septum screw caps for separate analysis of HANs, HNMs and HAAms. Head-space free conditions were employed. Chlorination experiments were conducted at a temperature of 21 °C for 72 hours in the dark.

2.8 Extraction of N-DBPs and detection by GC-MS

Seven N-DBPs in three different classes were analysed, including three HANs (monochloroacetonitrile, dichloroacetonitrile, trichloroacetonitrile), two HNMs (dichloronitromethane, trichloronitromethane) and two HAAms (chloroacetamide, trichloroacetamide). Chlorinated samples in triplicate were quenched with ascorbic acid (50 mg L^{-1}) for each analytical method. For HAN analysis, a headspace solid-phase microextraction (SPME) method with GC-MS was applied [19]. For HNMs and HAAms, a simplified liquid-liquid extraction (LLE) method was utilized, followed by simultaneous GC-MS detection [20].

3 RESULTS AND DISCUSSION

3.1 Total organic carbon and nitrogen characterisation of AOM

Water quality parameters of the AOM in the stationary growth phase are shown in Table 1. For TOC, concentrations increased rapidly during the early growth phase and continued to increase at a slower rate during the stationary phase. For example, the TOC of *C. vulgaris* increased from 2.9 mg L^{-1} in Jaworski media to 19.51 mg L^{-1} at day 25. The changes in TOC for *C.vulgaris* in this study are comparable with increases (18 and 20 mg L^{-1})observed during the same growth period by Henderson, et al. [1] and Hoyer, et al. [21], respectively. In contrast to TOC, TN initially decreased due to algal consumption of DIN, specifically nitrate, from the growth media. After stabilisation of the DIN concentrations during the early stationary phase (days 15-20), TN concentrations began to increase, although DIN concentrations remained steady. Therefore, the increases of both TOC and TN during the stationary phase were attributed to newly released AOM. Furthermore, constant TN/TOC ratios during the stationary phase (days 15-30) were observed for all species, suggesting that the exuded AOM probably comprised a constant molecular composition at this stage. Interestingly, *A. circinalis* had the highest TN/TOC ratio (0.13 ± 0.01 mg L^{-1}), which is almost twice the median or average DON/DOC ratios (~0.067) previously found in drinking source waters during algal blooms.[5, 22] The ratios of TN/TOC in *C. vulgaris* and *M. aeruginosa* were much more comparable to ratios measured in algal bloom-impacted source waters.[5, 22]

3.2 LC-OCD-OND analysis and MW distribution of AOM using ultrafiltration

LC-OCD-OND chromatograms of AOM from the three test species at day 20 (<0.45 μm) are shown in Figure 1. For all algal species, the main DON components were the biopolymer and building block fractions, which are likely to result from algal carbohydrates and proteins.[1] To note, the biopolymer fraction was the dominant component in all algal AOM, particularly in *C.vulgaris*. The biopolymer fraction of AOM (> 50kDa) of *C.vulgaris* in this study occupied 58% of total C (data not shown). The biopolymer proportion of C content of *C.vulgaris* AOM (> 30 kDa) was previously reported of 62%.[1] The OND/OCD ratios of fractional and bulk AOM for each algal species are shown in Table 2. The OND/OCD values calculated from the entire chromatograms for each species match the TN/TOC ratios reported in Table 1 very well. Therefore, the increasing TN concentrations of all algal species during the stationary phase (Table 1) can be attributed to increasing concentrations for these two DON fractions. However, there were variations in fractional DON/DOC ratios for all species (Table 2). It is clear that both cyanobacteria species had much larger DON/DOC ratios in their biopolymer fractions than *C. vulgaris*. However in the building blocks fraction around 1 kDa, *M. aeruginosa* had the lowest OND/OCD ratio (0.13), while *C. vulgaris* had the highest OND/OCD ratio (0.35).

Table 1 *Water quality parameters of the AOM extract solutions from stationary growth phases at 21 °C, ±STDEV, n≥3 except for DIN measurement (n=2), n.a.=not analysed, Lod=limits of detection.*

Media+ species	Growth phases	Cell Conc. (x10^7 cells mL^{-1})	pH	TOC (mg L^{-1})	TN (mg L^{-1})	TN/TOC	DIN (mean, mg L^{-1}), n=2		
							ammonia	Nitrite	nitrate
				Green algae in Jaworski media					
media	Jaworski (d0)	0.00	7.38	2.9 ±0.12	15.9 ±0.11	5.46 ±0.19	0.67	0.06	16.52
C.vulgaris	Early Stat. (d15)	2.82 ±0.11	10.01 ±0.17	17.79 ±1.07	1.31 ±0.11	0.07 ±0.01	0.33	0.01	0.00
	Middle Stat. (d20)	2.88 ±0.51	9.83 ±0.47	19.51 ±1.46	1.20 ±0.20	0.06 ±0.01	0.20	0.02	0.01
	Late Stat. (d25)	2.80 ±0.26	9.55 ±0.12	24.17 ±1.95	1.48 ±0.18	0.06 ±0.01	n.a.	n.a.	n.a.
				Cyanobacteria in MLA media					
media	MLA (d0)	0.00	7.6-7.8	4.43 ±0.29	30.51 ±1.33	7.61 ±1.34	<Lod	<Lod	28.45
M.aeruginosa	Early Stat. (d20)	6.35 ±0.15	9.07 ±0.15	49.36 ±4.99	3.43 ±0.81	0.07 ±0.02	<Lod	<Lod	0.47
	Middle Stat. (d25)	6.34 ±0.09	8.77 ±0.67	74.00 ±7.50	4.70 ±0.90	0.06 ±0.01	<Lod	<Lod	0.40
	Late Stat. (d30)	6.35 ±0.10	8.29 ±0.27	97.06 ±12.20	5.41 ±1.53	0.06 ±0.02	<Lod	<Lod	0.20
A.circinalis	Early Stat. (d20)	(5.08 ±0.42)x10^{-2}	9.53 ±0.26	25.90 ±2.37	3.34 ±0.50	0.13 ±0.02	n.a.	n.a.	n.a.
	Middle Stat. (d25)	(5.69 ±0.15)x10^{-2}	9.81 ±0.12	39.01 ±0.06	4.52 ±0.25	0.12 ±0.01	n.a.	n.a.	n.a.
	Late Stat. (d30)	(4.78 ±0.32)x10^{-2}	10.03 ±0.26	45.05 ±1.61	5.66 ±0.40	0.13 ±0.01	n.a.	n.a.	n.a.

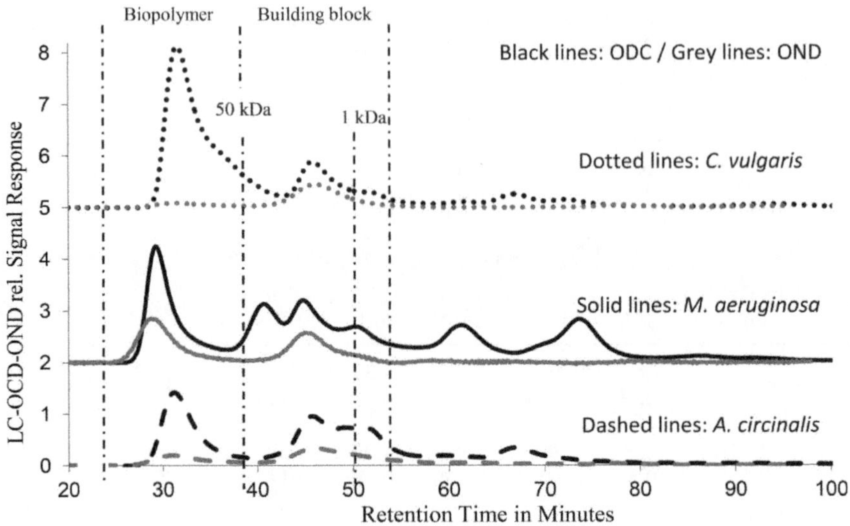

Figure 1 *LC-OCD-OND signal responses and MW distribution of AOM samples extracted during the stationary phases (d20). Molecular size of 50 kDa at ~38 min and 1 kDa at ~51 min were determined by analysing the chromatograms of fractional AOM after ultrafiltration.*

Ultrafiltration experiments separated the bulk AOM (<0.45 μm) into a further three fractions: >50 kDa, 1-50 kDa and <1 kDa. LC-OCD-OND chromatograms of these AOM fractions with known size ranges allowed the associated retention time for positions of 50 k and 1 kDa to be identified (Figure 1). It is obvious that the DON components of the extracts from the three species were mainly located in fractions of >50 kDa and 1 k-50 kDa. In comparison, the AOM extract from *M. aeruginosa* had very high DON content in both fractions of >50 kDa and 1-50 kDa. In addition, the OCD/OND signals of the AOM extract from *M. aeruginosa* also had the shortest retention times. This implies the DON components of the AOM extract from *M. aeruginosa* were of higher MW than those of the other two species.

Table 2 *Comparison of OND/OCD ratios of the AOM extracts from each species at stationary growth phase (day 20), derived from the MW distribution of N and C during LC-OCD-OND analysis at 21 °C,± STDEV, n≥3.*

	Biopolymer fraction	Building blocks fraction	Entire chromatograms
C.vulgaris	0.02 ±0.00	0.35 ±0.04	0.07 ±0.00
M.aeruginosa	0.10 ±0.01	0.13 ±0.02	0.08 ±0.00
A.circinalis	0.08 ±0.01	0.26 ±0.01	0.13 ±0.01

3.3 Formation of N-DBPs from chlorination of AOM fractions

HANs were the only N-DBP class detected from chlorination of all the AOM fractions with the initial Cl_2 and TOC concentration of 8 mg L^{-1} and 2 mg L^{-1}, respectively. The chlorinated HNMs and HAAms were not detected. The formation of different HAN species was found to vary depending on the algal species and their DON content in the different size fractions. Both *C.vulgaris* and *A.circinalis* only formed dichloroacetonitrile (DCAN), while *M.aeruginosa* produced both DCAN and monochloroacetonitrile (MCAN). While trichloroacetonitrile (TCAN) was not detected, it is noted that the detection limit of TCAN in this study (0.3 μg L^{-1}) is similar to reported concentrations of TCAN formation from NOM.[3]

A comparison of DCAN formation from *C. vulgaris* and *A. circinalis* is shown in Figure 2. *A. circinalis* had greater DCAN formation than *C. vulgaris* for all size fractions. However, the fractional N/C ratio was always higher in *C. vulgaris* except for the 1-50 kDa size fraction, where the N/C ratios were the same for both algae. This suggests that the N/C ratio might not correlate well with DCAN formation. The maximum yields of DCAN from the bulk fractions of <1.2 μm, and the 'dissolved' fraction of <0.45 μm of *C.vulgaris* and *A.circinalis* were about 1.6 and 2.4 μg L^{-1}, respectively. However, there was little difference in the DCAN formation that was observed between these two fractions, with respect to each species. This similarity suggests that the algal precursor for DCAN can be considered dissolved and more likely to persist in water sources. In addition, the 1-50 kDa size fraction formed more DCAN than the >50 kDa size fraction for both species, despite varying N/C ratios. Again, this suggests that the N/C ratio does not well reflect potential DCAN formation. In particular, the highest DCAN formation (2.4 μg L^{-1}) was produced by *A. circinalis* in this size fraction. Finally, it is clear that HMW AOM can produce more DCAN than LMW AOM. The total DCAN concentrations from AOM > 1 kDa were 1.35 and 2.93 μg L^{-1} for *C.vulgaris* and *A.circinalis*, respectively. The total DCAN concentrations from AOM < 1 kDa were 1.31 and 1.58 μg L^{-1}. For the bulk AOM (<0.45 μm) from *C. vulgaris* and *A. circinalis* in this study, the mean DCAN concentrations were 1.59 and 2.02 μg L^{-1}, respectively.

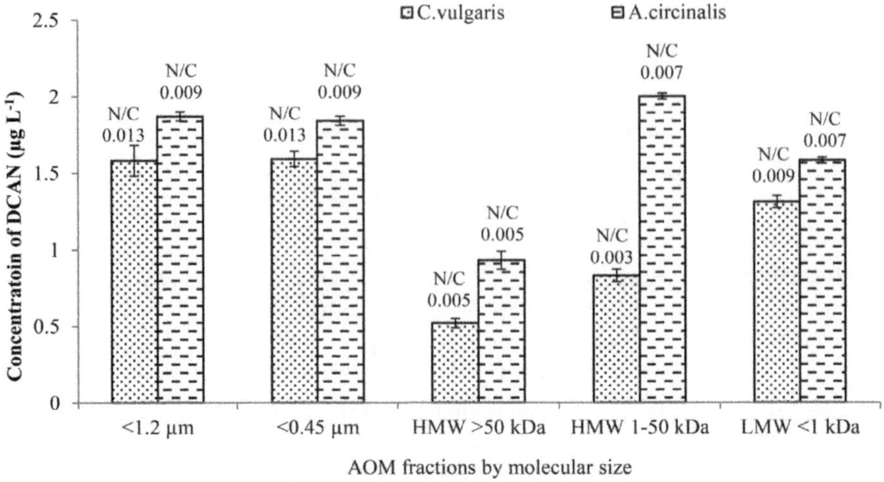

Figure 2 *Comparison of DCAN formed from various AOM fractions from day 20. Fractional N/C ratio was based on TOC/TN analysis.*

4 CONCLUSIONS

The AOM exuded from green algae (*C.vulgaris*) and cyanobacteria (*M.aeruginosa* and *A.circinalis*) was characterised using a combination of conventional TOC/TN analysis, ultrafiltration and advanced techniques (LC-OCD-OND). DON was found predominantly in the HMW size fractions. Chlorination experiments using fractionated AOM indicated that HANs were the only measured class of N-DBPs to form. All three species tested formed DCAN, while *M.aeruginosa* also formed MCAN. When initial TOC concentrations of all the AOM fractions were constant at 2 mg L^{-1}, the normalized maximum yields of HANs from the bulk or dissolved AOM extracts of *C.vulgaris, M.aeruginosa* and *A.circinalis* were about 0.8, 0.8 and 1.2 µg mg^{-1} as C, respectively. Cyanobacteria investigated in this study, whose AOM extracts contain larger fractions of HMW DON, can form higher concentrations of DCAN than the AOM extracts from green algae (*C. vulgaris*). More importantly, for all species tested with the same TOC concentration (2 mg L^{-1}), the formation of DCAN from HMW AOM was greater than that from LMW AOM. One interesting observation in this study was that the N/C ratio did not correlate well with DCAN formation. In summary, the difference of algal species and the DON content in the HMW fraction of AOM could be a dominant factor controlling DCAN formation in this study.

5 ACKNOWLEDGEMENTS

This research was supported under the Australian Research Council Linkage Project funding scheme (LP110100548), including support from Water Research Australia and the Water Corporation of Western Australia. Xiang Li is also supported by an Australian Postgraduate Award (APA) and a UNSW Engineering Research Excellence Award (EREA). The authors also thank Deborah Liew at Curtin University for analysis of HANs.

References

1. R. K. Henderson, A. Baker, S. A. Parsons and B. Jefferson, *Water Research*, 2008, **42**, 3435-3445.
2. L. Li, N. Gao, Y. Deng, J. Yao and K. Zhang, *Water Research*, 2012, **46**, 1233-1240.
3. J. Fang, X. Yang, J. Ma, C. Shang and Q. Zhao, *Water Research*, 2010_b, **44**, 5897-5906.
4. T. Berman and D. A. Bronk, *Aquatic Microbial Ecology*, 2003, **31**, 279-305.
5. P. Westerhoff and H. Mash, *Journal of Water Supply Research and Technology-Aqua*, 2002, **51**, 415-448.
6. S. W. Krasner, M. J. Sclimenti, W. Mitch, P. Westerhoff and A. Dotson, 2007.
7. J. Huang, N. Graham, M. R. Templeton, Y. Zhang, C. Collins and M. Nieuwenhuijsen, *Water Research*, 2009, **43**, 3009-3018.
8. S. D. Richardson, M. J. Plewa, E. D. Wagner, R. Schoeny and D. M. Demarini, *Mutation research*, 2007, **636**, 178-242.
9. W. H. Chu, N. Y. Gao, M. R. Templeton and D. Q. Yin, *Chemosphere*, 2011, **83**, 647-651.
10. W. Lee, P. Westerhoff and M. Esparza-Soto, *Journal American Water Works Association*, 2006, **98**, 102-110.
11. R. M. W. Amon and R. Benner, *Limnology and Oceanography*, 1996, **41**, 41-51.
12. A. D. Shah and W. A. Mitch, *Environ Sci Technol*, 2012, **46**, 119-131.
13. L. Guo, J. K. Lehner, D. M. White and D. S. Garland, *Water Research*, 2003, **37**, 1015-1022.
14. A. Dotson and P. Westerhoff, *Journal American Water Works Association*, 2009, **101**, 101.
15. W. A. Mitch, S. W. Krasner, P. Westerhoff and A. Dotson, *Occurrence and Formation of Nitrogenous Disinfection By-Products.*, Water Research Foundation, Denver, CO, USA, 2009.
16. C. J. S. Bolch, S. I. Blackburn, B. A. Neilan and P. M. Grewe, *Journal of Phycology*, 1996, **32**, 445-451.
17. S. A. Huber, A. Balz, M. Abert and W. Pronk, *Water Res*, 2011, **45**, 879-885.
18. Y. S. Lui, H. C. Hong, G. J. S. Zheng and Y. Liang, *Journal of Hazardous Materials*, 2012, **209-210**, 278-284.
19. I. Kristiana, C. Joll and A. Heitz, *Journal of Chromatography A*, 2012, **1225**, 45-54.
20. D. Liew, K. L. Linge, C. A. Joll, A. Heitz and J. W. A. Charrois, *Journal of Chromatography A*, 2012, **1241**, 117-122.
21. O. Hoyer, B. lusse and H. Bernhardt, *Z. Wasser. Abwas. For.*, 1985, **18**, 76-90.
22. B. Xu, T. Ye, D.-P. Li, C.-Y. Hu, Y.-L. Lin, S.-J. Xia, F.-X. Tian and N.-Y. Gao, *Science of The Total Environment*, 2011, **409**, 1116-1122.

OCURRENCE OF POLYCYCLIC AROMATIC HYDROCARBONS (PAHs) AND THEIR CHLORINATION BY-PRODUCTS IN DRINKING WATER AND THE COATINGS OF WATER PIPES

J.Tillner, C.Bach, C. Hollard, C. Rosin, J.F. Munoz, X. Dauchy*

ANSES, Nancy Laboratory for Hydrology, Water Chemistry Department, 40 rue Lionnois, 54000 Nancy, FRANCE
*Corresponding author: xavier.dauchy@anses.fr

1 INTRODUCTION

Polycyclic aromatic hydrocarbons (PAHs) are a group of aromatic organic compounds that can be generated during the incomplete combustion of organic materials but also they can occur naturally in products derived from fossil fuels such as coal tar or asphalt. Several PAHs have shown genotoxicity *in vitro* and are suspected to be endocrine disrupting chemicals[1]. In France, the quality limits for some PAHs in drinking water are set at a total of 0.1 μ/L for benzo[b]fluoranthene, benzo[k]fluoranthene, benzo[g,h,i]perylene, and benzo[1,2,3-c,d]pyrene[2] and at 0.01 μ/L for benzo[a]pyrene, according to the European Council Directive 98/83/EC on the quality of water intended for human consumption[3]. The presence of PAHs in surface waters and water reservoirs is frequently an indication of an industrial waste contamination. In contrast, PAHs detected in drinking water rarely originate from the raw water source but from the water distribution systems[4]. In Europe until the 1970s, steel and cast iron water pipes were commonly lined with coal tar pitch to prevent corrosion. Although coal tar was then replaced by other materials, many coal tar lined systems are still in use today. Coal tars contain PAHs which have been shown to migrate into drinking water[5].

Water chlorination reduces PAHs levels but can also form chlorinated, hydroxylated and oxidised by-products[6]. Many chlorinated organic compounds exhibit greater toxicity than their precursors due to their increased lipophilicity and thus higher capacity to permeate biological membranes[7-9]. Several studies have been performed to assess the PAHs by-products formed during water chlorination[10-19]. It has been reported that PAHs form several by-products (ketones and /or quinones) with both chlorine (hypochlorite/hypochlorous acid) and chloride dioxide disinfection agents in aqueous solutions. Many of these investigations, however, were not conducted at realistic conditions of the water treatment process, working at pH levels below or above those of drinking water and applying relatively higher PAHs concentrations and/or desinfection agents.

The aim of this study was to assess the potential exposure to PAHs and their by-products due to drinking water. Two sampling campaigns were conducted in French communities with a drinking water distribution network known for recurrently showing substantial PAHs concentrations. Seventeen PAHs and 11 potential chlorinated by-products were analysed in drinking water samples from two distribution networks using different disinfection agents (ClO_2 and chlorine). The analysis of coal tar-based pipe coatings of two pipe sections was also performed. The analytical method was based on solid-phase microextraction-gas chromatography-mass spectrometry (SPME-GC-MS^n). The limit of quantification was 10ng/L for all the target compounds.

2 METHOD AND RESULTS

2.1 Reagents and standards

A solution of 17 PAHs, all at 1 g/L in dichloromethane, was obtained from Restek (Lisses, France). As not all the PAH chlorinated products identified in the literature were commercially available, standards were selected so as to best cover the range of substances targeted, regarding both hydrophobicity and the number of aromatic rings (see Table 1). All chlorinated PAH were purchased from Chiron AS (Trondheim, Norway) either as pure standards (99.8%) or as solutions in isooctane at different concentrations (50 mg/L, 1 g/L). A mixed solution of 6 deuterated standards (1,4-dichlorobenzene-d_4, naphthalene-d_8, acenaphthene-d_{10}, phenanthrene-d_{10}, chrysene-d_{12}, perylene-d_{12}) all at 2 g/L in dichloromethane was obtained from Restek (Lisses, France). Intermediate stock solutions were prepared at 2 and 5 mg/L in ethyl acetate (EtOAc) (Pestipur® quality) and stored in amber vials at 4 °C.

2.2 Water and pipe samples

A sampling campaign was conducted in January 2013 in two drinking water distribution networks known for showing not negligible PAH concentrations. Four samples were obtained from one (water distribution network B) and 8 from the other (water distribution network C). The samples from the distribution network C were collected from 3 different sections (C1, C2, and C3). The inner coating of two pieces of piping from the section C1 was also analysed.
The water samples were collected in amber glass vials with sodium thiosulfate in order to remove chlorine. Water samples were analysed within 6 days of sampling.

2.3 Solid-phase microextraction (SPME) procedure

SPME was performed with a 85 µm PA fibre (Supelco, Bellefonte, USA) following a protocol previously reported for water and pipes samples[20].

2.3.1 Water Samples

Briefly, 90 µL of a solution of 6 deutered standards (see Section 2.1) at 2 µg/L were added to 18 mL of water samples spiked with 0.5% of EtOAc. The fibre was then immersed in the liquid phase (Figure 1A). The extraction was carried out at 60 °C for 20 min. under continuous agitation set at 300 rpm. When the extraction was completed, the fibre was immediately desorbed for 7 min in the inlet in the pulsed splitless mode (37.5 psi, 7 min.) and temperature programmed as follows: 140-290 °C at 600 °C/min and 290 °C held for 13 min.

2.3.2 Pipes Samples

The top most layer of the interior coatings of the pipes obtained from the section C1 was scraped with an inox spatula. Ten milligrams of the powder was extracted with 10 mL of EtOAc for 48 h without agitation at room temperature. Ten microliters of the extract were diluted in 10 mL of EtOAc. An aliquot (usually 90 μL) of the diluted solution was used to spike 18 mL of ultra-pure water, which were treated like the water samples (Figure 1B) (see Section 2.3.1).

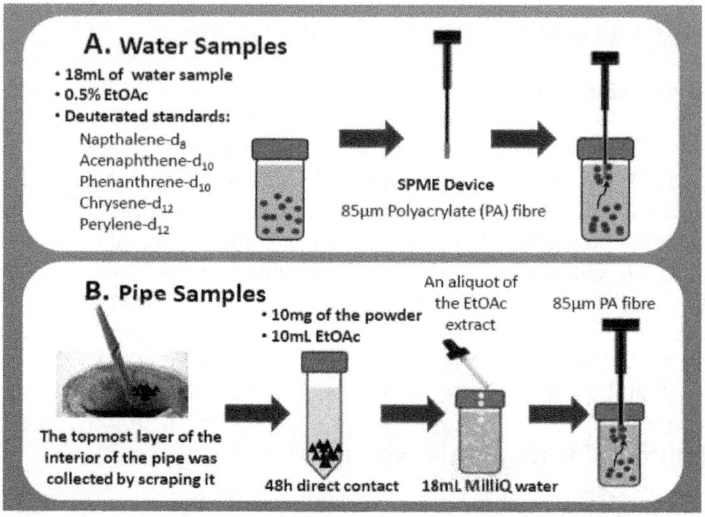

Figure 1 *Extraction procedure for drinking water (A) and pipe samples (B)*

2.4 GC-MS^n analysis

The analyses were performed with an Agilent Technologies 7890A gas chromatograph (GC) coupled to an Agilent Technologies ion trap 240 mass spectrometer (MS) (Palo alto, CA, USA). SPME extractions were automated by using a GC autosampler 120 from Agilent Technologies. Helium was the carrier gas set at a flow rate of 1.3 mL/min. Analytes were separated using a Rxi-5SilMS column (30 m × 0.25 mm; 0.25 μm film thickness) purchased from Restek (Lisses, France). The initial column temperature was 50°C and was programmed at 30 °C/min to 240 °C, then at 10 °C/min to 320 °C and held for 1.7 min. MS was operated in external electron impact. The source, ion trap, transfer line and manifold temperatures were 250 °C, 200 °C, 320 °C and 50 °C respectively. Optimised MS^n conditions for all target compounds are listed in Table 1.

2.5 Performances of the analytical method

The optimisation of analytical conditions and the performances of the method were carried out in a previous publication[20].

Blank samples (18 mL of ultra-pure water + 0.5% EtOAc) were carried out at an interval of every two samples to check for potential contamination and/or carryover.

The limit of quantification for an 18 mL water sample was 10 ng/L for each target compounds. The six-point standard calibration was conducted before each set of analyses. The recoveries in different tap water samples spiked at a concentration of 25 ng/L were between 82-111%.

Table 1 *Mass spectrometry conditions for the PAHs and their chlorination by-products.* [a] *Collision-induced dissociation (CID) voltage*

Compounds (Abbreviations)	Parent ion (m/z)	CID [a] voltage (V)	Quantifiers (m/z)	Qualifiers (m/z)
Naphthalene (Nap)	128	0	128	-
2 Methylnaphthalene (Me-Nap)	142	1.10	141	115
Acenaphthylene (Acy)	152	1.10	152	151
Acenaphthene (Ace)	153	1.10	152	151
1,5-Dichloronaphthalene (Cl-Nap)	196	0.90	196	161
Fluorene (Flu)	165	0.90	165	164
9,10-Dihydroanthracene	179	1.10	178	165
Fluorenone	180	0.42	180	152
Phenanthrene (Phe)	178	0.60	178	-
Anthracene (Anthr)	178	0.60	178	-
Anthraquinone	208>180	0.42	152	151
2-Chloroanthracene (Cl-Anthr)	212	1.50	177	212,176,178
Cyclopenta[d,e,f]phenanthrenone	205	1.25	205	176
Fluoranthene (Fla)	202	1.10	202	-
Pyrene (Pyr)	202	1.10	202	-
1-Chloroanthraquinone	243	55	214	208,186
3-Chlorofluoranthene (Cl-Fla)	236	1.60	236	201
1-Chloropyrene (Cl-Pyr)	236	1.60	236	201
1,5-Dichloroanthraquinone	277	85	222+220+150	187,186,184,185
Chrysene (Chry)	228	1.10	228	-
Benzo[a]anthracene (BaA)	228	1.10	228	-
Benzo[b]fluoranthene (BbP)	252	0.60	252	-
Benzo[k]fluoranthene (BkP)	252	0.60	252	-
Benzo[a]pyrene (BaP)	252	0.60	252	-
6-Chlorobenzo[a]pyrene (Cl-BaP)	286	1.90	251	286
Indeno[c,d]pyrene (IcdP)	276	1.10	276	-
Dibenz[a,h]anthracene (DBahA)	278	1.10	278	-
Benzo[g,h,i]perylene (BghiP)	276	1.10	276	-

2.6 Interpretation of results for water distribution network B and C

In drinking water from network B, fluoranthene, phenanthrene, and benzo[a]pyrene were detected in all the samples (Table 2). Concentrations of PAHs were often below to 100 ng/L, except for fluoranthene and phenanthrene, which were found between 250 - 480 ng/L in water samples 1B, 2B, and 3B. With regards to the 11 chlorination products targeted, anthraquinone, chlorofluoranthene, and were detected in all the samples from network B. Chloropyrene and fluorenone were observed in 3 of the samples. The individual concentrations of these by-products were not exceeding 100 ng/L. This is the first time that chloropyrene and cyclopenta[d,e,f]phenanthrenone were detected in drinking water. This last compound is presumably an oxidation product of cyclopenta[d,e,f]phenanthrene which can be found in coal tar[21], but which was not targeted in this campaign.
In drinking water from network C, phenanthrene was detected in 6 samples in a concentration ranging from 14 to 51 ng/L. Naphthalene and fluoranthene in 3 samples, fluorenone in 2 samples, and acenaphtene and pyrene in only 1 sample. Of the 11 chlorination by-products, only anthraquinone, and fluorenone were found in 3 samples from the section C2 (4C2, 5C2, and 8C2). The highest individual concentrations observed were 21 ng/L for fluorenone and 17 ng/L for anthraquinone (Table 2).

Table 2 *Concentrations (ng/L) of PAHs and their chlorination by-products (*) in drinking-water networks B and C (2 Methyl-Naphthalene, Acenaphthylene, 1,5-Dichloronaphthalene, 9,10-Dihydroanthracene, 2-Chloroanthracene, 1-Chloroanthraquinone, 1,5-Dichloroanthraquinone, Chrysene, Benzo[b]fluoranthene, Benzo[k]fluoranthene, Benzo[a]pyrene, 6-Chlorobenzo[a]pyrene, Indeno[cd]pyrene, Dibenzo[ah]anthracene and Benzo[ghi]perylene were not detected in any samples, and therefore were not included in the table).*

	Drinking water network B				Drinking water network C							
Sample code	1B	2B	3B	4B	1C1	2C1	3C1	4C2	5C2	8C2	6C3	7C3
Used disinfectant	ClO_2				Chlorine			ClO_2			ClO_2	
Naphthalene	<10	<10	<10	<10	<10	<10	<10	41	30	38	<10	<10
Acenaphthene	<10	<10	<10	<10	<10	<10	<10	12	<10	<10	<10	<10
Fluorene	23	20	23	<10	<10	<10	<10	13	<10	14	<10	<10
Fluorenone*	40	41	40	<10	<10	<10	<10	21	11	19	<10	<10
Phenanthrene	340	250	480	18	<10	51	29	49	33	48	14	18
Anthracene	31	32	32	<10	<10	<10	<10	<10	<10	<10	<10	<10
Anthraquinone *	55	52	40	18	<10	<10	<10	17	15	17	<10	<10
Cyclopenta[def]phenanthrenone*	46	52	43	22	<10	<10	<10	<10	<10	<10	<10	<10
Fluoranthene	295	255	275	16	<10	23	<10	13	<10	14	<10	<10
Pyrene	20	28	21	<10	<10	10	<10	<10	<10	<10	<10	<10
3-Chlorofluoranthene*	34	35	34	14	<10	<10	<10	<10	<10	<10	<10	<10
1-Chloropyrene*	15	15	13	<10	<10	<10	<10	<10	<10	<10	<10	<10
Benzo[a]anthracene	39	45	39	18	<10	<10	<10	<10	<10	<10	<10	<10

Reproduced by kind permission of Elsevier from the Journal of Chromatography A, 2013, 1315, 36-46

A strong association between fluorene and fluorenone was observed in both networks. When fluorene was detected, even in relatively low concentrations (between 13 and 23 ng/L), fluorenone could also be found (between 11 and 49 ng/L). Only in drinking water from network B, other associations could be confirmed for pyrene and chloropyrene, anthracene and anthraquinone, and fluoranthene and chlorofluoranthene.

In both networks, the concentration of anthraquinone is above that of anthracene. The slightly higher hydrophobicity of anthracene ($logK_{ow}$= 4.55) compared to anthraquinone ($logK_{ow}$= 3.03) could be the explanation. The same observation was made for fluorene and fluorenone.

2.7 Interpretation of results of the pipe coatings from network C

As expected, all 17 PAHs targeted were detected in the pipe coatings (Figure 2). The most dominants were phenanthrene, fluoranthene and pyrene. However, the absolute concentration between the two pipes sections (C1A and C1B) differed greatly, probably due to differences in composition of the two coatings and/or an inhomogeneous sampling. Of the chlorination by-products, anthraquinone, fluorenone, cyclopenta[d,e,f]phenanthrenone, chloropyrene, and chlorofluoranthene were found in both pipe coatings from different sections (C1A and C1B).

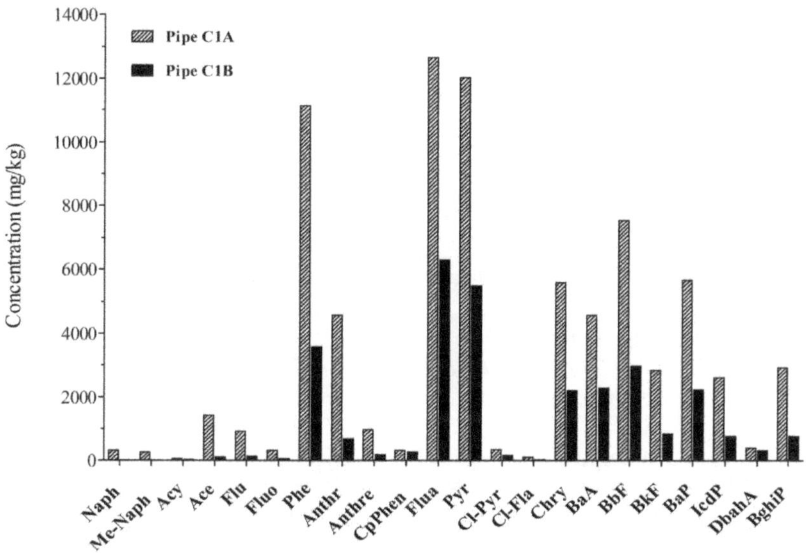

Figure 2 *Concentrations (mg/kg) of PAHs and their chlorination by-products in pipe coatings of the distribution network C (see Table 1 for the list of compounds analysed and their abbreviations)*
Reproduced by kind permission of Elsevier from the Journal of Chromatography A, 2013, 1315, 36-46

The ratios of PAHs/PAH by-products in the coatings were less than 5 for the oxygen-substituted PAHs (anthraquinone and fluorenone), whereas they ranged from 35 to 157 for chlorinated PAHs (chloropyrene and chlorofluoranthene). This observation suggests that PAHs form oxygen-substituted products more readily than chlorinated by-products with chlorine. The detection of chlorination by-products in the pipe coatings supports the hypothesis that such products are formed within the coating and not in the water after migration. Pipe coatings can therefore act as a reservoir for chlorination by-products, which slowly migrate into drinking water.

3 CONCLUSION

PAHs with less than five aromatic rings and thus with the lowest hydrophobicity were the most frequently detected in drinking water from both distribution networks. Variations in the occurrence of PAHs are probably due to differences in the composition of the coal tar linings in the two distribution networks investigated. In our experimental conditions, none of the examined distribution networks exceeded the French quality limits for PAHs in water intended for human consumption. It is worth noting that this is the first time that chloropyrene and cyclopenta[d,e,f]phenanthrene were found in drinking water and linked to the reactions of chlorine with PAHs of coal tar pipe coatings.

It is also noteworthy that coal tar can also contain other PAHs than those targeted in this study and other potentially toxicologically relevant molecules such as alkyl-substituted PAHs, nitrogen-substituted PAHs, sulphur-substituted PAHs and/or dibenzofuran. These compounds might not only migrate into water but also form disinfection by-products.

Further chemical compounds potentially present in coal tar linings than those found in this study, and their by-products should also be monitored. The investigation of further sites would provide more information about the different factors affecting the formation and the diffusion of chlorinated by-products of PAHs into drinking water.

References

1. B. f. Risikobewertung, *Markersubstanzen für polyzyklische aromatische Kohlenwasserstoffe (PAK) zur Lebensmittelüberwachung*, Bundesinstitut für Risikobewertung (BfR), 2009.
2. *Arrêté du 11 janvier 2007 relatif aux limites et références de qualité des eaux brutes et des eaux destinées à la consommation humaine mentionnées aux articles R. 1321-2, R. 1321-3, R. 1321-7 et R. 1321-38 du code de la santé publique* Ministère de la Santé et des Solidarités, France, 2007.
3. *Council Directive 98/83/EC of 3 November 1998 on the quality of water intended for human consumption*, European Council, 1998.
4. J. Kielhorn and A. Boehncke, *Guidelines for drinking-water quality, 2nd ed. Addendum to Vol. 2. Health criteria and other supporting information - Polynuclear aromatic hydrocarbons in Drinking-water - Background document for development of WHO Guidelines for Drinking-water Quality*, WHO, Geneva, 2003.
5. K. Alben, *Environ. Sci. Technol.*, 1980, **14**, 468-470.
6. R. M. Harrison, R. Perry and R. A. Wellings, *Environ. Sci. Technol.*, 1976, **10**, 1151–1156.
7. A. Hanberg, F. Waern, L. Asplund, E. Haglund and S. Safe, *Chemosphere*, 1990, **20**, 1161-1164.
8. A. Colmsjo, A. Rannug and U. Rannug, *Mutat. Res.*, 1984, **135**, 21-29.
9. A. L. Blankenship, K. Kannan, S. A. Villalobos, D. L. Villeneuve, J. Falandysz, T. Imagawa, E. Jakobsson and J. P. Giesy, *Environ. Sci. Technol.*, 2000, **34**, 3153-3158.
10. T. Merkel, M. Maier, F. Sacher and D. Maier, *Journal of Water Supply: Research and Technology - AQUA*, 1997, **46**, 289-303.
11. A. R. Oyler, D. L. Bodenner, K. J. Welch, R. J. Liukkonen, R. M. Carlson, H. L. Kopperman and R. Caple, *Anal. Chem.*, 1978, **50**, 837–842.

12. A. R. Oyler, R. J. Liukkonen, M. T. Lukasewycz, D. A. Cox, D. A. Peake and R. M. Carlson, *Environ. Health Perspect.*, 1982, **46**, 73-86.
13. R. J. Liukkonen, S. Lin, A. R. Oyler, M. T. Lukasewycz, D. A. Cox, Z.-J. Yu and R. M. Carlson, Water chlorination. Environment impact and health effects, Pacific Grove, California, 1983.
14. R. M. Carlson, R. Caple, A. R. Oyler, K. J. Welch, D. L. Bodenner and R. Liukkonen, *Water Chlorination, Environ Impact and Health Eff*, 1978, **2**, 59-65.
15. S. Johnsen, I. S. Gribbestad and S. Johansen, *Sci. Total Environ.*, 1989, **81-82**, 231-238.
16. J. C. Milano, D. Boubker and J. L. Vernet, *Environ. Technol.*, 1995, **16**, 189-195.
17. Y. Mori, S. Goto, S. Onodera, S. Naito and H. Matsushita, *Chemosphere*, 1991, **22**, 495-501.
18. S. Onodera, K. Igarashi, A. Fukuda, J. Ouchi and S. Suzuki, *J. Chromatogr.*, 1989, **466**, 233-249.
19. J. Hu, X. Jin, S. Kunikane, Y. Terao and T. Aizawa, *Environ. Sci. Technol.*, 2006, **40**, 487-493.
20. J. Tillner, C. Hollard, C. Bach, C. Rosin, J. F. Munoz and X. Dauchy, *Journal of Chromatography A*, 2013, **1315**, 36-46.
21. K. Alben, *Anal. Chem.*, 1980, **52**, 1825-1828.

IODO-THM FORMATION DURING CHLORINATION OF SOURCE WATERS WITH NATURALLY OCCURRING AMMONIUM AND HIGH SODIUM CONTENT

A.-M. Tugulea, R. Aranda-Rodriguez; M. Giddings, F. Lemieux, and J. Hnatiw

Health Canada, Ottawa, Canada

1 INTRODUCTION

Drinking water disinfection represents one of the greatest achievements of public health protection and since its advent in the 20-th century as it has virtually eliminated typhoid fever, cholera and many other waterborne diseases. Nevertheless, during water disinfection, various compounds, with known or suspected toxicity, are formed through reactions between disinfection agents and precursors present in the source water. The challenge is to minimize the formation of disinfection by-products (DBPs) without compromising disinfection efficacy. In the last 30 years, hundreds of DBPs have been identified,[1] including, more recently, the iodine-containing trihalomethanes (iodo-THMs).[2,3] The formation of iodo-THMs occurs when iodide is oxidized to hypoiodous acid (HOI) by the oxidant used in the disinfection process, and the subsequent reaction of HOI with the natural organic matter (NOM) present in the source water.[4,6] Iodide occurs naturally in waters, with higher concentrations found typically in the vicinity of sea water or in the presence of inland salt deposits and brines.[3] Iodine concentrations comparable to those typically found in sea waters have also been reported in some inland surface waters.[7] In the presence of free chlorine (hypochlorous acid-HOCl), HOI can be oxidised to iodate (IO_3^-),[5,6] the desired sink for iodine in drinking water due to its quick reduction to iodide *in vivo* and its consequent low toxicity to humans.[8] Free chlorine reaction with HOI reduces the potential for iodo-THM formation in drinking water disinfection processes using chlorine, although formation of iodo-THMs has been reported in such cases.[2,3] In contrast, monochloramine is a weaker oxidant than chlorine and HOI is much more stable in its presence, thus favoring the reaction of HOI with NOM,[4-6] resulting in a higher potential for iodo-THM formation in water treatment systems using chloramines. In fact, the formation of iodo-THMs in water treatment plants using chloramines has been reported in many countries,[2,3,9-13] in some cases leading to taste and odour issues[10] related to the formation of iodoform (TIM), which has a low odour (0.02 μg/L) and taste (5 μg/L) threshold.[2] In water treatment plants using chloramines, the formation of iodo-THMs has been predicted[4-6] and found[3,12,13] to be correlated with the free chlorine contact time allowed before the addition of ammonia. Treatment processes that allow longer free chlorine contact times have been reported to produce lower iodo-THM concentrations.[3, 12] In water plants using source water with naturally occurring ammonium, chloramines are formed immediately after the addition of chlorine and iodo-THMs formation is favored.[3, 13]

The National Survey of Disinfection By-Products in Canadian Drinking Water (2009-2010) (National Survey) investigated 65 water treatment plants and their distribution systems (WTPs) across Canada. It included the first survey of iodo-THM (dichloroiodomethane [DCIM], dibromoiodomethane [DBIM], bromochloroiodomethane [BCIM], chlorodiiodomethane [CDIM], bromodiiodomethane [BDIM], and TIM) occurrence in Canada[13] as well as determining bromide, iodide and total iodine concentrations. At least one iodo-THM congener was detected in the distributed water in 55% and 71% of the water systems tested in winter and summer, respectively. Total iodo-THM concentrations ranged from 0.02 μg/L (method detection limit [MDL], for DCIM) to 21.66 μg/L, with the maximum concentration being found at a site, where all six iodo-THMs were detected. In the National Survey, more than 90% of the total iodo-THM concentrations found in the investigated systems were under 1 μg/L. Although source waters with high bromide concentrations typically also had higher total iodine and iodide concentrations, no consistent correlations could be established between these parameters. Iodo-THMs concentrations found in treated water did not appear to have a consistent correlation with the concentration of bromide, iodide or total iodine in the source water.

Twelve of the 65 WTPs investigated during the National Survey used chloramines as secondary disinfectants.[13] The highest concentrations of total iodo-THM (>5 μg/L) were found in treated water from two WTPs classified as small systems (serving under 5,000 people) that used source water with sodium concentrations over 200 mg/L, naturally present ammonium and where chloramines were used as the secondary disinfectant. No significant concentrations of iodo-THMs were found in the treated water of the other 10 WTPs using chloramines. Based on the data, there appears to be no association between the use of chloramines alone and the formation of iodo-THMs at concentrations over 1 μg/L. Thus, other factors must be influencing the formation of iodo-THMs.

Since bromide, iodide or total iodine determinations, with the level of sensitivity required, are not generally part of routine water quality analysis in many water treatment systems, it is important to establish practical criteria that could be used to help identify water treatment systems with a high potential for iodo-THM formation, especially for small systems. How can we identify water systems with potential for high iodo-THM formation? Is the potential to form iodo-THMs equally relevant for all treatment systems using chloramines? And will the use of chloramine treatment as the main (or only!) criteria for screening WTPs for potential iodo-THM formation lead to overlooking water treatment systems that may, in fact, also have a high potential? These were the questions that lead to the design of the present study: "Targeted survey of emerging disinfection by-products in water systems using source waters with high sodium content" (Targeted Survey).

Based on data from the National Survey,[13] our hypothesis was that high sodium concentrations and the presence of ammonium in the source water could be used to predict water treatment systems with an elevated risk of iodo-THMs formation. Total iodo-THM concentration and speciation in such systems were also expected to be correlated to the free chlorine concentration[3, 5, 6] in the treated and distributed water. Therefore, the Targeted Survey focused on water systems using source waters with elevated sodium content (>200 mg/L) and naturally occurring ammonium. Sodium concentration was considered only a marker for the presence of high bromide and iodide concentrations in the source water, and at no time was sodium concentration considered to have a *de facto* contribution to the iodo-THM formation during water treatment. The study also investigated the occurrence of other emerging DBPs such as cyanogen chloride (CNCl); cyanogen bromide (CNBr); nitrosamines, bromate and perchlorate and the results will be published elsewhere.

2 SURVEY DESIGN

2.1 Site Selection

Sixteen WTPs with source water sodium concentrations ranging from 10 mg/L to 760 mg/L, including 10 systems with sodium concentrations >200 mg/L, were selected for the targeted survey. Twelve of the selected WTPs use groundwater sources, 3 use river water and one uses lake water. The WTPs with source water sodium concentrations >200 mg/L included in this survey were situated in two different geographic locations from the two WTPs with elevated sodium concentrations tested in the National Survey.

All WTPs in the study use ammonium in the treatment train and/or had naturally occurring ammonium, and used sodium hypochlorite (chlorine) as the disinfection agent. Many of the systems selected were also known to have comparatively high DOC and to form either very low or very high total THMs, with an important contribution of brominated congeners.

Figure 1 presents the geographic locations of the 16 sites selected for the Targeted Survey.

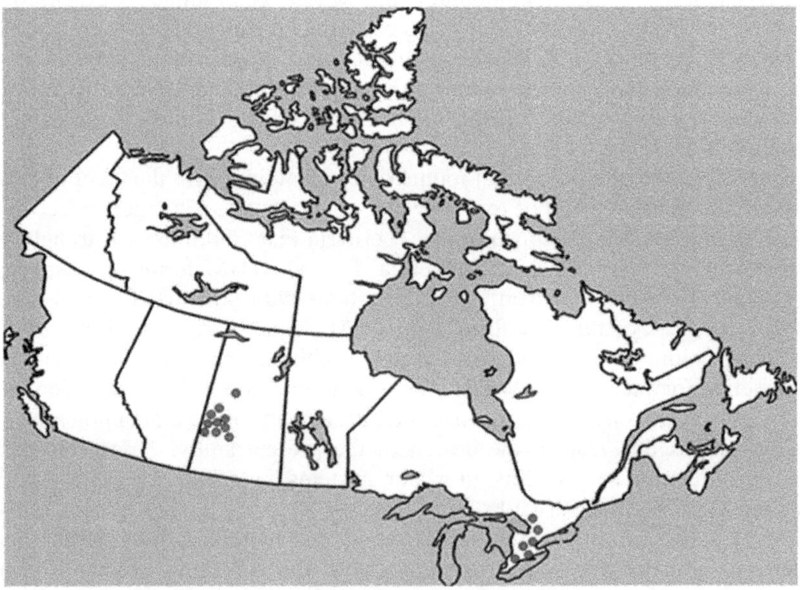

Figure 1 *Geographic location of survey sites*

2.2 Target DBPs and Information Collected

The six iodo-THMs (DCIM, DBIM, BCIM, CDIM, BDIM, and TIM) as well as 27 other neutral DBPs (THMs, haloacetaldehydes-including chloral hydrate, haloacetonitriles, haloketones, 7 nitrosamines, chloropicrin), oxyhalides (chlorite, chlorate, bromate, perchlorate) and some DBP precursors (bromide) and precursor characteristics (SUVA for natural organic matter) were determined in Health Canada laboratories. Source water characteristics were determined by a contract laboratory and included natural organic matter concentration (as DOC; TOC); ammonium and nitrite/nitrate concentrations; major ions, and a suite of 36 elements analyzed using ICP-MS. Free and total chlorine; water temperature; turbidity and water pH were determined on site. The characteristics of the water treatment process at the time of sampling, including disinfection procedures, were recorded using a questionnaire.

2.3 Sampling Schedule

Samples were collected from public water treatment plants and distribution systems twice in the same year, under winter and summer conditions, in most cases, in order to capture the extremes of water temperature during the year. For iodo-THM analysis, duplicate samples were collected at three locations from each WTP: in the raw water - before the treatment plant-(R); after treatment-in the treatment plant-(T); and at the mid-point of the distribution system-(D2) wherever access was available.

3 EXPERIMENTAL WORK

3.1 Water Sampling and Stabilization for Iodo-THM Analysis

The water sampling and stabilization methods used during the Targeted Survey were similar to the one used during the National Survey.[13] Duplicate water samples (sample A and sample B) were collected in 65 mL amber bottles pre-loaded with 0.3 mL fresh ascorbic acid solution (0.114 M) as a quenching agent and 1 mL of a 1 N HCl solution to achieve a final sample pH of 3-4. Due to the high total dissolved solids content at most sites, the 1N HCl solution was used to achieve the required pH while keeping the volume of acid solution low (<1 mL), to avoid diluting the sample. The amount of ascorbic acid used per sample was also higher than usual due to the possibility of high chlorine rates being used at some sites. The bottles were filled, with no headspace, from the raw and treated water taps at the water treatment plant or at the distribution sampling point, after running water for at least 5 minutes. Samples were shipped by overnight courier to the Health Canada laboratory. During shipping, water samples were kept at less than 10°C at all times and samples were analysed as soon as received at the laboratory.

3.2 Iodo-THM Analysis by Automated SPME_GC_ECD

Water samples were extracted and analysed by an automated SPME-GC-ECD method, using a 100 μm PDMS fibre, in direct immersion mode. The method is described in detail elsewhere.[13] A summary of the analytical methods performance parameters is presented in Table 1.

3.3 Bromide Analysis by Ion Chromatography

Bromide was determined in raw and treated water samples using an ion chromatography-conductivity detector. The method is described elsewhere.[13]

Table 1 *Analytical method performance parameters*

IodoTHM	*MDL (µg/L)*	*Dynamic Range (µg/L)*	*Average % Spike Recovery**	*Average Interday % RSD***
DCIM	0.05	0.05 - 5.0	101.0	6.58
BCIM	0.09	0.09 - 5.0	109.6	6.69
DBIM	0.06	0.06- 5.0	113.3	7.42
CDIM	0.12	0.12 - 5.0	106.1	7.18
BDIM	0.1	0.1 - 5.0	109.9	10.21
TIM	0.05	0.05 - 2.5	100.3	14.43

* Average % Recovery from raw, treated and distribution water samples spiked with the six iodo-THM mixture (0.5 µg/L iodoform; 1 µg/L for the other 5 congeners). Samples from 16 WTPs, in two Canadian provinces, collected over 2 sampling seasons were used in the calculation.

** Average Interday % RSD was calculated for each season, using the daily values for the second-lowest concentration from the daily calibration curve and averaged for both sampling seasons.

3.4 Neutral, Halogen-Containing DBPs

Water samples were analysed for 20 neutral DBPs including THMs, CNCl; CNBr; trichloroacetonitrile (TCAN); dichloroacetonitrile (DCAN); bromochloroacetonitrile (BCAN); dibromoacetonitrile (DBAN), and chloropicrin (CPK). Analysis was done using liquid-liquid extraction with MTBE followed by GC-ECD analysis, according to a method developed in Health Canada laboratories and described previously.[14] The Health Canada method is based on U.S. EPA-Method 551.2.

5 RESULTS AND DISCUSSION

4.1 Iodo-THM Concentration Range

One or more iodo-THMs congeners were found in treated water and distribution water from 15/16 WTP in the survey. The total iodo-THM concentrations (sum of six iodo-THMs) were obtained from T and D2. Maximum total concentration found in either T or D2 samples for each WTP ranged from 0.11 µg/L to 26.82 µg/L (Table 2).

The majority of the 16 water systems investigated in this survey (10) were selected to represent what we believed to be extremely favourable conditions for the formation of iodo-THMs during water disinfection: sodium concentrations over 200 mg/L and ammonium in the source water. The other 6 WTPs (with sodium concentrations between 10 mg/L and 80 mg/L, some having ammonium in the source water) were included for comparison purposes. The sodium concentration (routinely determined for most water treatment plants in Canada) was used as a marker for their potential to have higher concentrations of iodide and/or total iodine. Data from the National Survey[13] indicated that high sodium concentrations were better correlated with high iodide/total iodine than other cations (e.g. K^+; Ca^{2+}). However, iodide and total iodine were not determined in the Targeted Study, due to operational constraints.

Table 2 *Maximum total iodo-THM concentration and dominant iodo-THM congener determined in treated water from 16 WTPs.*

Water Treatment Plant (WTP)	*Max. Total Iodo-THM conc.in Ttreated Water (µg/L)*	*Dominant Iodo-THM*	*Ammonium conc.in Source Water (mg/L N)*	*Sodium conc. in Source Water (mg/L)*	*Free chlorine conc. in Treated Water (mg/L)*	*Bromide in Source Water (µg/L)*
1	0.66	DCIM	2.7	620	1.55	Not sampled
2	10.91	TIM	1.9	580	0.38	930
3	10.48	TIM	1.1	530	0.05	1140
4	3.99	DBIM	1.1	410	0.64	670
5	26.82	BCIM	0.87	420	2.5	730
6	21.54	TIM	nd	440	0.06	1490
7	1.48	BCIM	1.1	390	0.08	420
8	2.03	TIM	1.7	760	0.03	2550
9	1.69	DBIM	1.1	420	1.38	840
10	0.70	DCIM	0.37	240	0.19	590
11	0.40	DBIM	0.11	54	0.11	80
12	0.80	DCIM	0.16	52	2.08	90
13	0.34	BDIM	nd	41	1.32	120
14	0.16	DCIM	nd	10	1.6	50
15	0.11	DCIM	0.09	29	1.5	50
16	nd	Nd	0.08	74	0.72	240

The presence of naturally occurring ammonium in the source water was postulated to create favourable conditions for the formation iodo-THMS, due to the early formation of chloramines in the treated water.[3,5] The majority of the WTPs included in the study (13/16) had detectable naturally occurring ammonium concentrations ranging from 0.08 mg/L to 2.7 mg/L (calculated as nitrogen). Determined source water DOC ranged from 0.62 mg/L to 17 mg/L.

The iodo-THM concentration data collected during the Targeted Survey appear to support our hypothesis that the presence of high sodium concentrations (over 200 mg/L) and the presence of ammonium in the source water are good markers for identifying WTPs with a high potential to form iodo-THMs after treatment. Data presented in Table 2 show that all maximum total iodo-THM concentrations >1 µg/L were found in treated water from WTPs using source water with sodium concentrations >200 mg/L and all but one of those source waters also had detectable ammonium levels.

In the National Survey,[13] total iodo-THM concentrations found in the vast majority of the water treatment systems studied were <1 µg/L (>90%) and only one sample over 10 µg/L was found, in contrast, in the Targeted Survey, the maximum total iodo-THM concentrations were >1 µg/L, in 8/16 WTPs and >10 µg/L in 4/16 WTPs investigated. A comparison of iodo-THM occurrence frequency, average and median total concentrations for the National Survey and for the Targeted Survey is presented in Table 3. For WTPs with high sodium source water (>200 mg/L), 8 of 10 WTPs had maximum total iodo-THMs concentrations of >1 µg/L and 4/10 WTPs the maximum total iodo-THMs concentrations were >10 µg/L. Maximum total iodo-THMs concentrations found in treated water from individual WTPs supplied by source water containing >200 mg/L sodium and <200 mg/L sodium are presented in Figure 2, while the average and medium total iodo-THM concentrations for the two categories of WTPs can be found in Figure 3.

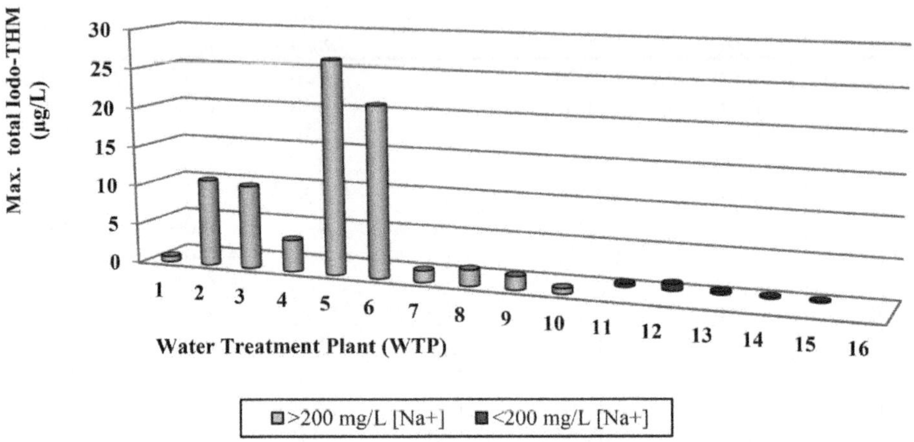

Figure 2 *Maximum iodo-THM concentration in Targeted survey for WTPs using source waters with sodium concentrations under and over 200mg/L.*

Table 3 *Comparison of iodo-THMs occurrence frequency, average and median maximum total concentrations (calculated only for sites where iodo-THM were found) determined as part of the National Survey and of the Targeted Survey.*

	2009/2010 National Survey	*2012/2013 Targeted Survey*
Number of WTPs investigated	65	16
% of WTPs with Na^+ concentration>200 mg/L	6.2	62.5
Maximum total iodo-THM concentration - (Average value)	1.2 µg/L	5.47 µg/L
Maximum total iodo-THM concentration - (Median value)	0.17 µg/L	1.48 µg/L

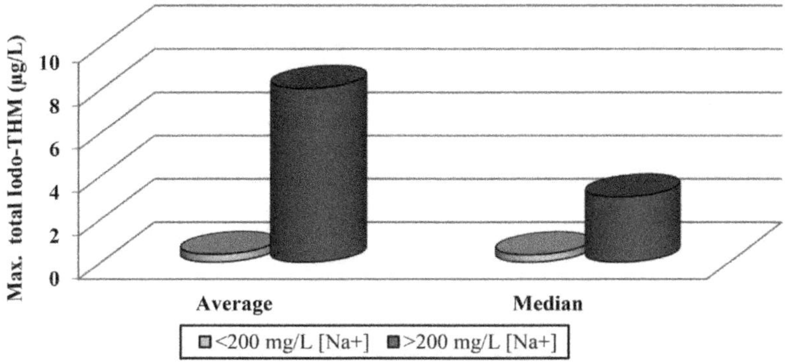

Figure 3 *Average and median maximum iodo-THMs concentrations determined in Targeted Survey for WTPs using source water with sodium concentrations under/over 200 mg/L.*

4.2 Iodo-THM Speciation and Free Chlorine Residual

The frequency of occurrence of individual iodo-THM congeners in the treated water was DCIM (13 WTPs) > DBIM (10 WTPs) > BCIM (8 WTPs) > CDIM (7 WTPs) > BDIM (6 WTPs) and TIM (5 WTPs).

All six iodo-THM congeners were present in the treated and distributed water from three of the four WTPs that had maximum total iodo-THM concentrations of >10 µg/L with the exception of WTP6, where DBIM was absent (Figure 4). The dominant species formed in waters was TIM. At three of these sites, ammonium was detectable in the source water and chloramines were formed at the first addition of chlorine. The highest TIM concentration (16.28 µg/L), was detected in treated water from WTP6, during summer, where it was observed that the source water had no detectable ammonium, low total chlorine residual (0.4 mg/L) and no detectable free chlorine residual (Table 4). However, this source water had high DOC (12 mg/L) and high bromide concentration (1490 µg/L). At WTP8, the dominant iodo-THM was also TIM, but much lower total iodo-THMs were found (2.03 µg/L), no free chlorine was detected in the treated water and the total residual chlorine concentration was 0.3 mg/L, thus suggesting that the dose of chlorine needed to form chloramine (for a source water concentration of 1.7 mg/L ammonium as N) was the limiting factor in the formation of iodo-THMs. It is very likely that if breakpoint chlorination was performed, the amount of iodo-THMs formed would have been considerably higher, in the range of those observed for WTP2 and WTP3.

WTP5 was the only WTP where maximum total iodo-THM concentrations greater than 10 µg/L were produced, and where large amounts of chlorine were used to achieve a free chlorine residual of 2.5 mg/L in the distribution system. BCIM was the dominant congener (5.4 µg/L), in the treated water at WTP5 and the data suggest that different reactions are responsible for iodo-THMs speciation at this WTP when compared to WTPs where free chlorine residuals were low and the dominant congener was TIM. BCIM was also the dominant congener in the treated water at WTP7.

At the other 9 WTPs where iodo-THMs were found, either DCIM (5 WTPs) or DBIM (3 WTPs) was the dominant congener. Typically at these WTPs, the free chlorine residual in the distribution system was higher than in the four systems for which TIM was the dominant congener formed.

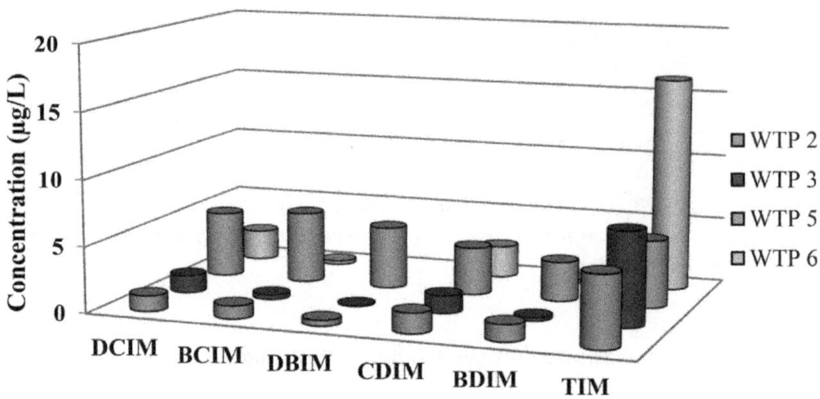

Figure 4 *Iodo-THM speciation in water sampled from the mid-point of the distribution system from the four WTPs where maximum total iodo-THM concentrations > 10µg/L were determined.*

Table 4 *Bromide, free chlorine and iodo-THM congeners concentration in water sampled from the mid-point of the distribution system for WTP2, WTP3, WTP5 and WTP6.*

WTP	*Bromide (µg/L)*	*Free Chlorine (mg/L)*	*DCIM (µg/L)*	*BCIM (µg/L)*	*DBIM (µg/L)*	*CDIM (µg/L)*	*BDIM (µg/L)*	*TIM (µg/L)*
2	930	0.38	1.19	0.99	0.46	1.56	1.29	5.42
3	1140	0.05	1.32	0.36	0.06	1.35	0.25	7.14
5	730	2.5	4.96	5.4	4.71	3.62	3.01	5.12
6	1490	0.06	2.3	0.28	nd	2.42	0.26	16.28

4.3 Iodo-THM Speciation and Bromide Concentration in the Source Water

Bromide concentration in the source waters (MDL = 5 µg/L) ranged from 50 µg/L to 2550 µg/L, with the higher bromide concentrations found in groundwater sources from Mid-Western Canada.

Similarly to results from the National Survey,[13] bromide concentrations do not appear to have a significant influence on iodo-THMs speciation at the four sites where TIM was the dominant congener (see Table 2). Notably, at WTP6, where the source water had a high bromide concentration (1490 μg/L), no DBIM, and very little BCIM (0.28 μg/L) and BDIM (0.26 μg/L) were found and the dominant congeners found were TIM (16.28 μg/L), CDIM (2.42 μg/L) and DCIM (2.3 μg/L). At the WTPs where either DCIM or DBIM was the dominant congener detected, no consistent positive correlation was found between the bromide concentration and DBIM as the dominant mono-iodo-THM congeners. This finding indicates that factors other than bromide availability may determine iodo-THM speciation at WTPs with higher free chlorine residuals.

4.4 Seasonal Variation of Iodo-THM concentrations

Ten of the WTPs were sampled in both summer and winter. Although total iodo-THM concentrations in the treated summer water appear to be higher at most WTPs sampled during both seasons, no consistent trend was observed (Figure 5). This is unlike regulated THMs, where higher concentrations are typically found in summer treated water samples. The small number of WTPs sampled during both seasons may not allow for trends to be observed. The variety of factors influencing the formation and speciation of iodo-THMs under different treatment conditions (see also sections 5.3 and 5.5) may further complicate the understanding of seasonal variability of iodo-THMs.

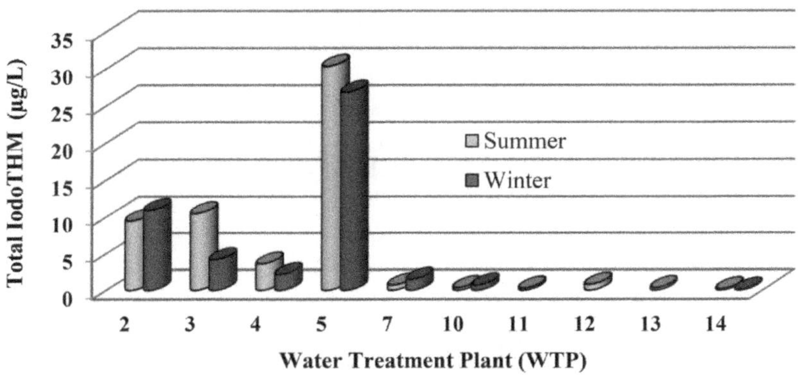

Figure 5 *Maximum total Iodo-THM concentrations in summer and winter treated water of the 10 WTPs sampled during both seasons.*

4.5 Total iodo-THM concentration and total THM concentrations

Total THMs concentrations in distribution system samples ranged from 0.27 μg/L at WTP3 to 363 μg/L at WTP5 in the winter and from <MDL to 423 μg/L at WTP5 in the summer. The confidence in the results for summer treated and distribution system water samples from WTP5 was low because the amount of free chlorine (8 mg/L) exceeded the quenching capacity of the ascorbic acid solution present in the sampling bottles. The results for iodo-THM from WTP-5 summer samples (up to 30.35 μg/L) were also excluded from the iodo-THM occurrence and speciation discussion for this reason. This issue was addressed in the winter sampling through the addition of a larger volume of ascorbic acid, so the winter data for WTP5 are considered to be accurate.

Treated and distribution water samples collected from WTP2; WTP3; WTP6 and WTP8 contained under 1 μg/L of regulated total THMs. However, total THMs concentration in treated and distribution water samples from WTP4, WTP5, WTP10 and WTP13 were greater than 100 μg/L. All other WTPs had regulated total THMs within the acceptable limits developed by Health Canada (MAC of 100 μg/L).[15] The results for total THMs follow the expected trend (i.e., higher THM concentrations in the summer samples vs winter samples and higher THM concentrations at the mid-point distribution sampling point than at the water plant).

Treated water from all four WTPs where TIM was found to be the dominant iodo-THM congener (WTP2, 3, 6 and 8) and where higher total iodo-THM concentrations were determined, had only trace amounts of regulated THMs (see Figure 6). Three of these four WTPs also had maximum total iodo-THM concentrations of >10 μg/L (WTP 2, 3, 6).

At four WTPs, total iodo-THM concentrations in treated water were higher than total regulated THMs. Similar situations have been reported before[13]. Higher Total iodo-THM concentrations than total regulated THMs concentrations were determined at WTPs 2 and 3 in both the winter and in the summer samples and at WTPs 6 and 8 only in the summer samples, as winter samples were not collected. The WTPs 2, 3, 6 and 8 are the WTPs where the dominant iodo-THM congener found was TIM. At these WTPs the free residual chlorine at the time of sampling was low (see Table 2). The use of breakpoint chlorination appears to be the cause of the low free chlorine residual at WTPs 2, 3 and 8, where ammonium was found in the raw water. At WTP 6, other factors (e.g. high DOC, high bromide concentration) may explain the low free chlorine residual. At WTPs 4, 5, 7 and 9 total iodo-THM concentrations over 1 μg/L were found, but the total regulated THM concentrations were considerably higher during both seasons (with the exception of WTP9 for which winter samples were not collected). In all these cases, the free chlorine residual determined at the time of sampling was higher than for WTPs 2, 3, 6 and 8 (see Table 2) and the dominant iodo-THM congener was either DBIM (WTP4 and WTP9) or BCIM (WTP5 and WTP7).

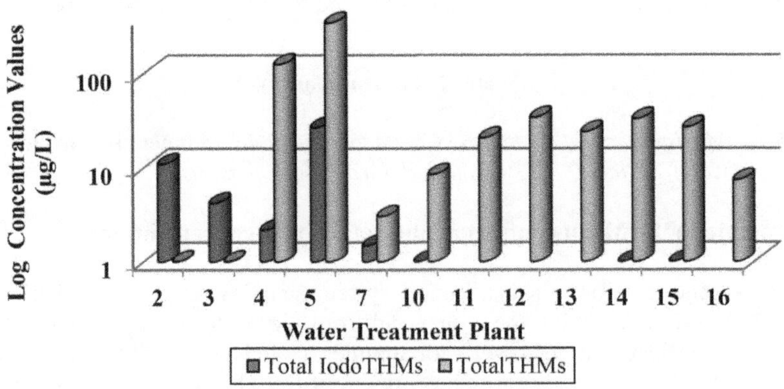

Figure 6 *Total Iodo-THM concentrations vs maximum total regulated THM concentrations in treated winter water samples (T-winter).*

5 CONCLUSIONS

Iodo-THMs results from this study show that source waters with sodium concentrations in excess of 200 mg/L and in which naturally-occurring ammonium is present have a considerably increased probability of producing total iodo-THMs concentrations exceeding 1 μg/L. Under specific treatment conditions, maximum total iodo-THMs concentrations can be in the 1 to 30 μg/L range and can exceed the concentrations of regulated THMs formed in the same systems, under the same treatment conditions. Due to the potential for iodo-THMs to be of higher toxicity than the currently regulated THMs[3,16] the potential formation of iodo-THMs should be taken into account when making decisions about the disinfection of source waters with high sodium content and naturally present ammonium.

This study indicates that when high total iodo-THMs concentrations are formed, iodoform tends to be the dominant iodo-THMs congener and, by far, the highest contributor to the total iodo-THMs concentration. Even at very low concentrations, the formation of iodoform creates aesthetic challenges (odour and taste issues) that may lead to lower use of distributed drinking water by the population. In the systems tested, the main factor favouring the formation of high total iodo-THMs, with iodoform as the main congener, in addition to the presence of ammonium and sodium (as a marker for increased available iodide), appears to be low free chlorine residual. Iodo-THM speciation in such systems does not appear to be directly correlated to the bromide concentration in the source water.

Based on results from the National Survey[13] and the current Targeted Survey, which were conducted in different geographic regions, we believe that sodium concentrations over 200 mg/L and the presence of ammonium in the source water are good screening criteria for the selection of water treatment systems that may have a high potential for the formation of iodo-THMs. This further supports the need for water treatment systems to test for naturally occurring ammonium, particularly for those systems supplied with source waters containing high sodium concentrations. Where source waters with both high sodium concentrations and naturally present ammonium are used, testing for the presence of iodo-THMs in treated and distributed water should be considered.

6 ACKNOWLEDGEMENTS

Data presented here were collected as part of the Targeted Survey of Selected Disinfection By-Products in source waters with high saline concentrations, a 3 year study funded by the Health Canada Monitoring and Surveillance Fund as part of the Chemical Management Plan (CMP) within Health Canada.

The authors would like to thank the members of the Canadian Federal-Provincial-Territorial Committee on Drinking Water for their gracious help in providing data for the site selection and the water plant operators for agreeing to participate in the study.

The authors would like to thank Zhiyun Jin for technical support.

References

1. S.D. Richardson, *Global Nest Journal*, 2005, **7**, 43.
2. B. Cancho, F. Ventura, M. Galceran, A. Diaz, S. Ricart, *Water Res*., 2000, **34**, 3380.
3. S.D. Richardson, F. Fasano, J.J. Ellington, F.G. Crumley, K.M. Buettner, J.J. Evans, B.C. Blount, L.K. Silva, T.J. Waite, G.W. Luter, A.B. McKague, R.J. Miltner, E.D. Wagner, M.J. Plewa, *Environ. Sci. Technol.*, 2008, **42**, 8330.
4. Y. Bichsel, and U. von Gunten, *Environ. Sci. Technol.,* 2000, **34**, 2784.
5. Y. Bichsel, and U. von Gunten, *Environ. Sci. Technol*., 1999, **33**, 4040.

6. G. Hua, D.A. Reckhow, J. Kim, *Environ. Sci. Technol.*, 2006, **40**, 3050.
7. J.E. Moran, S.D. Oktay, P.H. Santschi, *Water Resour. Res.*, 2002, **38**, 1149.
8. A. Taurog, E.M. Howells, H.I. Nachimson, *J. Biol. Chem.*, 1966, **241,** 4686.
9. S.W. Krasner, H.S. Weinberg, S.D. Richardson, S.J. Pastor, R. Chinn, M.J. Sclimenti, G.D. Onstad, A.D. Jr. Thruston, *Environ. Sci. Technol.*, 2006, **40**, 7175.
10. R.C. Hansson, M.J. Henderson, P. Jack, R.D. Taylor, *Water Res.*, 1987, **21**, 1265.
11. H.S. Weinberg, S.W. Krasner, S.D. Richardson, A.D. Jr. Thruston, http://www.epa.gov/athens/publications/reports/EPA_600_R02_068.pdf
12. E.H. Goslan, S.W. Krasner, M. Bower, S.A. Rocks, P. Holmes, L.S. Levy, S.A. Parsons, *Water Research,* 2009, **43,** 4698.
13. A.M. Tugulea, R. Aranda-Rodriguez, D. Berube, L. Dabeka, F. Breton, J. Hnatiw, *Environ. Sci. Technol.*, (submitted for publication).
14. D.T. Williams, G.L. LeBel, F.M. Benoit, *Chemosphere*, 1997, **34**, 299.
15. Health Canada, 'Guidelines for Canadian Drinking Water Quality: Guideline Technical Document – Trihalomethanes' Water Quality and Health Bureau, Healthy Environments and Consumer Safety Branch, Health Canada, Ottawa, Ontario. 2006.
16. Y.T. Woo, D. Lai, J.L. McLain, M.K. Manibusan, V. Dellarco, *Environ. Health Perspect.,* 2002, **110** , 75.

NDMA FORMATION AFTER COAGULATION WITH FERRIC CHLORIDE AND DECARBONATION

M. Sgroi[1,2], P. Roccaro[1,2], G.L. Oelker[3], S.A. Snyder[2]

[1]Department of Civil Engineering and Architecture, University of Catania, viale A. Doria 6, 95125 Catania, Italy
[2]Department of Chemical & Environmental Engineering, University of Arizona, 1133 E. James E. Rogers Way, Tucson, AZ 85721, USA
[3]United Water, Edward C. Little Water Reclamation Facility, 1935 South Hughes Way, El Segundo, CA 90245, USA

1 INTRODUCTION

Nitrosamines are a group of emerging disinfection by-products (DBPs)[1]. They are potent rodent carcinogens and most are classified as probable human carcinogens by the International Agency for Research on Cancer[2]. Of all the nitrosamines, N-Nitrosodimethylamine (NDMA) has been most commonly detected in drinking water and wastewater[3] and it is mutagenic with a life time cancer risk of 10^{-6} for a concentration of 0.7 ng/L[4]. The presence of NDMA is of concern in locations where wastewater effluents are used for potable reuse because disinfection of wastewater effluents may lead to formation of relatively high concentrations of NDMA[5,6]. Indeed, wastewater often contains significant concentrations of NDMA and other nitrosamines[7,8] as well as considerable concentration of NDMA precursors[9].

Several studies have shown that NDMA can be generated in water and wastewater treatment system by chlorine-based disinfection process[10,11,12,13]. More recently, researchers have reported that alternative disinfectants including chlorine dioxide (ClO_2) and ozone (O_3) are also able to produce nitrosamines[14,15].

To date, there is a lack of knowledge about the nitrosamine precursor pool and formation mechanisms. In addition, treatment processes or treatment trains specifically aimed at removing/transforming nitrosamine precursors within water treatment plants or post-formed nitrosamines need to be investigated. In this context, Sgroi et al.[16] showed the complexity of NDMA control by advanced technologies at an indirect potable reuse facility due to the wide variety of NDMA precursors.

Objectives of this study were (i) to investigate the removal of ozone-related NDMA precursors by coagulation process, and (ii) to evaluate the effect of decarbonation (i.e. reduction of carbon dioxide (CO_2) concentration in the water) in NDMA formation during chloramination. For this last point, it is noteworthy to observe that Lv et al.[17], in a theoretical study on NDMA formation, reported that at pH>7 NDMA can be formed from dimethylamine (DMA), a well known NDMA precursor, reacting with nitrite anion in the presence of carbon dioxide, which may catalyze the reaction.

2 MATERIALS AND METHODS

2.1 Materials

All purchased solvents (methanol, methylene chloride) were HPLC grade or superior, obtained from Fisher Scientific and used without further purification. Potassium indigotrisulfonate, sodium phosphate monobasic monohydrate, sodium phosphate dibasic heptahydrate and phosphoric acid (ACS reagent grade, ≥85 wt. % in H_2O) were purchased at >98% purity from Sigma-Aldrich. Sodium sulfate anhydrous (Granular, ACS reagent grade), hydrochloric acid, ferric chloride hexahydrate and sodium thiosulfate crystalline (> 99% purity) were obtained from Fisher Scientific. N-Nitrosodimethylamine (d6, 98%) from Cambridge Isotope Laboratories (Andover, MA) was used as a surrogate, N-Nitroso-Di-N-Propylamine (d14, 98%) from Cambridge Isotope Laboratories (Andover, MA) was used as an internal standard, and nitrosamine mix from Supelco (Bellafonte, PA) was used as the standard for NDMA analysis. Lime solution was provided by the facility's lab staff. Whatman (Clifton, NJ) borosilicate microfiber filters with a nominal pore size of 0.7 μm were used for filtration.

2.2 Sampling

To evaluate the role of coagulation in NDMA precursors removal, composite samples (24-h) of the final effluent of a municipal wastewater treatment plant (WWTP) placed in the south-western United States were collected. This WWTP serves about four million people and treats wastewater characterized by about 90% from domestic discharge. The unchlorinated final effluent is in part sent to a downstream indirect potable reuse (IPR) system.

The downstream IPR process scheme includes microfiltration (MF), reverse osmosis (RO), UV advanced oxidation (UV/H_2O_2), decarbonation, lime stabilization (pH 8), and secondary disinfection using sodium hypochlorite (NaClO) (dose of 3-4 mg/L as chlorine (Cl_2)). NaClO (10 mg/L as Cl_2) is also added before MF for membranes fouling control; chloramines are formed because of the elevated ammonia load present in water (Table 1). Decarbonation towers strip carbon dioxide (CO_2) raising the pH and reducing the amount of lime to add for stabilizing the water and reaching pH 8.

Table 1 *Water quality parameters*

Water parameter	*Wastewater effluent*	*IPR effluent*
Ammonia (mg/L as N[a])	40.6	3.0
pH	7.7	8.0
DOC[b](mg/L)	14.3	0.2

[a] *N, nitrogen.* [b] *DOC, dissolved organic carbon*

During the collection at the IPR facility, samples were collected before decarbonation tower (undecarbonated samples) (denoted as "Pre decarbonation") and immediately after decarbonation tower (completely decarbonated samples) (denoted as "Post decarbonation"). Samples collected in 20 liters containers were promptly amended by sodium hypochlorite and lime in order to reach a final chlorine concentration of around 4 mg/L and pH 8 (i.e. the final stabilization and disinfection of the IPR facility were simulated). During this sampling campaign, additional 1 L amber bottles were collected directly from the process before decarbonation tower and after decarbonation tower, and immediately quenched in situ by 80 mg/L of sodium thiosulfate without further manipulation ("day 0 not adjusted" samples).

All the collected samples were shipped overnight to The University of Arizona [Tucson, AZ (USA)], where they were stored at room temperature (22 – 23 °C) and covered with aluminum foil to protect the samples from light. Sub-sampling from the containers were performed after 1 and 4 days of chlorine contact time (denoted as "day 1" and "day 4" respectively). Duplicates sub-samples were collected in 1 L amber bottles and amended with 80 mg/L of sodium thiosulfate to quench any residual chlorine. With each daily aliquot, residual total chlorine and pH were always monitored. Furthermore, sub-samplings from the containers were also performed at the IPR facility and the sub-samples were quenched within 3 hours of the collection event (denoted as "day 0").

2.3 Bench-scale experiments

The ozonation system used for bench-scale ozonation is described in Sgroi et al.[18]. Residual dissolved ozone concentration was measured by the indigo method[19]. In all the wastewater ozonation tests in this work, the dissolved ozone residual measured immediately after ozone application was zero due to the high instantaneous ozone demand (IOD) common for wastewater[20], and the utilized reactor configuration and experimental set-up. Ozonation was performed after a simulated bench-scale coagulation process to evaluate the possible removal of NDMA precursors.

A Jar Test (Phipps & Bird six paddle stirrer) experiment was performed to simulate a coagulation-sedimentation process. Two different ferric chloride doses (20 mg/L and 50 mg/L) were used. The experimental procedure included a brief coagulation step with high stirring velocity, a longer flocculation step with low stirring velocity and a final sedimentation of the formed flocks. After the Jar Test experiment, samples were filtered at 0.7 μm before ozonation.

2.4 Analytical methods

Analysis of NDMA was conducted according to EPA method 521 using an Agilent 7000 Triple Quadrupole GC/MS-MS (Santa Clara, California). For dissolved organic carbon (DOC) analysis, samples were filtered through 0.45 μm hydrophilic polypropylene filter (GHP Membrane Acrodisc, Pall Life Sciences) and acidified to pH<3 with hydrochloric acid. A Shimadzu TOC-LCSH (Kyoto, Japan) total carbon analyzer was used for quantification. Analysis of inorganic water constituents was performed by the water reclamation facility's lab according to standard methods.

3 RESULTS AND DISCUSSION

3.1 Carbon dioxide effect on NDMA formation

All the samples collected at the IPR facility had very similar NDMA results. No immediate NDMA increase was evident in all the samples after chlorine addition and pH correction (day 0 samples, Table 2). NDMA rose to 7 – 8 ng/L after 24 hours, but remained stable later and always lower than 10 ng/L. Thus, no impact by decarbonation was evident.

Sgroi et al.[16] reported that at the investigated IPR facility the concentration of NDMA in the product water is periodically monitored and NDMA concentrations higher than 10 ng/L were observed several times in the final effluent. Likely, a very low concentration of precursors was present during the present experiments and high NDMA concentrations were not observed. Indeed, precursors concentration in water can be highly variable and may also be related to RO membrane conditions and performance. Probably, the low precursors concentration and water quality may also have affected the carbon dioxide catalyzing effect on NDMA production.

Table 2 *NDMA formation after 0, 1 and 4 days of chlorine contact time. Pre decarbonation, sample collected before decarbonation tower. Post decarbonation, sample collected after decarbonation tower. Day 0 not adj., day 0 not adjusted (samples not amended with chlorine and lime)*

Constituent	*Day 0 not adj.*	*Day 0*	*Day 1*	*Day 4*
Pre decarbonation				
NDMA (ng/L)	*2*	*3 (0.1)*[a]	*7 (0.9)*	*7 (0.1)*
Total Chlorine (mg/L)	1.1	4.5	2.9	1.7
pH	5.8	8.0	8.1	8.4
Alkalinity (mg/L as $CaCO_3$)	19	73	na	na
Total CO_2 (mg/L)	77	65	na	na
Post decarbonation				
NDMA (ng/L)	*3*	*4 (0.2)*	*7 (0.9)*	*8 (1.1)*
Total Chlorine (mg/L)	1.1	4.3	3.1	1.7
pH	7.8	8.2	8.6	8.6
Alkalinity (mg/L as $CaCO_3$)	21	24	na	na
Total CO_2 (mg/L)	19	21	na	na

na=not available; [a] *(± one standard deviation of the measurement derived from the standard curve).*

At pH 8 carbon dioxide is mainly present in the form HCO_3^-. The difference in carbon dioxide concentration between pre-decarbonation and post-decarbonation water is evident from the different measured values of alkalinity and total carbon dioxide (Table 2).

3.2 Role of coagulation in NDMA precursors removal

Results obtained from Jar Test experiments are reported in Figure 1 where ozone dose of 11 mg/L were applied.

The presented results confirm the ineffectiveness of the ferric chloride on NDMA precursor removal during a coagulation process.

4 CONCLUSIONS

The presence or generation of NDMA in drinking water is of concern for public health. This issue is noticed worldwide and especially in wastewater impacted potable supplies. Because of this, it was felt necessary to investigate treatment processes or treatment trains specifically aimed at removing/transforming nitrosamine precursors within water treatment plants or post-formed nitrosamines. In this study, NDMA formation was evaluated after ferric chloride coagulation of a wastewater effluent and decarbonation of the final effluent of an IPR system.

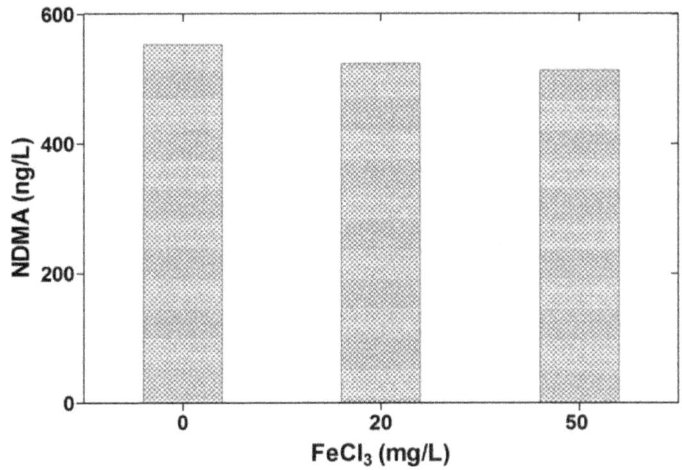

Figure 1 *NDMA formation during ozonation after coagulation with ferric chloride ($FeCl_3$). Ozone dose 11 mg/L*

The following conclusions can be summarized:

- coagulation with ferric chloride ($FeCl_3$) is ineffective in removing NDMA precursors;
- the concentration of carbon dioxide (CO_2) in water, at the concentrations tested, does not appear to affect NDMA formation during chloramination.

References

1. S.D. Richardson, *Anal. Chem.*, 2007, **79**, 4295.
2. IARC monographs on the evaluation of the carcinogenic risk of chemicals to humans, *Some N-nitroso compounds*, 1978, **17**, 365.
3. USEPA, Unregulated Contaminant Monitoring Rule 2, 2012. http://water.epa.gov/lawsregs/rulesregs/sdwa/ucmr/data.cfm#ucmr2010
4. USEPA, Integrated Risk Information System (IRIS), N-nitrosodimethylamine. Office of Research and Development (ORD), National Center for Environmental Assessment, 1987 www.epa.gov/iris/subst/0045.htm.
5. W.A. Mitch and D.L. Sedlak, *Environ. Sci.Technol.*, 2004, **38**, 1445.
6. K. Kosaka, M. Asami, Y. Konno, M. Oya and S. Kunikane, *Environ. Sci. Technol.*, 2009, **43**, 5236.
7. D.L. Sedlak, R.A. Deeb, E.L. Hawley, W.A. Mitch, T.D. Durbin, S. Mowbray and S. Carr, *Water Environ. Res.*, 2005, **77**, 32.
8. M. Krauss, P. Longrée, F. Dorusch, C. Ort and J. Hollender, *Water Res.*, 2009, **43**, 32.
9. S.W. Krasner, P. Westerhoff, B. Chen, B.E. Rittmann, S.-N. Nam and G. Amy, *Environ. Sci. Technol.*, 2009, **43**, 2911.
10. J. Choi and R.L. Valentine, *Environ. Sci. Technol.*, 2003, **37**, 4871.
11. W.A. Mitch and D.L. Sedlak, *Environ. Sci. Technol.*, 2002, **36**, 588.
12. I.M. Schreiber and W.A. Mitch, *Environ. Sci. Technol.*, 2005, **39**, 3811.

13. I.M. Schreiber and W.A. Mitch, *Environ. Sci. Technol.*, 2006, **40**, 6007.
14. I. Najm and R.R. Trussel, *J. Am. Water Works Assoc.*, 2001, **93**, 92.
15. P., Andrzejewski, B., Kasprzyk-Hordern and J. Nawrocki, *Water Res.*, 2008, **42**, 863.
16. M. Sgroi, P. Roccaro, G.L. Oelker and S.A. Snyder, *Water Res.*, 2015, **70**, 170.
17. C.L. Lv, Y.D. Liu, R. Zhong and Y. Wang, *J. Mol. Struct. Theochem*, 2007, **802**, 1.
18. M. Sgroi, P. Roccaro, G.L. Oelker and S.A. Snyder, *Environ. Sci. Technol.*, 2014, **48**, 10308.
19. K.L. Rakness, E. C. Wert, M. Elovitz and S. Mahoney, *Ozone: Sci. Eng.*, 2010, **32**, 33.
20. E.C. Wert, F.L. Rosario-Ortiz, D.D. Drury and S.A. Snyder, *Water Res.*, 2007, **41**, 1481.

AQUEOUS EQUILIBRIUM CONSTANTS OF FORMATION FOR BROMAMINES, CHLORAMINES, AND BROMOCHLORAMINES: QUANTUM CHEMISTRY ESTIMATES.

Daniela Trogolo,[1] *Urs von Gunten,*[2,3] *J. Samuel Arey* [1,3]

[1] Environmental Chemistry Modeling Laboratory, EPFL, Station 2, 1015 Lausanne (CH)
[2] Laboratory for Water Quality and Treatment, EPFL, Station 2, 1015 Lausanne (CH)
[3] Eawag, Swiss Federal Institute of Aquatic Science and Technology 8600 Dübendorf (CH)

1 INTRODUCTION

During disinfection of drinking and swimming pool waters, reactive halogenated species such as HOCl and HOBr undergo substitution reactions with ammonia, leading to the formation of chloramines and bromamines:[1–5]

$$NH_3 + HOX \leftrightarrow NH_2X + H_2O \quad (1)$$

$$NH_2X + HOX \leftrightarrow NHX_2 + H_2O \quad (2)$$

$$NHX_2 + HOX \leftrightarrow NX_3 + H_2O \quad (3)$$

where X =Br, Cl. Bromochloramines can additionally form, resulting from the reactions of mixed halogenated species with the hypohalous acids.[6,7]

As an unintended consequence of water treatments, numerous disinfection by-products (DBPs) may be formed from the reactions of halamines with naturally occurring inorganic and organic water constituents.[8–11] The kinetics and the thermodynamics of halamine formation in treated water may influence the rates of disinfection by-product formation. However, currently available information precludes a reliable and well-defined estimation of the specific reactivity of individual halamines in aquatic systems. Existing experimental work in these areas is limited, due to the experimental challenges of measuring some of these kinetically unstable and extremely reactive species in the laboratory.[5,12]

Quantum chemistry could offer more tractable estimates of the thermochemistry of chloramines, bromamines and bromochloramines. We developed and applied a computational method, named TA14,[13] to predict the electronic structures and the thermodynamic properties of halamines.

2 COMPUTATIONAL METHODS

Free energies of formation of halamines were computed using the Gaussian09[14] and CFOUR[15] program suites. The TA14 composite method combines a systematic sequence of wavefunction theory calculations, including the evaluations of dynamical and static electron correlation (CCSDTQ), relativistic effects, and anharmonic vibrations.

Free energies of reaction were computed as follows:

$$\Delta G_{rxn,aq} = \Delta G_{rxn,gas}\,(TA14) + \Delta\Delta G_{rxn,solv} \tag{4}$$

where $\Delta G_{rxn,gas}$ (TA14) is the gas phase free energy of reaction computed with the TA14 method and $\Delta\Delta G_{rxn,solv}$ is the free energy of solvation.

To assess the effects of aqueous solvation, one explicit water molecule was introduced and the SMD[16] implicit continuum solvation model was applied with the B2PLYPD[17,18] electronic structure method. The aug-cc-pVQZ[19–21] basis set was employed in all the calculations in solvation phase, except when NBr_3 and $NBrCl_2$ were involved in the reaction: in this case the aug-cc-pVTZ basis set was used, due to convergence difficulties with the larger basis set.

Free energies of solvation at 1 M standard state[22] for each compound A were calculated taking advantage of the cluster-continuum approach by Bryantsev et al.:[23]

$$\Delta G_{solv}(A)^* = G_{gas}(A\text{-}H_2O) - G_{gas}(A) - G_{gas}(H_2O) + \Delta G_{smd}(A\text{-}H_2O) - \Delta G_{smd}(H_2O) - \Delta G^{0\to *} - RT\ln[H_2O] \tag{5}$$

where the term "$G_{gas}(A\text{-}H_2O) - G_{gas}(A) - G_{gas}(H_2O)$" is the gas phase binding energy as described in Bryantsev's paper, $\Delta G_{smd}(A\text{-}H_2O)$ and $\Delta G_{smd}(H_2O)$ represent the free energies needed to bring the A-H_2O and H_2O from gas-phase to aqueous phase, as evaluated with the SMD implicit solvation model. The terms $\Delta G^{0\to *}$ and $RT\ln[H_2O]$ correspond to the standard state and cluster correction contribution to the free energy of reaction.

3 RESULTS

Reaction free energies in gas phase were established based on the TA14 method, whereas the solvation contributions were taken into account with the cluster continuum approach.

These computational predictions (Table 1) have an estimated accuracy of 2 logarithmic unit in the prediction of aqueous equilibrium constants (Table 1). The major contribution to the error is the calculation of the solvation effects on the free energy of reaction.

Observed discrepancies between K_{eq}^{Comp} and K_{eq}^{Expt} indicate that some experimental values needed to be revised.

4 CONCLUSIONS

Aqueous equilibrium constants describing the formation of halamines were provided with an estimated accuracy of about 2 logarithmic unit.

The results of this study and the developed computational approach have significant implications for the interpretation and revision of existing equilibrium and kinetic data describing reactions of halamines. In particular, improved estimates of reaction thermodynamics will help us to properly assess the formation rate constants of these reactive halogenated species, allowing future models of water disinfection treatments to be more accurately formulated and tested.

Table 1. *Computed and Experimental Aqueous Free Energies of Formation of Chloramines, Bromamines, and Bromochloramines.*

Reaction	Log K_{eq}^{Comp}	Log K_{eq}^{Expt}
$NH_3 + HOCl \leftrightarrow NH_2Cl + H_2O$	11.81	13.04[3]
$NH_2Cl + HOCl \leftrightarrow NHCl_2 + H_2O$	11.75	(est. 10.00)[3]
$NHCl_2 + HOCl \leftrightarrow NCl_3 + H_2O$	9.22	(est. 6.4)[3]
$NH_3 + HOBr \leftrightarrow NH_2Br + H_2O$	10.06	(est. 12.22)[24]
$NH_2Br + HOBr \leftrightarrow NHBr_2 + H_2O$	9.87	(est. 10.41)[24]
$NHBr_2 + HOBr \leftrightarrow NBr_3 + H_2O$	11.51	(est. 8.44)[24]
$NH_2Br + HOCl \leftrightarrow NHBrCl + H_2O$	12.57	Not available
$NH_2Cl + HOBr \leftrightarrow NHBrCl + H_2O$	10.82	Not available
$NHBrCl + HOCl \leftrightarrow NBrCl_2 + H_2O$	12.00	Not available
$NHBrCl + HOBr \leftrightarrow NBr_2Cl + H_2O$	-7.77	Not available
$NHCl_2 + HOBr \leftrightarrow NBrCl_2 + H_2O$	11.65	Not available
$NHBr_2 + HOCl \leftrightarrow NBr_2Cl + H_2O$	-4.99	Not available

References

(1) Weil, I.; Morris, J. C. *J. Am. Chem. Soc.* **1949**, *71*, 1664–1671.

(2) Morris, J. C. In *Principles and Applications of Water Chemistry*; Faust, S. D.; Hunter, J., Eds.; 1967; pp. 23–53.

(3) Morris, J. C.; Isaac, R. A. In *Water Chlorination: Environmental Impact and Health Effects. Vol. 4*; Jolley, R. L.; Brungs, W. A.; Cotruvo, J. A.; Cumming, R. B.; Mattice, J. S.; Jacobs, V. A., Eds.; Ann Arbor Science, Ann Arbor, MI, 1981; Vol. 4, pp. 50–62.

(4) Margerum, D. W.; Gray, E. T. In *Organometals and Organometalloids, Occurance and Fate in the Environment*; 1978; pp. 278–291.

(5) Jafvert, C. T.; Valentine, R. L. *Environ. Sci. Technol.* **1992**, *26*, 577–586.

(6) Gazda, M.; Margerum, D. W. *J. Inorg. Chem.* **1994**, *33*, 118–123.

(7) Gazda, M.; Dejarme, L. E.; Choudhury, T. K.; Cooks, R. G.; Lafayette, W. *Environ. Sci. Technol.* **1993**, *27*, 557–561.

(8) National Research Council. *Drinking Water and Health, Volume 2*; The National Academies Press, Washington, DC, 1980.

(9) National Research Council. *Drinking Water and Health, Volume 7 Disinfectants and*

Disinfectants By-Products; The National Academies Press, Washington, DC, 1987.

(10) Diehl, A. C.; Speitel Jr., G. E.; Symons, J. M.; Krasner, S. W.; Hwang, C. J.; Barrett, S. E. *J. AWWA* **2000**, *92*, 76–90.

(11) Lei, H.; Minear, R. A.; Mariñas, B. J. *Environ. Sci. Technol.* **2006**, *40*, 2559–2564.

(12) Hery, M.; Hecht, G.; Gerber, J. M.; Gendre, J. C.; Hubert, G.; Rebuffaud, J. *Ann. occup. Hyg.* **1995**, *39*, 427–439.

(13) Trogolo, D.; Arey, J. S. *Phys. Chem. Chem. Phys.* **2014**, In Press.

(14) Frisch, M. J.; Trucks, G. W.; Schlegel, H. B.; Scuseria, G. E.; Robb, M. A.; Cheeseman, J. R.; Scalmani, G.; Barone, V.; Mennucci, B.; Petersson, G. A.; Nakatsuji, H.; Caricato, M.; Li, X.; Hratchian, H. P.; Izmaylov, A. F.; Bloino, J.; Zheng, G.; Sonnenberg, J. L.; Hada, M.; Ehara, M.; Toyota, K.; Fukuda, R.; Hasegawa, J.; Ishida, M.; Nakajima, T.; Honda, Y.; Kitao, O.; Nakai, H.; Vreven, T.; Montgomery Jr., J. A.; Peralta, J. E.; Ogliaro, F.; Bearpark, M.; Heyd, J. J.; Brothers, E.; Kudin, K. N.; Staroverov, V. N.; Kobayashi, R.; Normand, J.; Raghavachari, K.; Rendell, A.; Burant, J. C.; Iyengar, S. S.; Tomasi, J.; Cossi, M.; Rega, N.; Millam, M. J.; Klene, M.; Knox, J. E.; Cross, J. B.; Bakken, V.; Adamo, C.; Jaramillo, J.; Gomperts, R.; Stratmann, R. E.; Yazyev, O.; Austin, A. J.; Cammi, R.; Pomelli, C.; Ochterski, J. W.; Martin, R. L.; Morokuma, K.; Zakrzewski, V. G.; Voth, G. A.; Salvador, P.; Dannenberg, J. J.; Dapprich, S.; Daniels, A. D.; Farkas, Ö.; Foresman, J. B.; Ortiz, J. V.; Cioslowski, J.; Fox, D. J. Gaussian09 Revision D.01, 2009.

(15) J.F. Stanton, J. Gauss, M.E. Harding, P. G. S.; with contributions from A.A. Auer, R.J. Bartlett, U. Benedikt, C. Berger, D.E. Bernholdt, Y.J. Bomble, L. Cheng, O. Christiansen, M. Heckert, O. Heun, C. Huber, T.-C. Jagau, D. Jonsson, J. Jusélius, K. Klein, W.J. Lauderdale, D.A. Matthews, T. Metzroth, L, A. V. M. and C. van W. CFOUR, Coupled-Cluster techniques for Computational Chemistry, a quantum-chemical program package with integral packages MOLECULE, PROPS, ABACUS and ECP routines, 2010.

(16) Marenich, A. V.; Olson, R. M.; Kelly, C. P.; Cramer, C. J.; Truhlar, D. G. *J. Chem. Theory Comput.* **2007**, *3*, 2011–2033.

(17) Schwabe, T.; Grimme, S. *Phys. Chem. Chem. Phys.* **2007**, *9*, 3397–3406.

(18) Grimme, S. *J. Chem. Phys.* **2006**, *124*, 034108.

(19) Dunning Jr., T. H. *J. Chem. Phys.* **1989**, *90*, 1007–1023.

(20) Kendall, R. A.; Dunning Jr., T. H.; Harrison, R. J. *J. Chem. Phys.* **1992**, *96*, 6796–6806.

(21) Dunning Jr, T. H.; Peterson, K. A.; Wilson, A. K. *J. Chem. Phys.* **2001**, *114*, 9244.

(22) Ben-Naim, A. *Solvation Thermodynamics*; Plenum Press, New York, 1987; p. 246.

(23) Bryantsev, V. S.; Diallo, M. S.; Goddard III, W. A. *J. Phys. Chem. B* **2008**, *112*, 9709–9719.

(24) Hofmann, R.; Andrews, R. C. *Water Res.* **2001**, *35*, 599–604.

CHLORAMINATION AND BROMAMINATION OF AMINO ACIDS

Virginie Simon, Florence Berne and Hervé Gallard

IC2MP, UMR CNRS 7285, Equipe Eaux, Géochimie Organique, Santé, 1 rue Marcel Doré TSA 41105 86073 Poitiers cedex 9

1 INTRODUCTION

Chlorine is widely used as a disinfectant to treat drinking water and more generally to inactivate microorganisms in many human activities. But its high reactivity with natural organic matter leads to the formation of harmful disinfection by-products (DBPs) that are now regulated in many countries. Since the detection of chloroform in water disinfected by chlorine,[1] some countries have used monochloramine (NH_2Cl) as an alternative disinfectant because of its higher stability in system networks and its lower reactivity towards natural organic matter.[2]

Among all model compounds, the reactivity of chlorine and monochloramine with amino acids has been investigated because it partly explains the inhibition activity of microorganisms by these oxidants.[3] The reactions of chlorine or chloramine with amino acids produced N-chloroorganic amines and DBPs such as cyanogen chloride, chloropicrine, dichloroacetonitrile, trihalomethanes and haloacetic acids.[4-6]

In the presence of bromide ion and ammonia, other haloamines such as monobromamine NH_2Br can be formed during disinfection of water by chlorine and ozone. Compared to chlorine the reactivity of haloamines has been less studied and their reactivity with organic compounds is not well known. Few kinetic studies determined rate constants for reactions between NH_2Cl and model organic compounds.[7-10] No data are available for bromamines, which can be explained by their low stability,[11] which makes their investigation challenging. The chemistry between such nitrogeneous compounds and chlorine and chlorine derivatives is also of interest in swimming pools because it is a source of toxic volatile DBPs.

The aim of this work was to study the bromamination and chloramination of amino acids. The kinetic studies of monochloramination and monobromamination of several amino acids were conducted in a continuous flow reactor in an excess of amino acids to simulate pseudo-first order kinetics. Also, the formation potentials of trihalomethanes (THMs) and haloacetic acids (HAAs) were determined in a batch reactor in excess of oxidant with a contact time of 72h.

2 EXPERIMENTAL SECTION

2.1 Chemicals and preparation of solutions

All solutions were prepared from reagent-grade chemicals and purified water delivered by a Millipore system (Milli RX). A solution of monochloramine (2 mM) was prepared daily by progressively mixing a solution of chlorine into an ammonium chloride solution in a stirred reactor and using a nitrogen to chlorine ratio equal to 2. The solution was maintained at pH 8 using 10 mM phosphate buffer. The concentration of monochloramine was determined iodometrically with sodium thiosulfate 0.01 M and by measuring the absorbance at 244 nm (Epsilon = 450 $L.mol^{-1}cm^{-1}$).

A solution of monobromamine (0.4 mM) was prepared by ozonation of a bromide solution to form hypobromous acid,then ammonium chloride solution was added with a nitrogen to bromine ratio equal to 100. The solution was maintained at pH 8.5 using 10 mM borate buffer to reduce the decomposition of monobromamine and to avoid dibromamine formation. The concentration of monobromamine was determined by measurement of the absorbance at 278 nm (Epsilon = 380 $L.mol^{-1}cm^{-1}$). In the presence of borate buffer, monobromamine was sufficiently stable to perform kinetic experiments for 1 hour after preparation.

2.2 Experimental procedures

2.2.1 Kinetic experiments. All kinetic experiments were conducted at 20°C in a continuous flow reactor with an excess of amino acid compared to the concentration of oxidant (R=20 for chloramination and R=10 for bromamination) to obtain the pseudo-first order rate constant. The continuous-flow reactor consisted of a dual-channel syringe pump KDS model 200 series (KDScientific, Holliston, MA, USA) used to deliver the haloamine and amino acid solutions at the same flow rate. Two 50-mL syringes containing each reactant solution were connected with Peek tube to a 1/16" mixing tee. The mixing tee was then connected to a reaction tube (0.88 mm-ID Peek tube) where the reaction took place. The volumes of the reactor were equal to 0.602 mL and 0.228 mL respectively for chloramination and bromamination experiments. The other extremity of the reaction tube was connected to a spectrophotometer cell placed in a SAFAS UV mc2 spectrophotometer. The whole set-up was in a thermostated room at 20°C. The UV spectra were obtained between 200 and 400 nm for different flow rates i.e. different reaction times. The experiments were performed with 12 aminoacids. Aromatic aminoacids were not studied because they absorb in the UV like haloamines and N-haloaminoacids.

2.2.2 Disinfection by-product formation potentials. Experiments were conducted in a batch reactor for a contact time of 72 h with an excess of haloamines compared to amino acids (see Table 1 for experimental conditions). Three aromatic amino acids i.e. phenylalanine, tryptophan and tyrosine were tested. Reaction was initiated by adding an appropriate volume of amino acid solution into a buffered solution of haloamine. 10 mL aliquots of the mixing solution were immediately transferred into a series of 20 mL headspace vials which were then hermetically sealed with PTFE-faced-silicone septa. Vials were kept in a thermostated room at 20°C for 72h. The reaction was stopped by adding 70 μL of HCl, 1M and a slight excess of ascorbic acid. Trihalomethanes were analysed using a gas chromatograph (Agilent 7890A) equipped with an automatic headspace sampler (Combi Pal CTC Analytic) and coupled to a mass spectrometrer

(Agilent 5975C). The capillary column was a HP-5MS 30m × 0.25 mm (1 μm internal diameter). Separation was carried out with nitrogen as carrier gas at a flow rate of 1 mL/min. Operating temperatures were 120 °C, 40 °C and 280 °C for the injector, the oven and the detector respectively. Detection was carried out using Single Ion Monitoring. Vials were introduced into the headspace sampler for 10 min at 50 °C. Calibration curves were constructedfrom standard THM solutions from 4 to 200 μM. The analytical conditions are reported in table 2.

Table 1 *Experimental conditions for DBP formation potential*

Conditions	**NH_2Cl**	**$NH_2Cl + Br^-$**	**NH_2Br**
[Oxidant] (mM)	2	2	0.4
[AA] (μM)	100	100	20
pH	8	8	8.5
[Br^-] (mM)	0	2	0
Buffer	10 mM phosphate	10 mM phosphate	10 mM borate
T (°C)	20	20	20

Table 2 *Analytical conditions for the analysis of THMs by headspace-GC/MS*

Temperature gradient		**40 °C (3 min)** 5 °C/min **100°C (1 min)** 50 °C/min **220°C (2 min)**
GC column		HP-5MS (30 m x 0,25 mm I.D., thickness : 1 μm)
Injector	Volume Temperature Split	2500 μL 120 °C Pulsed Split 1:10
Detector	Transferline EI Source Quadrupole	280 °C 230 °C 150 °C
SIM acquisition mode	**Group** 1 2 3 4	**m/z** — **Analyte** 47-83-85 — Trichloromethane 83-85-127-129 — Bromodichloromethane 127-129 — Dibromochloromethane 171-173 — Tribromomethane

Table 3 *Analytical conditions for the analysis of haloacetic acids*

Temperature gradient		35 °C (3 min) — 2 °C/min — 55°C (1 min) — 5 °C/min — 85°C (2 min) — 10 °C/min — 150°C — 40 °C/min — 200°C (2 min)
GC column		HP-5MS (30 m x 0,25 mm I.D.; thickness : 0,25 μm)
Injector	Volume	1 μL
	Temperature	150 °C
	Split	Pulsed Splitless
Detector	Transferline	280 °C
	EI Source	150 °C
	Quadrupole	230 °C

	Group	**m/z**	**Analyte**
SIM acquisition mode	1	59 – 77	MCAA
	2	59 – 93	MBAA
	3	59 – 83	DCAA
	4	59 – 117	TCAA
	5	59 – 127	BCAA
	6	59 – 107	DBAA
	7	59 – 141	BDCAA
	8	59 - 128	CDBAA
	9	171-251-253	TBAA

Haloacetic acids were analysed following EPA method 552.2. 10 μL of surrogate standard (2-bromopropanoïc acid at 20 mg/L in MeOH), 3g of sodium sulfate anhydrous, 0.8 mL sulphuric acid (96%) and 3 mL methyl tert-butyl ether (MTBE) were added with 10 mL sample to a 15 mL conical-bottom tube. The mixture was vigorously shaken for 1 min. After 2 min the organic phase was transferred into another 15 mL conical-bottom tube containing 1 mL sulphuric acid at 10 % v/v in MeOH. The tube was closed and heated at 50 °C for 2h. After cooling, 4 mL of a saturated solution of sodium carbonate was added to quench the reaction. The organic phase was transferred into a 2 mL vial for analysis by GC-MS using the same system as detailed previously for THMs, except that the MTBE extract was injected in pulsed splitless mode. The analytical conditions are listed in table 3.

3 RESULTS AND DISCUSSION

3.1 Determination of rate constants for chloramination and bromamination

Figure 1 shows as an example the evolution of UV spectra for a 40 mM glycine solution in presence of 2 mM chloramine for different reaction times.

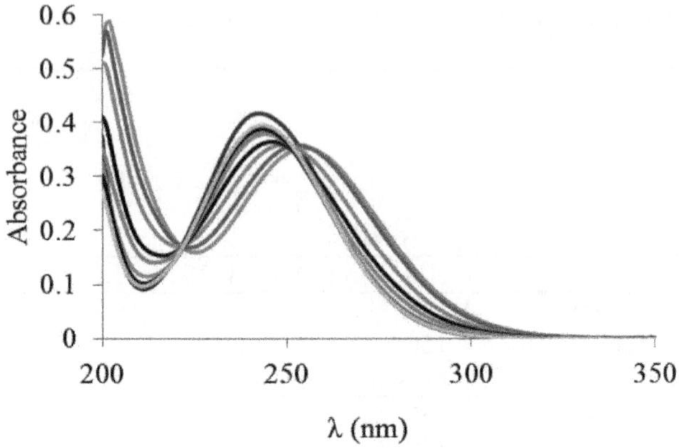

Figure 1 *Evolution of UV spectra during chloramination of a solution of glycine ($[NH_2Cl]_0$=2 mM, $[Glycine]_0$ =40 mM, pH 8, 10 mM phosphate buffer, 20°C)*

The results clearly show the decay of monochloramine at a wavelength of 244 nm and a shift of absorbance to higher wavelengths due to the the formation of the N-chloroglycine. In our conditions i.e. with an excess of aminoacids, the absorbance was the sum of the absorbance of NH_2X and N-haloaminoacid (N-X-AA). The concentration of NH_2X was determined from the absorbances at two wavelengths by solving the two equations 1 and 2.

$$A(\lambda_1) = l(\varepsilon_1[NH_2X] + \varepsilon'_1[N\text{-}X\text{-}AA]) \quad (1)$$
$$A(\lambda_2) = l(\varepsilon_2[NH_2X] + \varepsilon'_2[N\text{-}X\text{-}AA]) \quad (2)$$

The wavelenghts λ_1 and λ_2 were chosen as 244 and 280 nm for chloramination and 278 and 290 nm for bromamination. The extinction coefficients of N-X-AA were determined for complete consumption of NH_2X and assuming that the final concentration of N-X-AA was equal to the initial concentration of NH_2X.

The concentrations of haloamines were calculated for all reaction times and the logarithmic decay of haloamines showed a linear curve for our conditions (see Figure 2 for an example). The pseudo-first-order rate constant k_{obs} was obtained from the slope of the linear curve. We verified that k_{obs} increased linearly with the concentration of amino acid (see Figure 3).

Thus, the reaction of haloamine could be considered as a second-order kinetic law, first-order with respect to each reactant and the apparent second-order rate constant k_{app} was determined from k_{obs} according to the following equations:

$$-d[NH_2X]/dt = k_{app}[AA][NH_2X]$$
$$-d[NH_2X]/dt = k_{obs}\,[NH_2X] \text{ and } k_{app} = k_{obs}/[AA]$$

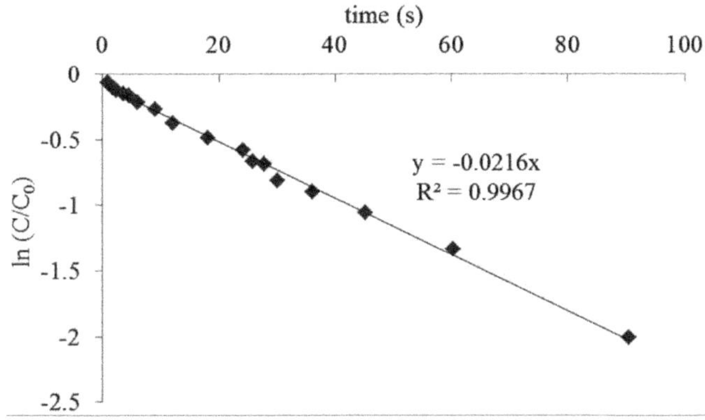

Figure 2 *Logarithmic decay of NH_2Cl in presence of glycine ($[NH_2Cl]_0$=2 mM, $[glycine]_0$=40 mM, pH 8, 10 mM phosphate buffer, 20°C)*

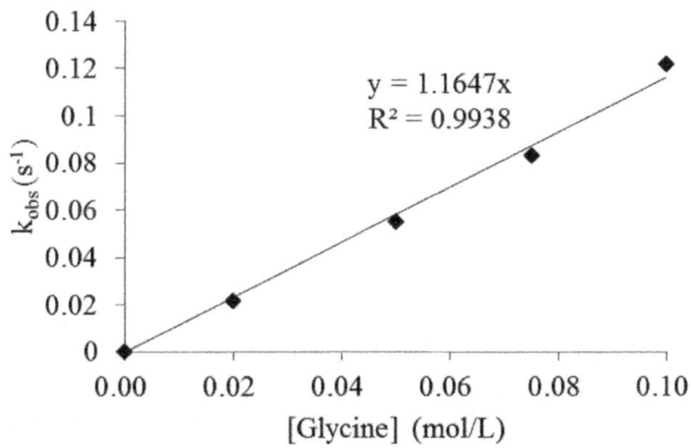

Figure 3 *Evolution of k_{obs} as a function of the concentration of glycine ($[NH_2Cl]_0$=2 mM, $[Glycine]_0$=40-200 mM, pH 8, 10 mM phosphate buffer, 20°C)*

The apparent second-order rate constants of the initial reaction of NH_2Cl at pH 8 and NH_2Br at pH 8.5 with 12 aminoacids are reported in table 4.

Table 4 *Kinetic rate constants for the chloramination and bromamination of selected amino acids.*

Amino acids		k with NH_2Cl in $M^{-1}s^{-1}$ at 20°C	k with NH_2Br in $M^{-1}s^{-1}$ at 20°C
diacid	Glutamic acid	1.25 ± 0.11	509 ± 21
dibasic	Histidine	2.13 ± 0.10	681 ± 29
	Arginine	1.53 ± 0.05	551 ± 17
	Lysine	1.09 ± 0.05	477 ± 34
amide	Glutamine	2.11 ± 0.11	659 ± 26
hydroxylated	Serine	1.94 ± 0.11	704 ± 40
	Threonine	1.85 ± 0.07	690 ± 12
hydrophobic	Glycine	1.08 ± 0.01	368 ± 11
	Valine	0.95 ± 0.01	155 ± 20
	Leucine	0.92 ± 0.02	153 ± 24
	Isoleucine	0.90 ± 0.04	138 ± 5
	Alanine	0.71 ± 0.09	132 ± 8

For monochloramine, rate constants varied from 0.71 $M^{-1}s^{-1}$ for alanine to 2.13 M^{-1} s^{-1} for histidine. The values were similar to the values of 0.55, 1.52 and 2.55 M^{-1} s^{-1} obtained by Snyder and Margerum for alanine, glycine and threonine, respectively,[7] and values of 0.88 and 1.30 $M^{-1}s^{-1}$ obtained by Issac and Morris for alanine and glycine respectively at pH 8 and 25°C.[8] Values of monobromamine determined in our study were much higher since the rate constants varied between 132 M^{-1} s^{-1} for alanine and 704 M^{-1} s^{-1} for serine. Figure 4 shows that a relatively good linear correlation exists between both sets of rate constants. The reactivity of monobromamine with amino acids is about 400-fold more important than the reactivity of monochloramine. Thus, bromine transfer is faster than chlorine transfer, which is explained by the higher electrophilic nature of the bromine atom compared to the chlorine atom.

Rate constants were plotted against the pKa values of the amine group in Figure 5. For both NH_2Cl and NH_2Br, three areas could be distinguished: 1) amino acids with hydrophobic side chains showing the lowest rate constants and pKa values of 9.7 – 9.9, 2) glutamic acid with an intermediate reactivity and 3) amino acids with polar side chains with the highest rate constants. For the dibasic amino acids, the reactivity increased with the number of nitrogen atoms in the side chain and the highest rate constant was obtained for histidine with the imidazole group.

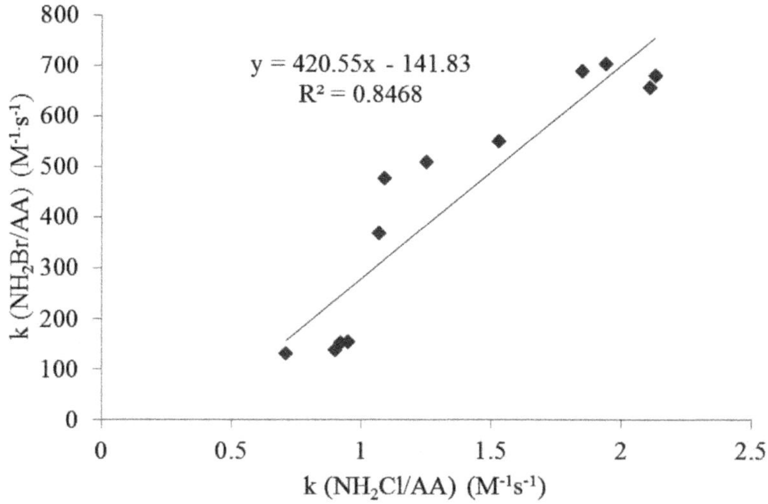

Figure 4 *Relation between kinetic rate constants obtained for chloramination and bromamination of amino acids*

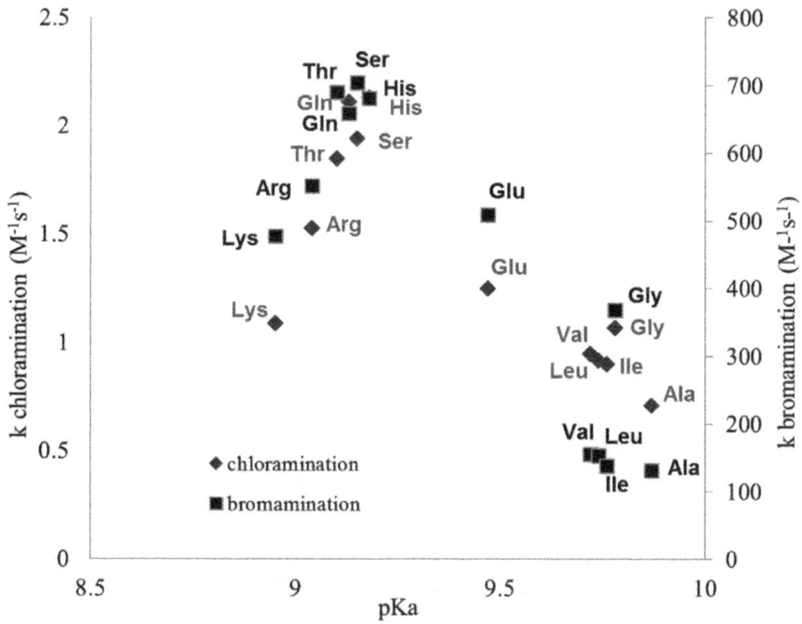

Figure 5 *Relation between kinetic rate constants and the nature of the side chain of amino acids*

3.2 Disinfection by-product formation potentials

3.2.1 THM-FP Figures 6a), b) and c) presents THM yields obtained by chloramination, bromamination and chloramination in presence of bromide.

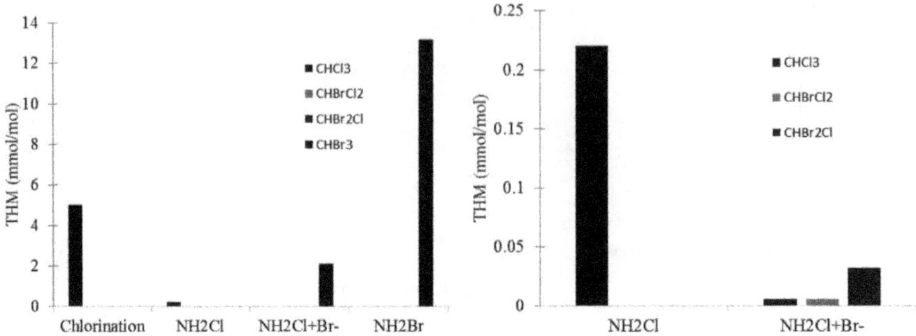

Figure 6a

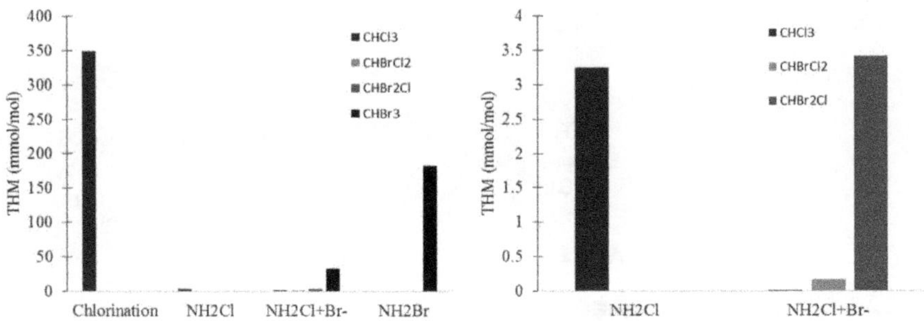

Figure 6b

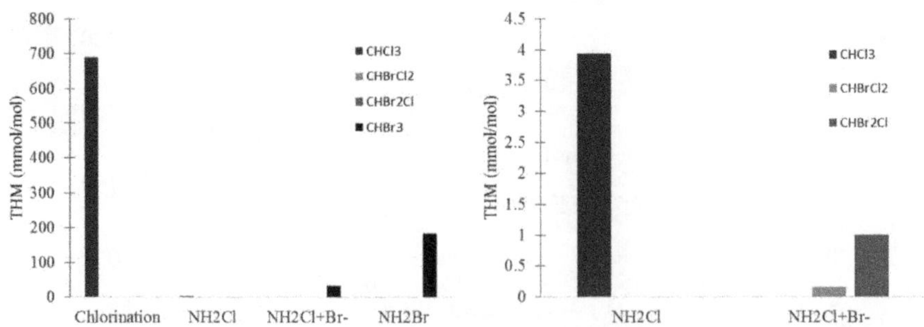

Figure 6c

Figures 6 *THM formation during chloramination, bromamination and chlorination of a) phenylalanine, b) tyrosine and c) tryptophan (Results of chlorination are from Hureiki et al.*[12] *Figures on the right side are zooms of figures on the left side).*

Results obtained for chlorination[12] were plotted for comparison. Compared to tyrosine and tryptophan, phenylalanine produced less THMs whatever the oxidant tested. The yields of chloroform from chloramination were 0.22, 3.25 and 3.94 mmol/mol for phenyalanine, tyrosine and tryptophan compared to 5, 348 and 690 mmol/mol, respectively for chlorination. Thus, chloroform formation was much lower during chloramination than chlorination, as is already known for model organic compounds and natural organic matter.

Bromoform molar yields produced by bromamination were up to 50 times higher than yields of chloroform obtained by chloramination. Finally, the experiments performed with chloramine in presence of bromide showed a mixture of the 4 THMs. Bromoform was the main THM formed and was produced by NHBrCl. These results are in agreement with fast Br rate transfer from bromamine compared to Cl transfer from chloramine as illustrated by rate constants determined previously.

3.2.2 HAA-FP Figure 7 shows the HAA yields obtained by chloramination and bromamination of the 3 aromatic amino acids.

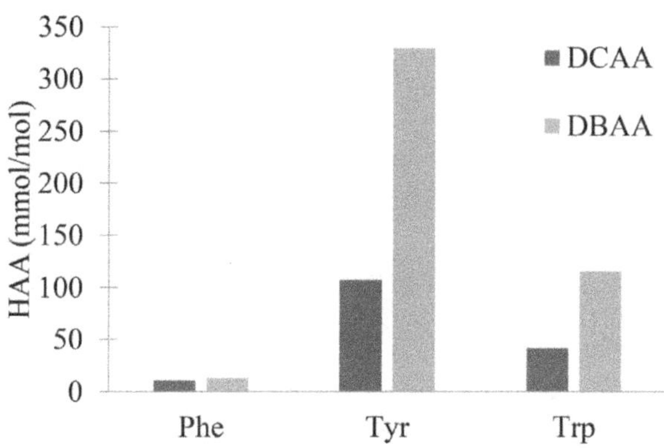

Figure 7 *Formation of dichloroacetic acid and dibromoacetic acid from chloramination and bromamination of aromatic amino acids*

DHAA was the major HAA and DBAA produced by bromamination was up to 3 times higher than DCAA produced during chloramination. As for THMs, phenylalanine formed a low amount of HAAs compared to tyrosine and tryptophan.

3 CONCLUSIONS

Monochloramine and monobromamine react with amino acids producing N-halo amino acids as the initial step. Kinetic rate constants ranged from 0.71 $M^{-1}s^{-1}$ for alanine to 2.13 $M^{-1}s^{-1}$ for histidine for monochloramination and were 400-fold higher for NH_2Br with rate constants between 132 $M^{-1}s^{-1}$ for alanine and 704 $M^{-1}s^{-1}$ for serine.

Monobromamination of aromatic amino acids produced much higher THM and HAA yields than chloramination.

Bromine transfer from haloamines to amine compounds such as amino acids is faster than Chlorine transfer. Higher disinfection efficiency is expected in presence of NH_2Br but it will generate more DBPs than NH_2Cl.

References

1 J.J. Rook, *Water Treatment and Examination*, 1974, **23**, p 234-243
2 G. Hua, D.A. Reckhow, *Water Research*. 2007, **41**, p 1667–1678
3 M.M. Donnermair, E.R. Blatchley III, Water Research, 2003, **37,** p 1557-1570
4 X. Yang, C. Fan, C. Chang, Q. Zhao *Water Research*. 2010, **44**, p 2691-2702
5 H. Chang, C. Chen, G. Wang, *Water Research*. 2011, **45**, p 3753-3764
6 X. Yang, Q. Shen, W. Guo, J. Peng, Y. Liang, C*hemosphere*, 2012, **88**, p 25-32
7 M.P. Snyder, D.W. Margerum, *Inorganic Chemistry*, 1982, **21**, p 2545-2550
8 R.A. Issac, J.C. Morris, *Environmental Science & Technology*. 1983, **17**, p 738-742
9 N. Cimetière, F. Dossier-Berne, J. De Laat, *Environmental Science & Technology,* 2009,**43**, p 9380-9385
10 Y. Wan, A. Jia, Z. Zhu, J. Hu, *Chemosphere,* 2013, **90**, p 1427-1434
11 H. Lei, B. J. Mariñas, R. A. Minear, *Environmental Science & Technology*, 2004, **38**, p 2111-2119
12 L. Hureiki, J.P. Croué, B. Legube B. *Water Research*. 1994, **28**, p 2521-253

FATE OF THE ANTIDIABETIC DRUG METFORMIN DURING CHLORINE DISINFECTION OF WATER

D. Armbruster[1], O. Happel[1*], M. Scheurer[1], K. Harms[3], T. C. Schmidt[2], H.-J. Brauch[1]

[1] DVGW-Technologiezentrum Wasser (TZW), Karlsruher Str. 84, 76139 Karlsruhe, Germany; eMail: oliver.happel@tzw.de
[2] University of Duisburg-Essen, Faculty of Chemistry, Universitätsstr. 5, 45141 Essen, Germany
[3] Philipps-University of Marburg, Faculty of Chemistry, Hans-Meerwein-Str., 35032 Marburg, Germany

1 INTRODUCTION

Metformin is the current gold-standard drug for treatment of *Diabetes mellitus Type 2*[1] and prescribed at annual quantities of more than 1100 t within Germany.[2, 3] It is not metabolized by the human body and only partially degraded during biological wastewater treatment.[4] As an environmental pollutant of anthropogenic origin, metformin is present in the range of several hundred ng/L up to one µg/L in most surface waters.[5, 6, 7]

The production of drinking water from surface waters containing metformin is expected to lead to the formation of disinfection byproducts during chlorine treatment. Scheurer *et al.* reported that aqueous solutions of metformin rapidly develop an intensive yellow colour upon treatment with sodium hypochlorite (NaOCl; Figure 1).[8] Investigations on the chemical reactions taking place during the transformation of metformin by NaOCl led to the discovery of two transformation products **Y** and **C**, which are specifically formed during the chlorination of metformin. The primary *yellow* transformation product **Y** is a cyclic triazole-derivate with an absorption maximum at 385 nm, carrying an active chlorimino-moiety. It undergoes slow decomposition into the chloroorganic nitrile **C**, a secondary *colourless* and stable transformation product.

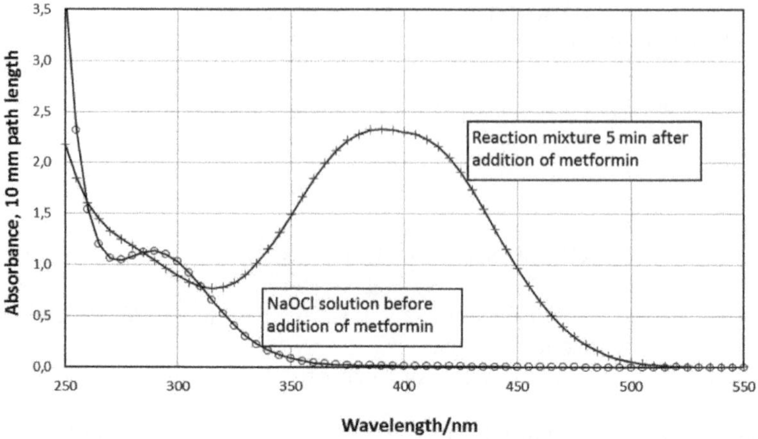

Figure 1 *Development of an absorption maximum at 385 nm during 5 min after addition of metformin (4 mmol/L) to a NaOCl solution (16 mmol/L; 4 Eq)*

2 AIMS AND METHODS

The aim of this study was to characterize the chemical processes taking place during and after chlorine treatment of aqueous metformin solutions. Reaction pathways, kinetics and stoichiometry, as well as the determination of the influence of pH and temperature were investigated. An important aspect was to determine whether the transformations take place in tap water matrix and at trace concentrations. The transformation products **Y** and **C** were purified by column chromatography, recrystallized and characterized by HPLC-DAD, GC-MS, HPLC-ESI-TOF, ^{1}H-NMR and single-crystal X-ray structure determination. Formation rates were observed by HPLC-MS analyses and determination of active chlorine using flow injection analysis (FIA) with DPD-reagent.

3 FORMATION OF **Y** AND DEGRADATION OF **Y** TO **C**

Rapid formation of the primary transformation product **Y** was observed in the ranges between 10 °C and 30 °C. At initial concentrations of 100 µmol/L metformin and 500 µmol/L NaOCl (35 mg/L; 5 Eq) the reaction was completed within 20 min (at 30 °C) and 60 min (at 10 °C), respectively. A decrease of pH and an increase of the absorption at 385 nm were observed (Figures 1 and 2). The reaction proceeded between pH 3 and pH 10, in both ultrapure and tap water, even at trace quantities of reactants (µg/L-range for metformin, mg/L-range for free chlorine). Thus, it is expected to take place at the conditions applied during drinking water disinfection.

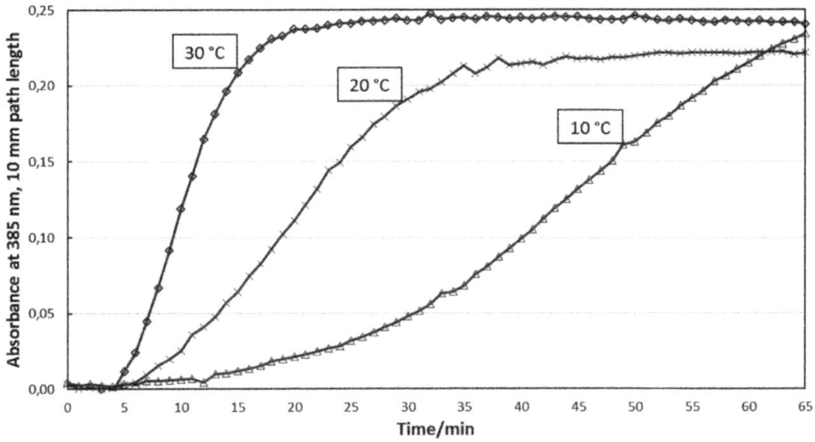

Figure 2 *Formation of **Y** during treatment of metformin (100 µmol/L) with NaOCl (500 µmol/L): observation of UV absorption at 385 nm at 10 °C, 20 °C and 30 °C*

Y is not stable in aqueous solution or in crystalline state at ambient conditions. In ultrapure water, at low temperature and absence of light its decomposition rate is low, whereas decomposition proceeds within days in the presence of light, elevated temperature and tap water (Figure 3). **Y** decomposed within hours after dissolution in organic solvents. Trace amounts of **Y** in tap water were found to decompose within minutes due to the influence of matrix factors. However, a secondary colourless and refractory degradation product **C** was discovered. It has a low photolysis rate at ambient day light, while being resistant to heat and not readily degraded in the presence of organic solvents. It proved to be stable in tap water at 8 °C and absence of light, suggesting that the substance is persistent in the drinking water distribution network. Under these conditions no significant degradation was observed at an initial concentration of 10 ng/L over 40 days.

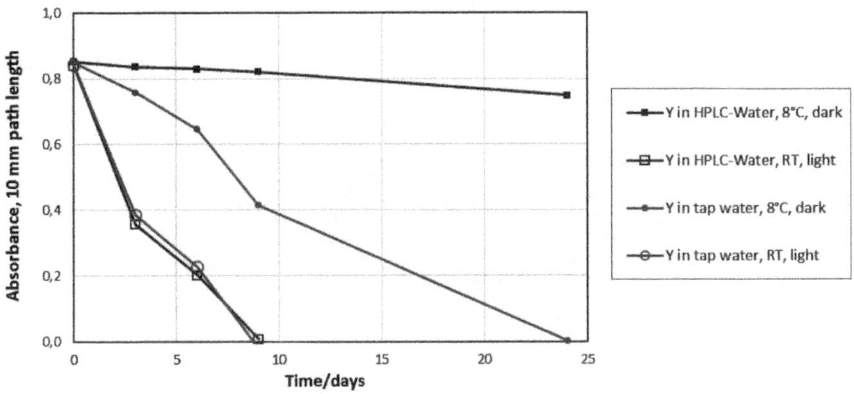

Figure 3 *Decomposition of **Y** (observed at 385 nm) during 25 d in ultrapure water and tap water at ambient light and room temperature or at 8 °C and in darkness. Initial concentration: 20 mg/L (125 µmol/L)*

4 STRUCTURE ELUCIDATION

Both substances were isolated and purified by column chromatography and recrystallization. **Y** was found to deflagrate if heated rapidly beyond 61 °C, whereas **C** proved to be stable towards heat. Due to the heat-sensitivity of **Y** mass-spectrometric analyses were insufficient for structural elucidation. The molecular structures of **Y** and **C** were derived through single-crystal X-ray structure determination (Figure 4). Both molecules are fully planar and yield exactly two signals of identical intensity during ^{1}H-NMR analyses. These properties indicate a positive charge of the dimethylamino-nitrogen along with the linkage to the molecular residue by a double bond, prohibiting free rotation. Although the molecules may be formulated as neutral molecules, they are present in zwitterionic state both in crystalline form as well as in solution.

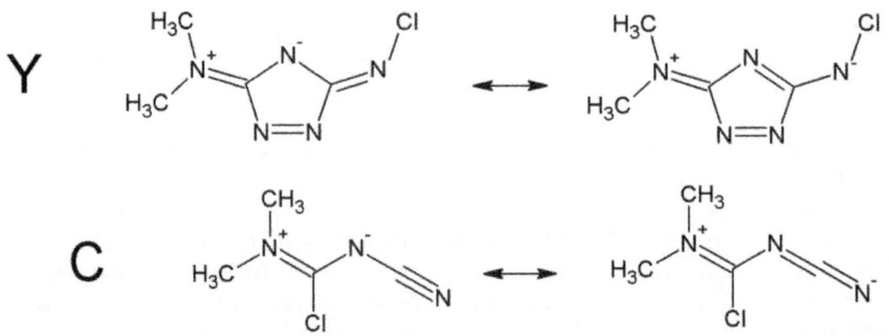

Figure 4 *Mesomeric structures of* ***Y*** *(above) and* ***C*** *(below)*

The immediate major chlorination product **Y** is a cyclic dehydro-1,2,4-triazole derivate ($C_4H_6ClN_5$; Mw = 159.59 g/mol). In its pure state the active chlorimine is a solid substance of intensive yellow colour, crystallizing in long needles (Figure 5). Decomposition of **Y** leads to the formation of a colourless and stable chloroorganic nitrile **C** ($C_4H_6ClN_3$; Mw = 131.57 g/mol), crystallizing in compact needles. Both substances were not described in previous literature and are not listed in the Chemical Abstract Service *Scifinder*.[9]

Figure 5 *Crystalline needles of* ***Y*** *after recrystallization from* H_2O

5 SUGGESTION OF A TRANSFORMATION PATHWAY

The transformation of metformin to **Y** is likely to take place via a sequence of three chlorination steps along with formation of five equivalents of HCl. It involves cyclization and consecutive oxidation of metformin, leading to the metastable product **Y** (Figure 6). Decomposition of **Y** to **C** involves elimination of molecular nitrogen and shift of the chlorine substituent from a nitrogen atom to a carbon atom, which transforms the active chlorimine **Y** into the chloroorganic nitrile **C**.

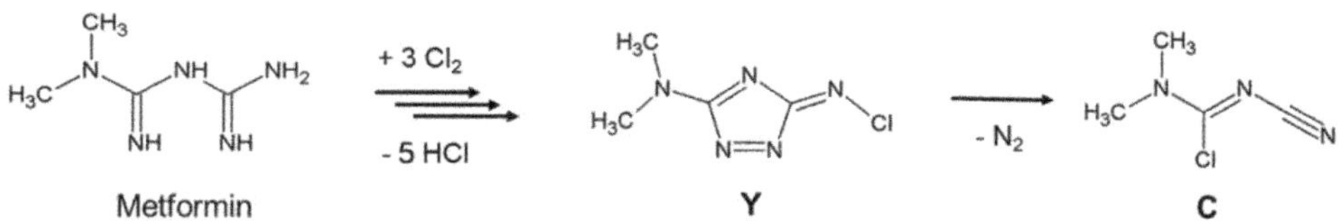

Figure 6 *Transformation of metformin into **Y** by NaOCl and decomposition of **Y** into **C***

6 HPLC-DAD-MS ANALYSES

HPLC-DAD analyses using a reversed phase column (Gemini C18; Phenomenex, Aschaffenburg, Germany) facilitated baseline separation of **Y** and **C**. **C** has an absorption maximum at 242 nm and **Y** at 385 nm, allowing a photometric differentiation of the compounds (Figure 7).

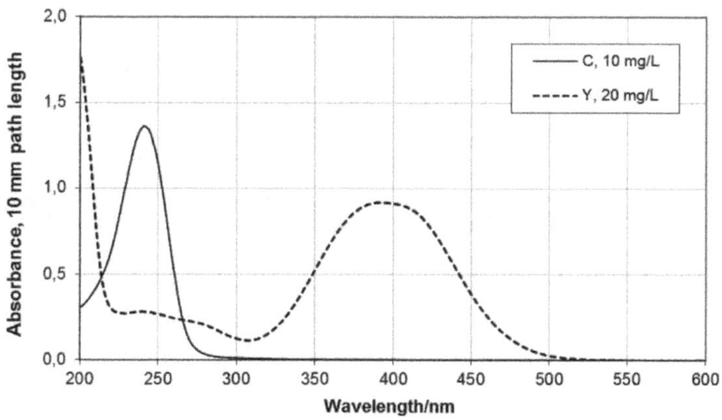

Figure 7 *UV-VIS absorption of purified **Y** and **C***

Due to the instability of **Y** to heat, complete decomposition to **C** was observed during GC-MS analyses. Both analytes yielded identical signals at an identical retention time. **Y** decomposed nearly completely to **C** during HPLC-MS techniques. At a concentration of 1 mg/L it was possible to detect **Y** as molecular ion (m/z = 160.0383 $[M+H]^+$) with a weak signal intensity by HPLC-ESI-TOF. The signal allowed the calculation of the correct empirical formula. At lower concentrations HPLC-MS/MS analyses allowed a differentiation of the two analytes according to their specific retention times (Figure 8). Identical mass signals (m/z = 132.0326 $[M+H]^+$) were detected at 6.3 min (**C**) and 6.8 min (**Y**).

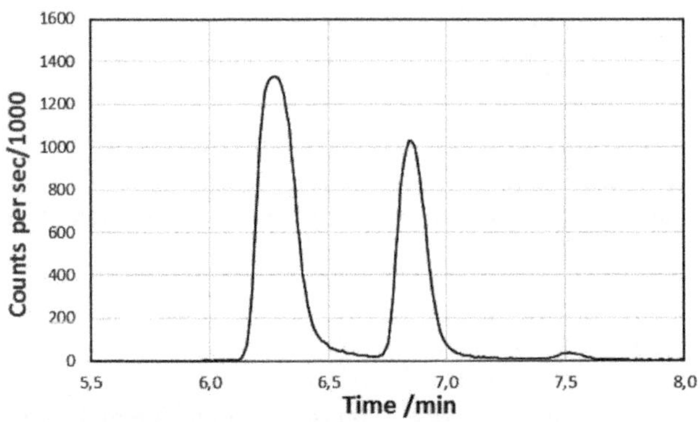

Figure 8 *Detection of* ***Y*** *and* ***C*** *by HPLC-ESI-TOF at a concentration level of 1 mg/L: extracted ion chromatogram for m/z = 132.03 to 132.04*

7 TRACE ANALYSES

After transformation of metformin (100 ng/L) with NaOCl (0.4 mg/L) in tap water, mass signals attributable to **C** were derived by HPLC-MS/MS analysis without prior analyte enrichment. Quantification tests on an AB SCIEX Triple Quad™ 5500 LC-MS/MS system revealed a LOQ below 10 ng/L for **C** by direct injection (S/N-ratio = 15) according to the specific mass transition m/z = 131.9 to m/z = 61.9. Due to the ephemerality of **Y** no reproducible enrichment or detection within the ng/L-range was possible. Enrichment of **C** from the ng/L-range was successfully achieved by solid-phase extraction using Bond Elut PPL cartridges (Agilent, Waldbronn, Germany) with recovery rates beyond 80% at an enrichment factor of 1000.

8 CONCLUSIONS

Two previously unknown chlorine-transformation products of metformin, **Y** and **C**, were discovered and characterized. The substances were purified and crystallized, which allowed structural elucidation by single-crystal X-ray structure determination. Metformin is primarily transformed into a metastable triazole-derivate **Y**, which further decomposes to **C**, a stable chloroorganic nitrile. The reaction was observed at trace concentrations of reactants and is expected to take place during chlorine disinfection of water containing residual metformin. Whereas trace quantities of **Y** rapidly decompose in tap water, the persistence of **C** in the drinking water distribution network is likely.

References

1 M. L. Reitman and E. E. Schadt, *J. Clin. Invest.*, 2007, **117**, 1226.

2 WHO (2014): http://www.whocc.no/atcddd/ (accessed: 31.10.2014).

3 K. Mengel, *Antidiabetika.* In: *Schwabe U, Paffrath D (Hrsg.) Arzneimittel-verordnungs-Report*, Springer, Berlin Heidelberg, 2011, 351.

4 M. Scheurer, F. Sacher and H.-J. Brauch, *J Environ Monit*, 2009, **11**, 1608.

5 D. W. Kolpin, E. T. Furlong, M. T. Meyer, E. M. Thurman, S. D. Zaugg and L. B. Barber, *Environ Sci Technol.*, 2002, **36**, 1202.

6 C. Trautwein, J.-D. Berset, H. Wolschke and K. Kümmerer, *Environment International*, 2014, **70,** 203.

7 E. Vulliet and C. Cren-Olivé, *Environmental Pollution*, 2011, **159**, 2929.

8 M. Scheurer, A. Michel, H.-J. Brauch, W. Ruck, F. Sacher, *Water Research*, 2012, **46,** 4790.

9 CAS Scifinder: http://www.cas.org/products/scifinder (accessed: 31.12.2014).

DBP FORMATION FROM THE CHLORINATION OF ORGANICS IN TEA AND COFFEE

T. Bond, S. Tang and M.R. Templeton

Department of Civil and Environmental Engineering, Imperial College London, South Kensington campus, London SW7 2AZ, UK.

1 INTRODUCTION

Tea is produced from the leaves of *Camellia sinensis* and is classified into three types, depending on the amount of fermentation: green tea, oolong and black tea[1]. The composition of tea leaves is dominated by polyphenols, especially flavonoids, which can comprise up to 30% of the dry weight of tea leaves[2]. Other components include amino acids, sugars and caffeine. Coffee beans are the fruit of coffee plants, with *Coffea Arabica* and *Coffea Robusta* the commonest species[3]. Polysaccharides form more than 50% of green coffee beans, with a large percentage as cellulose. Polyphenols in coffee include chlorogenic acid and caffeic acid. The brown colour of coffee is attributed to melanoidins, brown nitrogenous polymers of uncertain structure[4].

Tea and coffee are the most consumed beverages in the world, and over 80% of tap water consumption is associated with making tea, coffee and other hot drinks[5]. Since tea and coffee are rich in organic matter, they can potentially act as disinfection by-product (DBP) precursors. Nonetheless, DBPs generated from reactions between residual chlorine in tap water and organic matter in tea and coffee have received limited attention in literature[6]. In this study, model compounds were selected to represent the different types of organics found in tea and coffee (Table 1). Their chlorine demand and DBP formation was measured and the potential for DBP formation during real-life preparation of tea and coffee assessed.

2 MATERIALS AND METHODS

Nine model compounds were selected to represent various components of tea and coffee (Table 1). Chlorine demand was measured after a 24 h contact time and chlorine dose of 35 mol/mol using the DPD-FAS titration method[7]. Boiling experiments were conducted on RO water spiked with set concentrations of chlorine. DBPs were measured by gas chromatography with electron capture detection (GC-ECD, Perkin Elmer Clarus 500 GC) with a Restek Rxi-5 Sil MS column of dimensions 30m x 0.25mm x 0.25μM and a modified verison of USEPA method 551.1[8]. DBPs reported were chloroform (trichloromethane), 1,1-dichloropropanone (1,1-DCP) and 1,1,1-trichloropropanone (1,1,1-TCP) (Sigma Aldrich). The chlorination conditions were as for chlorine demand experiments.

Table 1 *Model compounds*

Model compound (abbreviation)	Classification	Structure
(+) catechin hydrate (CH)	Phenolic	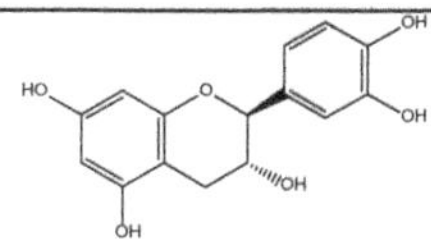
Caffeic acid (CFA)	Phenolic	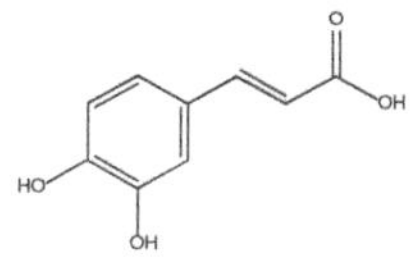
Caffeine (CF)	Alkaloid	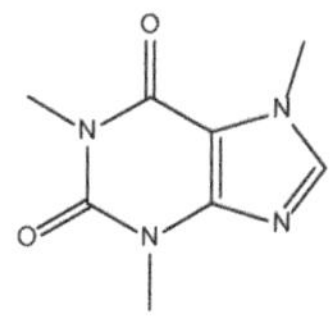
Chlorogenic acid (CGA)	Phenolic	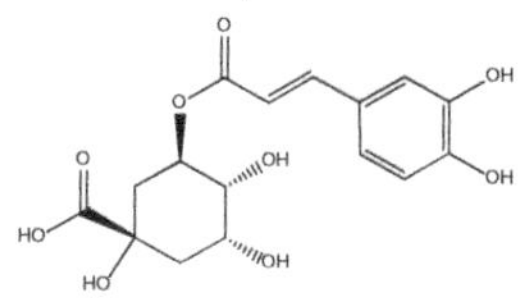
D (-) quinic acid (QA)	Carboxylic acid	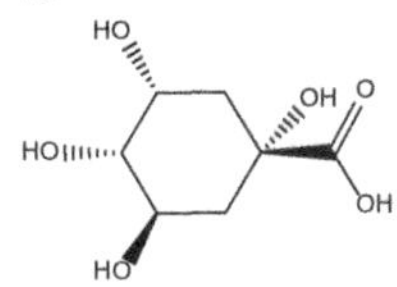
Epigallocatechin gallate (EGCG)	Phenolic	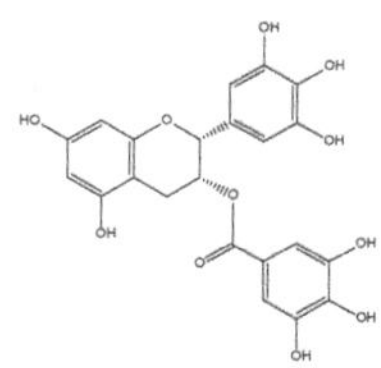
D (+) Galactose (GLA)	Monosaccharide	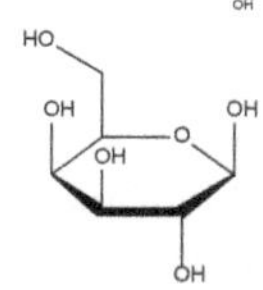
Gallic acid (GA)	Phenolic	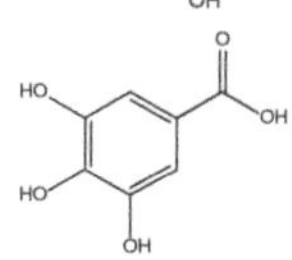
Pyrocatechol (PC)	Phenolic	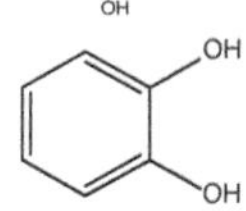

3 RESULTS

After boiling in a kettle, there was only a 5.1-19% reduction in initial chlorine concentration (Figure 1), indicating the majority of chlorine remained available for reaction. There was no significant difference between C/C_0 at different initial C_0 values, which implies that the extent of reduction during boiling was not a function of concentration. It is relevant to note here that in the UK water companies generally maintain chlorine at <0.5 $mg \cdot L^{-1}$ [9]; whereas in the USA up to 4 $mg \cdot L^{-1}$ is permitted[10].

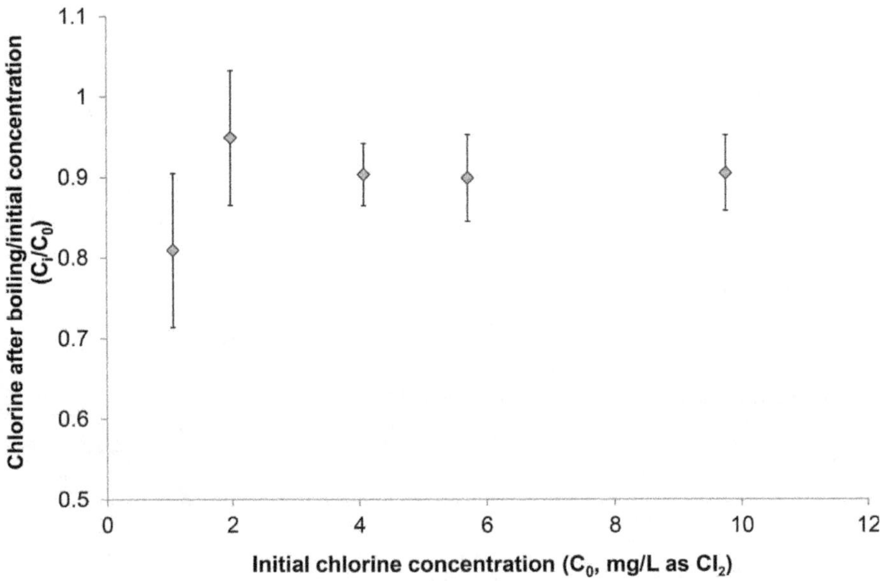

Figure 1 *Impact of boiling on concentration of aqueous chlorine spiked into ultrapure water*

The chlorine demand of the model compounds varied from 0.6 mol chlorine/mole model compound for epigallocatechin gallate to 11.3 mol/mol for catechin hydrate (Figure 2). These values are similar to those reported in literature for surrogates of natural organic matter found in drinking water source. For example, the chlorine demand of galactose and pyrocatechol was previously reported as 0.8 and 4.1 mol/mol, respectively[11-13].

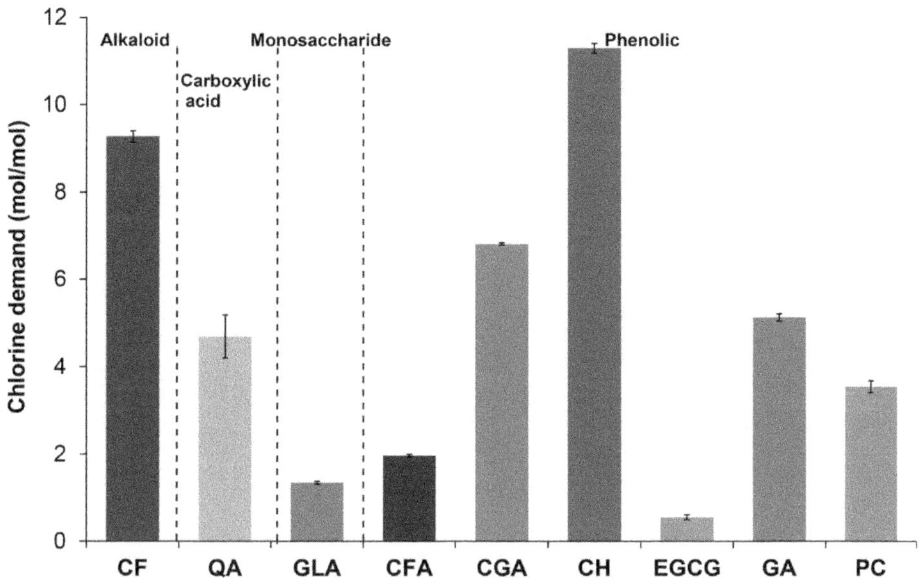

Figure 2 *Chlorine demand of model compounds*

Chloroform was the dominant DBP observed (Figure 3). The most potent DBP precursor was catechin hydrate, which generated 47.6 % mol mol^{-1} chloroform, whilst galactose produced the least amount of chloroform (0.5 % mol mol^{-1}). Only a limited number of model compounds have been reported in drinking water research which generate higher yields of THMs. Small quantities of 1,1-DCP and 1,1,1-TCP were observed: up to 0.34% mol mol^{-1} and 0.17% mol mol^{-1} respectively (Figure 3). The potency of catechin hydrate can be explained by this compound containing a resorcinol structure, well known for generating high yields of chloroform when reacted with chlorine[14].

4 CONCLUSIONS

- Chlorine at concentrations of 1-10 $mg{\cdot}L^{-1}$ was reduced by between 5-19% upon boiling, in other words the majority of aqueous chlorine remained available for reaction with organics in tea and coffee.
- Chlorine demand and chloroform formation of phenolic surrogates was high and is comparable with reactive DBP precursors found in natural waters. The most potent DBP precursor tested was catechin hydrate, which generated 48 % mol/mol of chloroform. This surrogate also had the highest chlorine demand recorded: 11.3 mol/mol.
- DBP formation during preparation of tea and coffee is predicted to be limited by kinetics and chlorine concentrations in tap water. In the UK water companies generally maintain chlorine at <0.5 $mg{\cdot}L^{-1}$ [9]; in the USA up to 4 $mg{\cdot}L^{-1}$ is permitted[10].

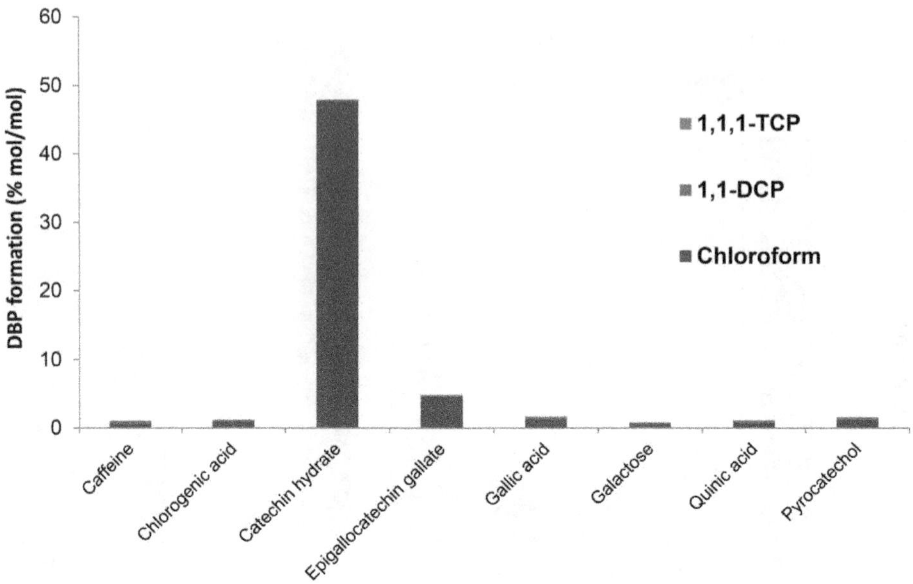

Figure 3 *DBP formation from model compounds*

References

1. D.A. Balentine, in *Phenolic Compounds in Food and Their Effects on Health I*, American Chemical Society, 1992, vol. 506, pp. 102-117.
2. A. Robertson, in *Tea*, eds. K. C. Willson and M. N. Clifford, Springer Netherlands, 1992, pp. 555-601.
3. Y. Wang and C.-T. Ho, *Journal of Agricultural and Food Chemistry*, 2009, **57**, 8109-8114.
4. S.I.F.S. Martins and M.A.J.S. van Boekel, *Food Chemistry*, 2003, **83**, 135-142.
5. DETR/DWI, *Aspects of Exposure Estimates of Disinfection By-Products from Water: A Review of Water Intake and Contact and the Uptake of Disinfection By-Products by Different Exposure Routes. Final Report to the Department of the Environment, Transport and the Regions* DETR/DWI report 4861, 2000.
6. W.W. Wu, P.A. Chadik, W.M. Davis, D.H. Powell and J.J. Delfino, *Journal of Agricultural and Food Chemistry*, 1998, **46**, 3272-3279.
7. APHA, AWWA and WEF, *Standard Methods for the Examination of Water and Wastewater*, 21st edn., American Public Health Association, Washington, DC, 2005.
8. USEPA, *Method 551.1. Determination of chlorination disinfection byproducts, chlorinated solvents and halogenated pesticides/herbicides in drinking water by liquid-liquid extraction and gas chromotography with electron capture detection. Revision 1.0* 45268, Cincinnati, Ohio, 1995.
9. DWI, Drinking Water Inspectorate. Chlorine, 2010. Available at: http://dwi.defra.gov.uk/consumers/advice-leaflets/chlorine.pdf (accessed 31 December 2014).

10. USEPA, United States Environmental Protection Agency.National Primary Drinking Water Regulations: Disinfectants and Disinfection Byproducts, USEPA. Fed. Reg. 63 (241), 69389.
11. S. Navalon, M. Alvaro and H. Garcia, *Water Res.*, 2008, **42**, 3990-4000.
12. J. de Laat, N. Merlet and M. Dore, *Water Research*, 1982, **16**, 1437-1450.
13. T. Bond, E. H. Goslan, S. A. Parsons and B. Jefferson, *Environmental Technology Reviews*, 2012, **1**, 93-113.
14. J. J. Rook, *Environmental Science & Technology*, 1977, **11**, 478-482.

NITROSAMINES, NOT ONLY DISINFECTION BY-PRODUCTS (DBPs)

C. Pallez, X. Dauchy, J.F. Munoz, C. Rosin

ANSES, Nancy Laboratory for Hydrology, Water Chemistry Department, 40 rue Lionnois, 54000 Nancy, FRANCE

1 INTRODUCTION

Treatment processes using disinfectants are essential to remove micro-organisms and provide safe water. Disinfection by products (DBPs) may be formed by reaction of disinfectants like chlorine, chlorine dioxide or ozone with organic matter in raw water. Among over 600 DBPs identified in drinking water only few of them are regulated, regularly monitored and subject to World Health Organization Guidelines[1].
Nitrogenous forms and more particularly N-nitrosamines in drinking water are of concern due to their high genotoxicity and cytotoxicity compared with regulated DBPs. The N-Nitrosamines identified as chlorination DBPs in 1989 had been classified as probable human carcinogens by the International Agency for Research on Cancer[2].

Nitrosamines are polar, low molecular weights (< 200 g/mol) and very water soluble compounds. Their partition coefficients in octanol/water are generally low and they are consequently difficult to extract with organic solvents or to remove by adsorption on a water treatment plant. Due to their small Henry constants these compounds cannot be removed from water by stripping.

Nitrosamines can originate from industrials sources such as cosmetics processing, leather tanning, food, dye and rubber. Consequently industrial wastewater may contain high levels of nitrosamines and contaminate natural water. In urban and industrial wastewater, the measured concentrations are generally high; for example 2 mg/L of NDMA were measured in downstream industrial water of a tyre factory from Ontario[3]. NDMA, NMOR, NDBA, NPIP and NDEA have been detected from 5 to 25 ng/L at the entrance of an urban wastewater treatment plant in Switzerland[4].
Some treatments used in drinking water production are also responsible for NDMA contamination: anion exchange resins (with quaternary amine function group) and flocculants like polyDADMAC (polydiallyldimethylammonium chloride) may prove to be NDMA precursors[5].

Currently, in Europe, no regulation exists for nitrosamines in drinking water. Nevertheless the addition of NDMA on the European directive for drinking water was recently proposed. This revision was supported by the French government suggesting a guideline value of 100 ng/L, in accordance with World Health Organization guideline. Due to the significant risk for human health, guidelines values on the presence of

NDMA and other nitrosamines have been recently implemented in different countries. According to Peto et al. studies [6 7], US-EPA and German Federal Agency[8] propose a guideline value of 10 ng/L. Based on the same studies, Californian Environment Agency proposed a maximum allowable concentration of 3 ng/L. Santé Canada proposed in January 2011 a parametric value for NDMA in drinking water at 40 ng/L. US-EPA recently included five nitrosamines (NDMA, NDEA, NDPA, NPYR, NDPhA) to the final Drinking Water Contaminant Candidate List (CCL 3). Moreover six nitrosamines are included as emerging pollutants in the screening survey list 2 within US-EPA's second unregulated contaminant monitoring regulation. All of these are included in the present study.

The main objectives of this study were to develop a sensitive method to analyze nine nitrosamines (Table 1) in drinking water and to evaluate their occurrence in raw water and treated water samples from public water systems.

Table 1 *Chemical Names*

Chemical Names		
N-Nitroso-dimethylamine	NDMA	$C_2H_6N_2O$
N-Nitrosomethylethylamine	NMEA	$C_3H_8N_2O$
N-Nitrosopyrrolidine	NPYR	$C_4H_8N_2O$
N-Nitrosodiethylamine	NDEA	$C_4H_{10}N_2O$
N-Nitrosopiperidine	NPIP	$C_5H_{10}N_2O$
N-Nitrosomorpholine	NMOR	$C_4H_8N_2O_2$
N-Nitroso-di-n-propylamine	NDPA	$C_6H_{12}N_2O$
N-Nitrosodibuthylamine	NDBA	$C_8H_{18}N_2O$
N-Nitroso-diphenylamine	NDPhA	$C_{12}H_{10}N_2O$

2 METHOD AND RESULTS

2.1 Reagents and standards

The mix standard solution containing Nine N-nitrosamines (NDMA, NMEA, NDEA, NMOR, NPIP, NPYR, NDBA, NDPA and NDPhA) at 1000µg/ml in methanol was purchased from ACSD. Isotope-labelled standards $NDMA_{d6}$, $NDPA_{d14}$ were provided by Restek, and $NDPhA_{d6}$ and $NPyr_{d8}$ were supplied by Cambridge Isotope Laboratories Inc (Andover, MA, USA). Methanol and water (LCMSMS-grade) were purchased from Biosolve (Valkenswaard Holland) and dichloromethane from VWR.

A 10 µg/mL standard solution was prepared in methanol and conserved in a freezer at temperature lower than –15°C for up to 3 months. Working mix solution (1 to 200 ng/mL) were diluted in water : methanol (95:5 v/v). These solutions were surrogated with isotope-labelled standard at 25 ng/mL (excepted $NDPhA_{d6}$ surrogated at 100 ng/mL) and stored at 4°C for up to one month.

Cartridges used for off-line Solid Phase Extraction (SPE) were Coconut Charcoal 2 g/6 mL from Supelco.

2.2 Sample collection

Sampling of raw and tap water samples was performed by personnel of the French Ministry of Health from October 2011 to May 2012. All the 100 French departments were included in this study. For each department, three sample sites were monitored: i) the drinking water source producing the greatest flow in the department, ii) a randomly selected drinking water source and iii) a drinking water of interest selected according to industrial environment (rubber, salting), presence of pesticides which may induce nitrosamines, treatment plants using resins or high DBP levels. Raw and treated water were systematically collected for each site. A total of 299 raw water samples and tap water samples were analyzed, representing about 25% of French population. All water samples were collected in polyethylene bottles containing sodium thiosulfate to neutralize free residual chlorine, shipped on cold packs in polystyrene boxes, and arrived at our laboratory within 24–48 h.

2.3 Analytical Procedure

Several sorbents have been evaluated and used for the extraction and concentration of nitrosamines. Activated carbons showed good extraction rate for all nine nitrosamines of interest, even for hydrophilic compounds like NDMA. Nevertheless some precautions had to be taken to avoid the formation of nitrosamines from their corresponding amines on the activated carbon.

The raw and tap water samples were spiked with internal standards $NDMA_{d6}$ (25 ng/L), $NPYR_{d8}$ (25 ng/L), $NDPA_{d14}$ (25 ng/L) and $NDPhA_{d6}$ (100 ng/L) and stored at 4 °C before analysis. In most cases, samples were extracted within 1–2 days of collection.

Pre-treatment SPE method was performed with a Gilson GX-274 automate. Cartridges were dried with a Chromabond vacuum chamber. Evaporation was performed with a Thermochem (Labortechnik LIEBISCH). Final extracts were conserved at 4°C.

A TSQ Quantum Ultra High Performance liquid chromatograph tandem mass spectrometer with an Accela pump (Thermo Scientific France) was used in Atmospheric Pressure Chemical Ionization (APCI) mode. The analytical column Hypersil Gold aQ (150mm x 2,1mm, 3 µm, Thermo Scientific) was chosen and allowed a good separation of nitrosamines.

2.4 Performances of the analytical method

Quantification of nitrosamines was performed using isotope-labelled standards added to each sample before the SPE step to check the overall recovery of target chemicals. Concentrations of target compounds were determined with a nine standards calibration curve with a quadratic fit. The limit of quantitation (LoQ) was defined as the smallest concentration that has been validated with acceptable accuracy by applying the complete analytical method. Limit of detection (LoD) was considered as LoQ/3. In each SPE sequence, a blank surrogated in NDMA at 5 ng/L (LoQ) and a natural sample surrogated at 50 ng/L were extracted and analysed to check the absence of contamination and the efficiency of the protocol even at low level of concentration. Identification of nitrosamines was validated according to different criteria defined in SANCO document (12495/2011).

Analytical uncertainty was between 25% and 40% according to the concentration measured and recovery better than 60%.

2.5 Results for national campaign

In raw water samples, no molecule was detected for 95% of the samples. Only N-nitrosodimethylamine (NDMA) and N-nitrosomorpholine (NMOR) were identified in some samples.
In treated water, none of the analytes were detected for 81% of the samples. Among the nine nitrosamines analysed, only NDMA, NMOR and N-nitrosodiethylamine (NDEA) were identified at a level higher than the limit of quantification. Among 299 treated water concentrations of nitrosamines were higher than 10 ng/L for only 3 samples (Table 2).

Table 2 *Maximal content found in Resources and in Treated water (ng/L)*

Molecules	*LoQ (ng/L)*	*Resources*		*Treated Water*	
		Maximal content found (ng/L)	*Frequency of detection (>LoD)*	*Maximal content found in (ng/L)*	*Frequency of detection (>LoD)*
NDEA	5	Traces	1.7 %	32	7.4%
NDMA	5	72	1.7 %	14	3.0%
NMOR	10	1200	1.0%	1220	0.7%

The higher frequency of detection of nitrosamines in treated water illustrates the role of water treatment plant in their formation.

2.6 Particular case: Bolbec contamination

This campaign also highlighted concentrations higher than 1000 ng/L for NMOR in raw and treated water of a municipality. Further investigations were performed in partnership with the departmental Ministry of Health to determine the extent and origins of this contamination. Concentrations of NMOR up to 3000 ng/L in nearby water supplies were measured (Figure 1). NMOR concentrations reached 17 500 ng/L in the nearby river downstream of the industrial discharge.

This pharmaceutical site uses annually several thousand tons of morpholine in its process of manufacturing and NMOR is produced unintentionally in the industrial wastewater treatment plant, under acidic conditions. The general scheme of morpholine's degradation is shown on Figure 2.

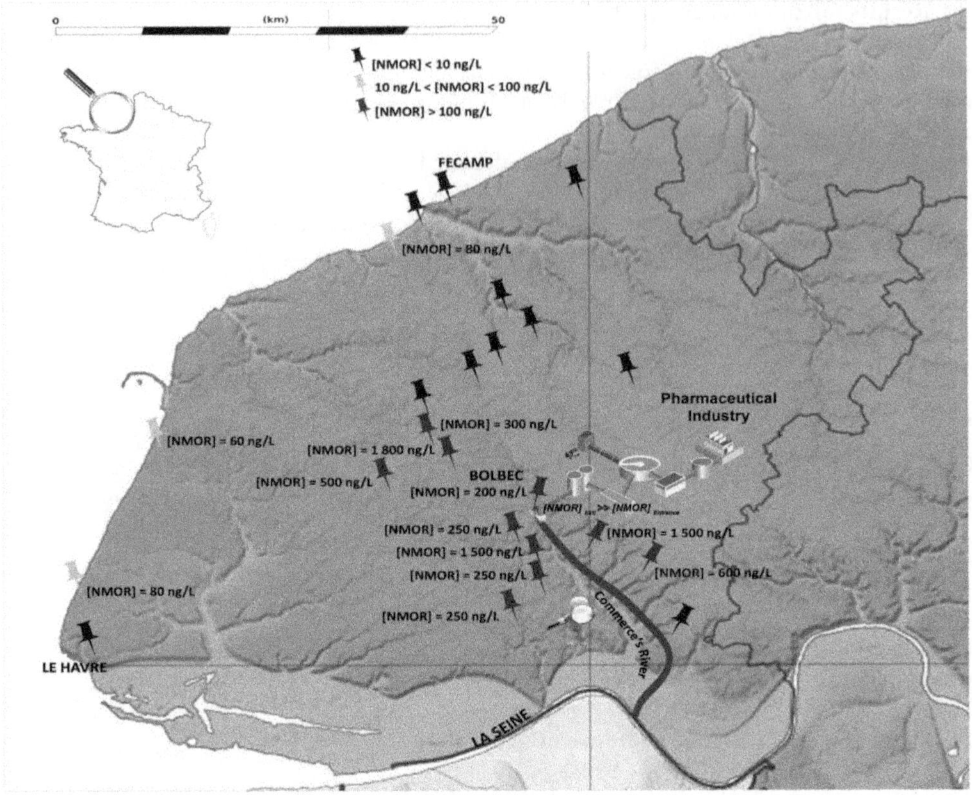

Figure 1 *Extent of the pollution in NMOR in Seine-Maritime department*

$$2\,HNO_2 \rightleftarrows N_2O_3 + H_2O$$

$$N_2O_3 \rightleftarrows NO^+ + NO_2^-$$

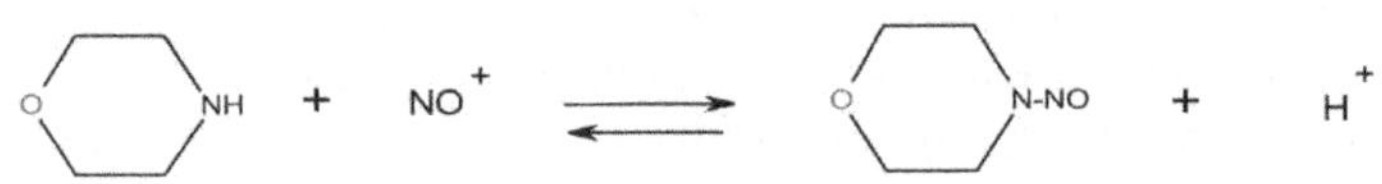

Morpholine N-Nitrosomorpholine

Figure 2 *Degradation of the Morpholine*

Further to these investigations, the wastewater treatment plant has been modified by adding UV treatment and drinking water resources were substituted.

The monitoring of the river contamination revealed a decrease of the contamination along the river from more than 15 000 ng/L just after the discharge point to lower than LoQ after Seine confluence (Figure 3). The contamination was monitored for several months. Results show a slow decrease after the cessation of the discharge probably due to exchanges with groundwater which acts as a buffer zone.

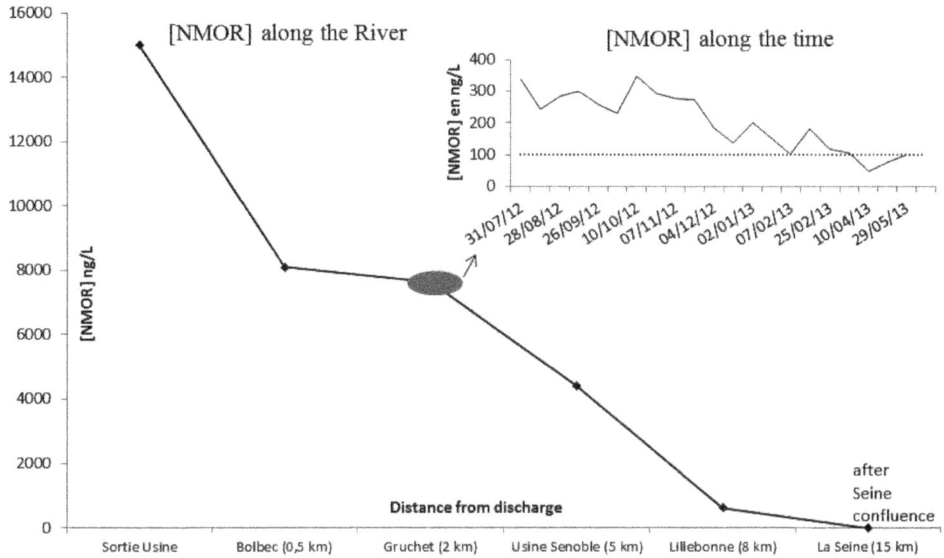

Figure 3 *Evolution of the concentration in NMOR (ng/L) along the river and with time at Gruchet*

As soon as the pollution was confirmed, citizens were advised not to consume tap water. Amounts of activated carbon in the water treatment plant were increased and interconnecting work was implemented.

Analysis in milk, eggs, meat, and vegetable were also planned to measure the extent of this contamination. No trace of nitrosamines was detected.

On the other hand, French Agency for Food, Environmental and Occupational Health & Safety (Anses) performed risk assessment and proposed 100 ng/L as a guide value for NMOR[9].

3 CONCLUSIONS

This sampling campaign has drawn up an inventory of the concentration levels of nitrosamines covering 25% of the drinking water produced in France. Three compounds NDMA, NDEA and NMOR were quantified. An industrial site was responsible for high concentrations of NMOR in drinking water and induced important work.

This study also highlighted the interest of having a good knowledge of industrial discharges and monitoring nitrosamines precursors which could explain phenomena of degradation and/or transformation of these compounds in the environment.

References

1. *Guidelines for Drinking Water Quality*, World Health Organization, International Agency for Research on Cancer, 2008.
2. J.M. Barnes and P.N. Magee, *British Journal of Industrial Medicine,* 1954, **11**, 167-174.

3. W.A. Mitch, J.O. Sharp and al., *Environmental Engineering Science*, 2003, **20,** 389-404.
4. M. Krauss, P. Longrée and al., *Water Research*, 2009, **43**, 4381-4391.
5. S.H. Mhlongo, B.B. Mamba, R.W. Krause, *Physics and Chemistry of the Earth*, 2009, **34**, 819–824.
6. R. Peto, R. Gray and al., *Cancer Research*, 1991, **51**, 6452-6469.
7. R. Peto, R. Gray and al., *Cancer Research*, 1991, **51**, 6415-6451.
8. J. Nawrocki and P. Andrzejewski, *Journal of Hazardous Materials*, 2011, **189**, 1-18.
9. *Avis de l'Anses relatif à l'évaluation des risques sanitaires liés à la présence de N-nitrosomorpholine dans l'eau destinée à la consommation humaine*, http://www.anses.fr/Documents/EAUX2012sa0172.pdf (accessed 31st Dec. 2014) , ANSES,2012.
10. A. Bruchet, Z. Do-Quang, *Journées Information Eaux*, 2008, Sous-produits de traitement de l'eau : Nitrosamines et NDMA, cas de la France. Poitiers (France).
11. J.W.A. Charrois, J.M. Boyd, and al., *Journal of Environmental Engineering and Science*, 2007, **6**, 103-114.
12. N.V. Huy, M. Murakami and al., *Water Research*, 2011 **45**, 8.
13. W.K. Lutz, *Genetic Toxicology and Environmental Mutagenesis*, 1999, **443**, 251-258.
14. B. Pfundstein, A.R. Tricker and al., *Food and Chemical Toxicology*, 1991, **29**, 733-739.
15. C. Planas, O. Palacios and al., *Talanta*, 2008, **76**, 906-913.

TOTAL TRIHALOMETHANE DISINFECTION BY-PRODUCT CONCENTRATIONS LEVELS OBSERVED IN DRINKING WATER FOR SOME LOCATIONS IN BAGHDAD CITY AND THE EFFECT OF ULTRAFILTRATION IN MINIMIZING THEM

O. G. Ameen[1], W. I. Abdulla [1], F. M. Mohamed[2], A. Qays Kaittan[3] E. Abdljabbar[4] and A.M.Rashed[5]

[1] Water Treatment Technology Centre, Directorate of Environment and Water, Ministry of Science and Technology, Baghdad, Iraq
2-5 Ministry of Science and Technology, Baghdad-Jaderia, Iraq

1 INTRODUCTION

In Jaderia-Baghdad drinking water treatment plant the Taiger river water is treated by using the conventional drinking water treatment process which is coagulation, flocculation, sedimentation, sand filtration and in the last stage disinfection using chlorine gas. The main purpose for drinking water treatment plants is to remove the organic, inorganic contaminants, colloidal and microorganisms such as viruses, bacteria and protozoans[1]. However, recently researchers have found that this treatment may not enough because of soluble organic contaminants or natural organic matter (NOM), both of which could be a source for the formation of disinfection by-products when water is disinfected by chlorine. Hence, NOM removal is one of the most important treatment requirements for the production of good quality drinking water [2]. The NOM reaction with chlorine gas generally forms halogenated compounds[3]. These compounds for example trihalomethanes (THMs) and haloacetic acids (HAAs) are regulated in some countries[4,5]. Recent studies identified new DBPs in disinfected drinking water by using most of chemicals employed for disinfection[6]. Enhanced coagulation is one of the methods used to minimize the organic materials in treated drinking water plants [7,8]. Also small particle size of granular activated carbon (GAC) can adsorb natural organic material (NOM) and this can be increased by acidification of raw water[9]. The ultrafiltration process using organic or inorganic membranes technology in water treatment plants has started to take its place in the world which prove their ability to minimize the organic material in the treated water[10]. This is achieved by using ultrafiltration hollow fibre membranes or flat sheets in submerged systems that are situated inside tanks filled with the raw water to be filtered. The driving force in this process is carried out by vacuum on the permeate side created by suction side of a pump which is usually self-priming.

1 MATERIALS AND METHOD

Samples were taken from Jaderia-Baghdad conventional drinking water treatment plant after the sand filter stage (before chlorine gas addition) and the results were compared

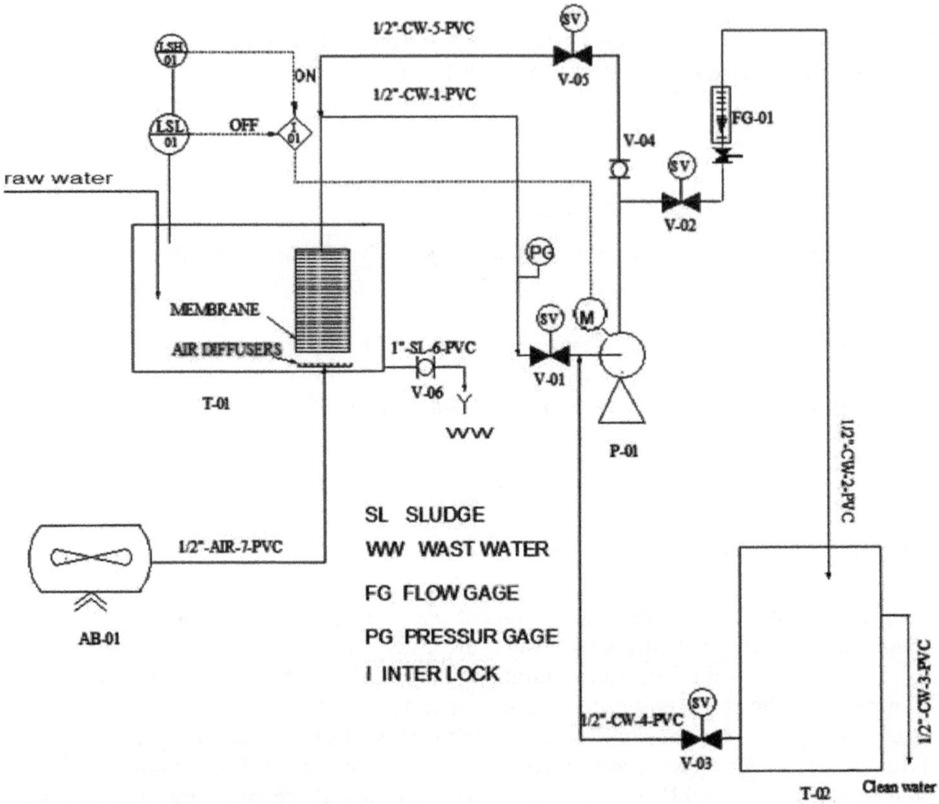

Figure 1 *Ultrafiltration membrane unit*

with water samples produced by a water ultrafiltration unit as shown in the Figure 1below and also tap water samples for THM analysis were taken from (13) thirteen different locations in Baghdad city.

The ultrafiltration membrane used in this unit was a 0.04 micron pore size PVDF (polyvinylidene fluoride) with 20 square metre surface area and a flux rate of 0.1 to 0.4 cubic metres per hour. The raw water was treated without alum addition.
The samples were sent directly for TOC analysis carried out by USEPA standard method 5910B Ultra violet absorption method[11], the equation below is used to calculate the TOC concentrations in mg/l.

$$UV_{254} = 0.0144\, KPH\,(mg/l) + 0.0018 \qquad (1)$$

After modification of the equation:-

$$KPH\,(mg/lit) = 69.4\, UV_{254} - 0.125 \qquad (2)$$

Where KHP (potassium hydrogen phthalate) presents as organic material to be equivalent to TOC (Total Organic Carbon) the measuring was carried out using a spectrophotometer BIOTECH Engineering UV - 9200. Ashery[12] has shown that the relationship between organic material concentrations and UV254 absorption is linear according to the equation below:-

$$NOM\ (mg/l) = 60.65(\ UV_{254}) - 1.84 \qquad (3)$$

The above equation is very similar to that of the USEPA standard method 5910B. Where the NOM concentration unit is mg/l

The percent removal of TOC can calculate by the equation

$$\%TOC\ removal = \frac{TOC\ in - TOC\ ont}{TOCin} \times \%100 \qquad (4)$$

Where: - TOC_{in} is the initial concentration (raw water)
TOC_{out} is the concentration after treatment

The water samples (both from the plant and the ultrafiltration unit) with the same volume were then treated by chlorine addition up to 5 mg/l final chlorine concentration in the water sample to allow the chlorine reaction to proceed for 1 hour and tested for total THMs by using GC (Gas Chromatography) SHIMADZU 2010 fitted with an electron capture detector (ECD) detector and a headspace sampler AOC – 5000. The principle of the method is that the sample is placed in a sealed vial and allowed to equilibrate with its headspace vapour at 70°C. Then the sample aliquot is taken from the vial headspace using a special syringe and injected into the GC using a 0.25 mm i.d. 60 metre length fused silica capillary column. This method is derived from the UK reference booklet entitled "Determination of very low concentrations of hydrocarbons and halogenated hydrocarbons in water" (1984-5), Methods for the Examination of Waters and Associated Materials, London Her Majesty's Stationary Office, ISBN 0117520047). The calibration graph was derived from a THM standard ampoule 1 ml mixture 2000µg/ml each THM in methanol from Sigma-Aldrich/SUPELCO.

2 RESULTS AND DISCUSSION

Figure 2 shows the TOC values from the river (raw water) 8.87 mg/l average value, drinking water treatment plant 3.39 mg/l average value and the ultrafiltration membrane unit 2.06 mg/l average value which clearly shows the minimum values came from the ultrafiltration membrane unit. There is no consistency in the TOC concentrations from day to day as the water source chemical characteristics are not the same every day. The remaining TOC depends on the pH, soluble organics, size and the molecular weight of organics in the raw water. It is readily apparent that the ultrafiltration plant is more efficient in removing TOC than the conventional water treatment plant.

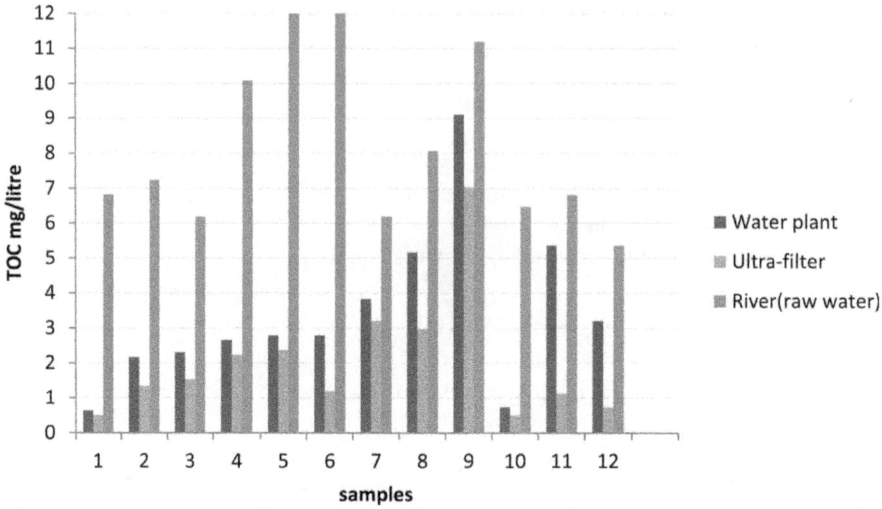

Figure 2 *TOC concentrations for samples from the river, the water plant and the ultrafiltration unit*

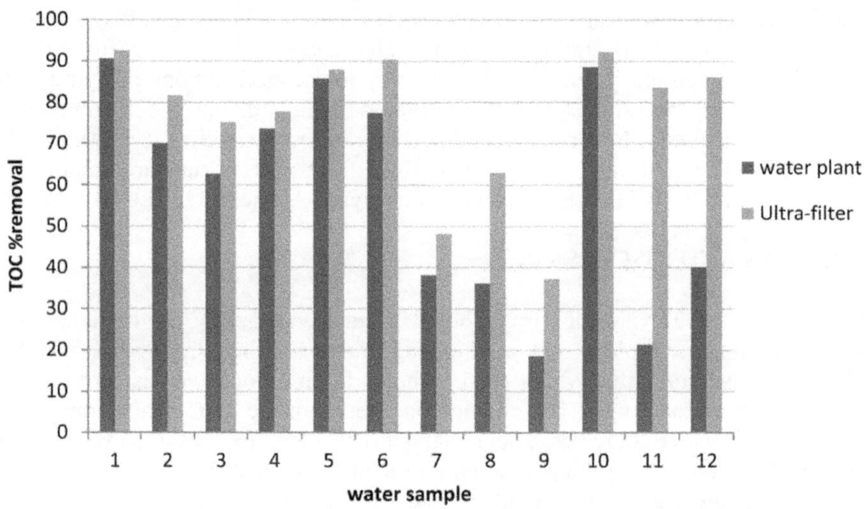

Figure 3 *TOC % removal in the water plant and the ultrafiltration unit*

Figure 3 shows the percentage removal of TOC in the water treatment plant. The average removal values were 58.5% for the conventional water plant and 76.3% for the ultrafiltration unit.

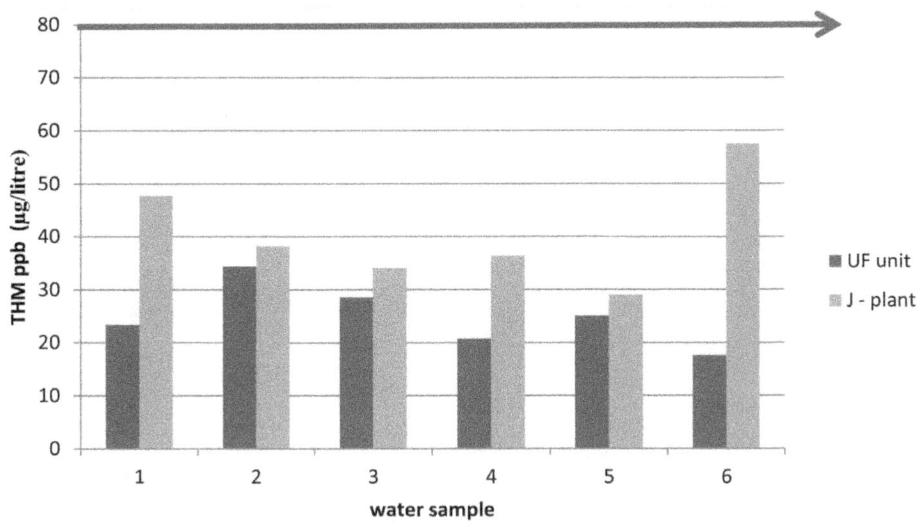

Figure 4 *THM concentrations in water plant and ultrafiltration unit*

Figure 4 shows that the THM levels in the water samples taken in September and October 2012 were less than 80 μg/l (USEPA regulation) and 150 μg/l (Iraqi water regulations) in both the drinking water treatment plant (40μg/l average value) and ultrafiltration unit (25 μg/l average value). However, in the water plant, the highest value was 57 μg/l corresponding to 71% of the maximum THM USEPA permitted level (80μg/l) whilst the highest THM level in the ultrafiltration unit was 35 μg/l corresponding to only 43.3% of THM permitted USEPA level. Thus, there is an advantage for better water quality by using an ultrafiltration unit to deal with potential high input concentrations of TOC.

4.1 Levels of THMs in some locations of Baghdad City

Fresh tap water sample were taken from different locations in Baghdad City as shown in Table 1 below. These were analysed to measure the level of THMs on the same day by the same method the samples were taken in the last quarter of 2012.

Table 1 *Concentrations of THMs in 13 locations in Baghdad city*

No.	Location	THM s (μg/l)	No.	location	THM (μg/l)
1	Hay Al Qahera	31.6	8	Al Karaada	41.7
2	Hay Tunes	34.7	9	Al Jaderia	42.8
3	Hay Al Shaa'b	29.0	10	Al Qadiseeia	23.5
4	Hay Al Bunook	34.0	11	Al Shuu'ala	15.8
5	Al Taleebia	23.1	12	Al Gazalia	9.6
6	Al Sadeer city	43.6	13	Mahmudia	24.2
7	Palastain Street (Al Muhandessen)	29.0			

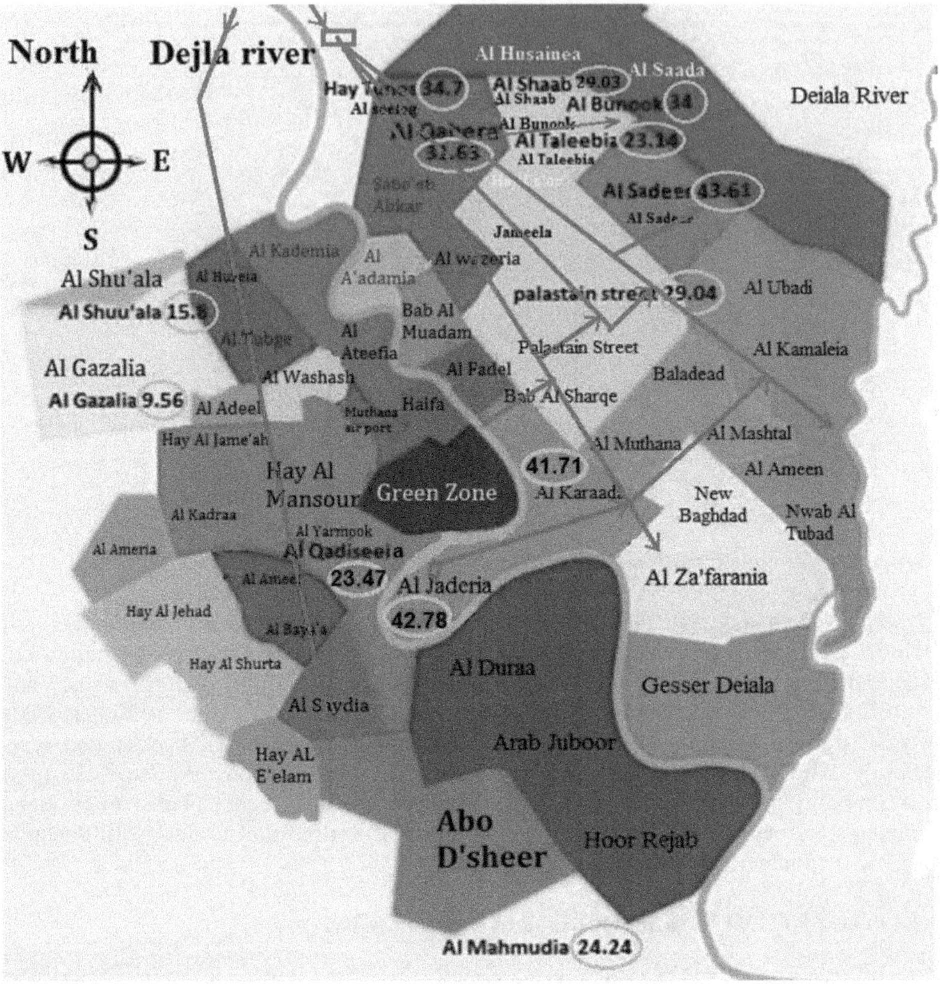

Figure 5 *Baghdad City map shows THM concentrations in tap water in different locations*

The drinking water supply for Baghdad city comes from two main drinking water treatment plants about 20 kilometre North of the Baghdad City (upstream on the river) the first one for the Karkh side and the second one for the Resafa side which serves about 70% of the Baghdad city population. The main pipe in the Resafa side is divided into three main lines connected to the distribution drinking water system as depicted in Figure 5.

Figures 5 and 6 and also Table 1 show the levels of THM in thirteen locations in Baghdad city. All the results were less than 80 μg/l and 150 μg/l the USEPA and Iraqi drinking water regulations respectively and reflect the good job done by the mayoralty of Bagdad. It should be noted that there were higher THM values in Al Sadeer 44 μg/1, Karada 42 μg/1 and Jaderia 43 μg/1. These sampling points are on the Resafa side and the last two are located almost at the end of the pipeline which means the water takes about 3-4 hours to reach that point. The Sadeer city result was ascribed to a water shortage (very high population area) that is supported by another source of water from a small drinking water treatment plant locate on the river bank as shown in Figure 5. The Karada and Jaderia results are also felt to be influenced by the fact that they were taken from the end of the distribution system (low pressure water pipeline). The most probable reason for high THM values is from the contaminated water sources which came from the city daily activity in addition to the old distribution system (cracks and leaks), using home suction pumps to overcome the water low pressure and the relatively long traveling time in the main pipeline.

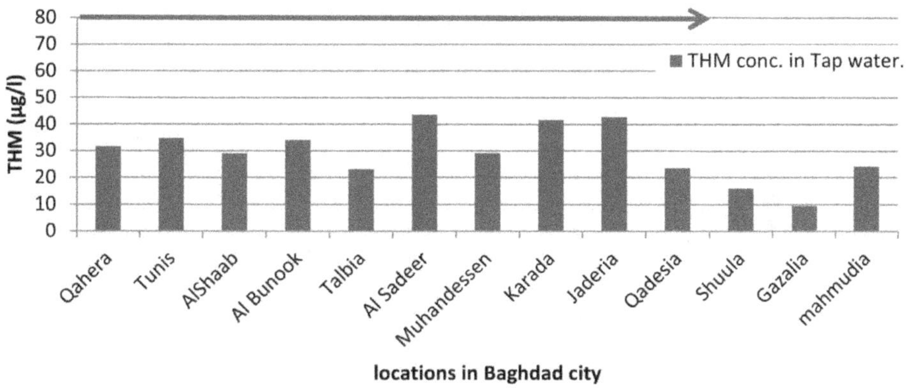

Figure 6 *THM concentrations in tap water in different locations in Baghdad city*

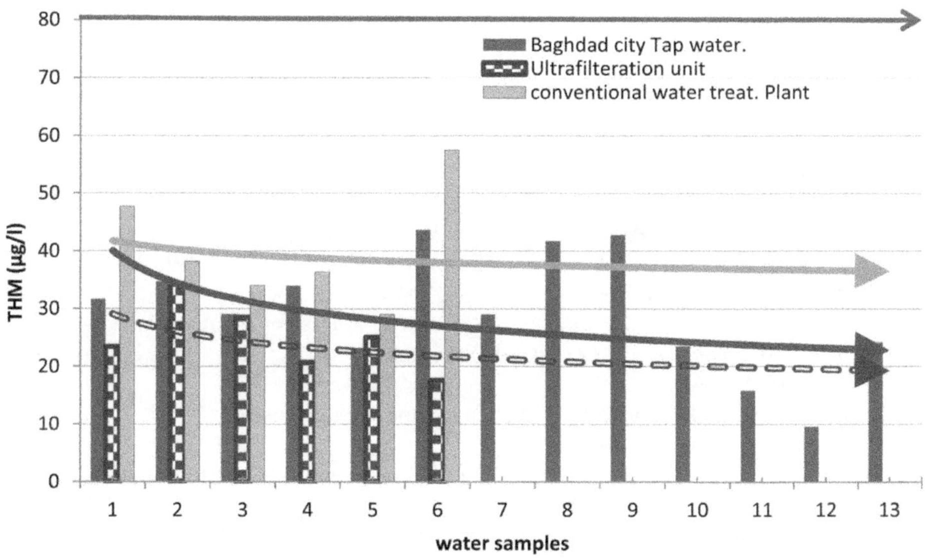

Figure 7 *Comparing the THM concentrations in tap water, the Jaderia plant and the ultrafiltration unit*

Figure 7 is a comparison chart between the measured levels of THMs for the Jaderia conventional water treatment plant, tap water in Baghdad city and the ultrafiltration unit. The average values are 40 μg/l, 29 μg/l and 25 μg/l respectively. Figure 7 also shows that the highest levels were in the Jaderia plant (after adding chlorine 5 mg/l), then in tap water from Baghdad city and the lowest values were achieved from the ultrafiltration unit. This is because this unit removes more organics from treated water than the conventional drinking water plant. Also, there is better water source quality (low TOC concentration) than that from the Jaderia water treatment plant where the water source is contaminated by higher levels of organic contaminants resulting from daily city activities.

3 CONCLUSIONS

The produced water after treatment by ultrafiltration has lower levels of TOC than from the conventional drinking water plant and this minimizes the undesired DBP chemicals, mainly THMs, after disinfection by chlorine gas.

Using an ultrafiltration unit membrane in the drinking water treatment plant could be a suitable choice for better drinking water quality in minimizing the TOC concentrations levels and controlling disinfection by products formation. It was found that by using this system, the total THM levels were well within the accepted and regulated limits for safe drinking water in distribution systems.

The THMs levels in tapwater for different locations of Baghdad City were well within the 80 μg/l and 150 μg/l the USEPA and Iraqi regulatory limits for drinking water respectively. This indicates the safe drinking water but still there are some higher than average values (44, 42 and 43 μg/l) in Al Sadeer City, Al Karaada and Al Jaderia respectively. These are probably caused by a different water source and this may need more attention from Iraqi authorities in the mayoralty of Baghdad.

References

1. J.Crittenden, et al., *Water treatment principles and design/MHW, 2*nd edition, 2005, ISBN 0-471-11018-3.
2. T.Leiknes, H.Odegaard, and H. Myklebust. *Journal of Membrane Science*, 2004, **242**.
3. J. J. Rook. *Water Treat. Exam.,* 1974, **23**, 234.
4. US Environmental Protection Agency.*Fed. Regist.*, 2006, **71,** 388.
5. T. Karanfil, S. Krasner, W. P. Westerhoff and Y. Xie. Washington, DC: American Chemical Society, 2008, pp. 2–19.
6. S. W. Krasner, *The formation and control of emerging disinfection by-products of health concern,* Philosophical Transactions of the Royal Society A ,2009, 367, 4077–4095.
7. A. F. Ashery, K. Radwan, I. Mohamed and G. A. Rashed. *International Water Technology Journal,* 2012**, 2**.
8. F. Vaezi, J. Nouri, A. Ghasri, A. Mohagheghian and M. R. Eshraghian, *Iranian J. Env. Health Sci. Eng.* 2005, **2**, 43.
9. A. H. Sulaymon, A. M. Ali and S. K. Al-Nasiry. *Desalination*, 2009, **245**, 155.
10. A. Chen, Q. Fan, and Q. Tian. *International Journal of Environmental Science and Development*, 2013, **4.**

11. USEPA *Standard method 5910B UV-absorbing organic constituents, Ultraviolet Absorption Method.*
12. A. F. Ashery, K. Radwan and I. Mohamed, G. A. Rashed. *Fifteenth International Water Technology Conference, IWTC*, 2010, 15, Alexandria, Egypt.

EVALUATING THE USEFULNESS OF SPRING WATER: PRODUCTION OF TRIHALOMETHANE FROM CHLORINATION

Yumi MANAKA, Masayuki GOTO and Chie SATO

Department of Health and Nutrition, Wayo Women's University, 2-3-1 Konodai, Ichikawa-shi, Chiba, Japan

1 INTRODUCTION

In Japan, spring waters have long been widely used for domestic purposes such as agriculture and consumption. Because spring waters are so closely intertwined in our everyday lives, many springs have been well maintained by the local residents. Depending on factors such as water quality and volume, a large number of these sources are designated by the Ministry of Environment as providing high-quality water. The names and locations of spring water sources can easily be found on-line.

In this study, we investigated whether chlorination of spring waters from five different sources would result in formation of trihalomethanes, a disinfection by-product of public health concern. The objective was to consider the possibility of utilizing spring water as a safe and useful water resource.

2 MATERIALS AND METHODS

2.1 Sample Collection

On May 28, 2014, spring water samples were collected in glass bottles at five locations in the Chiba Prefecture (Pond of Katsushika Shrine, Funabashi Daijingu, Taki Fudoson, Omachi Sizen Koen, and Rakannoi). At sampling locations, measurements were taken for water temperature and properties that are defined in the Japanese tap water quality standard (pH levels, turbidity, and chromaticity).

2.2 Chlorination Treatment of Spring Water

A sodium hypochlorite solution (produced by Kanto Kagaku) was added to the spring waters to achieve chlorine concentrations of 0.5 mg/L (approximately the same as tap water). The chlorine concentration was measured using a di-ethyl-p-phenylenediamine technique. This treatment was performed on the same day that the samples were collected.

2.3 Analysis of Trihalomethane and Total Organic Carbon (TOC)

The measurements of TOC were made using a total organic carbon measurement method in accordance with Ministerial Notification No. 261 (Clause No.30), which was introduced by the Ministry of Health, Labor and Welfare (MHLW) in 2003. Four trihalomethanes (chloroform, dibromochloromethane, bromodichloromethane, and bromoform) were simultaneously measured using a headspace-gas chromatographic-mass spectrometer (HS-GC-MS) in accordance with Ministerial Notification No. 261 (Clause No.15), which was introduced by the MHLW in 2003.

3 RESULTS AND DISCUSSION

The measurement results (pH, turbidity, chromaticity, and TOC) of five spring waters before chlorination treatment and a tap water sample collected from Chiba Prefecture were compared with the tap water quality standard values (Table 1). With the exception of the one spring water collected from the Pond of Katsushika Shrine, all of the properties were found to be satisfactory with regard to the tap water standards. The Pond of Katsushika Shrine had a turbidity of 7.9 and a chromaticity of 13.5, which were considerably higher than the standard values (turbidity: 2 or less, chromaticity: 5 or less). The pH was 7.0, which was less than that of the standard value. The TOC was 1.9 mg/L, which was within the range of the standard value (3 mg/L). However, it was high compared to the other four spring waters (TOC values of 0.4–0.5 mg/L). The four spring water sources that satisfied the standard values already have collection facilities in place. The public visits these locations to collect water, and some people already use it as drinking water. In the Pond of Katsushika Shrine, however, water collects in a pond that works as a reservoir and there is no facility for collecting water directly from the spring. Therefore, the turbidity and chromaticity are assumed to be higher because contamination accumulates due to run-off from the surrounding environment.

Trihalomethanes were not detected in the spring waters before chlorination treatment. After chlorination treatment (Table 2), the chloroform concentration was 0.007 mg/L and the bromodichloromethane concentration was 0.010 mg/L in the spring water from the Pond of Katsushika Shrine. In the Omachi Sizen Koen, the concentration of bromoform in the spring water was 0.013mg/L. Each of these spring waters had concentrations that were lower than the tap water quality standard values. Additionally, the total trihalomethane concentration was found to be lower than the total trihalomethane concentration of the tap water in the Chiba Prefecture. No trihalomethane was detected in the post-chlorination spring waters from Funabashi Daijingu, Taki Fudoson, and Rakannoi.

At much higher concentrations than typically found in drinking water, trihalomethanes have been shown to have carcinogenic properties in animals. In drinking water, trihalomethanes are produced during disinfection with chlorine to reduce the microbiological risk. In general, a higher concentration of organic carbon produces more trihalomethanes. The results of this study showed that the quantity of trihalomethanes in these spring waters after the chlorination process was very low and below the water quality standard values, which is consistent with the TOC values, which are generally lower than tap water (Table 1).

Table 1 *Results for (temperature, pH, turbidity, chromaticity, and TOC of water samples*

	Water temperature (℃)	*pH*	*Turbidity*	*Chromaticity*	*TOC* (mg/L)	*Chlorine concentration* (mg/L)
Pond of Katsushika Shrine	21.0	7.0	7.9	13.5	1.9	0.4
Funabashi Daijingu	19.0	6.8	0.2	1.0	0.5	0.5
Taki Fudoson	18.0	6.6	0.1	0.1	0.4	0.5
Omachi Sizen Koen	17.0	6.4	0.4	0.5	0.5	0.5
Rakannoi	18.0	6.4	0.2	0.5	0.5	0.5
Tap water	21.0	6.4	0.5	0.5	0.8	0.4
Standard value		5.8 or more and 8.6 or	2	5	3	
Target value		About 7.5	1			1

Table 2 *Concentrations of trihalomethane after chlorine treatment*

	Chloroform (mg/L)	*Dibromo chloromethane* (mg/L)	*Bromodichloro methane* (mg/L)	*Bromoform* (mg/L)	*Total trihalomethane* (mg/L)
Pond of Katsushika Shrine	0.007	ND	0.010	ND	0.017
Funabashi Daijingu	ND	ND	ND	ND	ND
Taki Fudoson	ND	ND	ND	ND	ND
Omachi Sizen Koen	ND	ND	ND	0.013	0.013
Rakannoi	ND	ND	ND	ND	ND
Tap water	0.010	ND	0.008	ND	0.018
Standard value	0.06	0.1	0.03	0.09	0.1

ND: Not detected

4 CONCLUSIONS

Although there are concerns related to the effects from microorganisms since there is no sterilization, the use of spring waters as sources of drinking water has gained more attention due to the naturally derived, high-quality minerals contained in them. In this study, four spring waters with water collection facilities satisfied the water quality standards for drinking water for pH, turbidity, and chromaticity levels. These spring waters were also found to have low concentrations of TOC, so only very small amounts of trihalomethane were produced from chlorination. On the basis of these results, we believe that in the future, the chlorination of spring water can potentially be used to secure safe, high-quality, and palatable drinking water.

Reference

1 Japan Water Works Association, Standard Methods for the Examination of Water 2011 Edition, General Review Papers and References, 2011.

EVALUATION OF CHLORITE AND CHLORATE PROPAGATION IN A DRINKING WATER DISTRIBUTION SYSTEM

S. Sorlini and M. Biasibetti

Department of Civil, Environmental, Architectural Engineering, and Mathematics, University of Brescia, Via Branze 43, 25123 Brescia, Italy

1 INTRODUCTION

Disinfection is applied in drinking water treatment plants (DWTPs) to ensure water quality and to avoid bacterial contamination in drinking water distribution systems (DWDSs). However, disinfectant concentration in DWDSs decreases with time, while disinfection by-product (DBP) formation increases; this is due to chemical reactions of the disinfectant with dissolved and particulate matter in the water, biofilm and pipe wall material.[1-4]

If chlorine dioxide (ClO_2) is used as disinfectant, chlorite (ClO_2^-) and chlorate (ClO_3^-) can be produced as DBPs. Therefore, it is important to evaluate spatial and temporal variation in residual chlorine, ClO_2^- and ClO_3^- within the DWDS. In fact, the World Health Organization (WHO) recommends a free residual chlorine concentration of 0.2 mg L^{-1} in the DWDS;[5] moreover, since ClO_2^- and ClO_3^- can cause oxidative damage to red blood human cells, each compound should not exceed the WHO guideline value (GV) of 700 μg L^{-1} in drinking water.[5]

Prediction of the disinfectant residual and DBP propagation in DWDS can be achieved using Water Distribution Modelling (WDM), which allows simulation and evaluation of a DWDS under different operating conditions. For example, the software Epanet 2.0 is a WDM in the public domain, developed by the United States Environmental Protection Agency (USEPA) that models the hydraulic and water quality behaviour of water distribution piping systems.

In the literature, several studies are available on the measurement and modelling of chlorine in DWDSs[4,6-13], while the kinetic reactions of ClO_2, ClO_2^- and ClO_3^- and the simulation of their propagation in DWDSs by means of WDM must be further investigated. Some researchers modelled chlorine residual propagation in DWDSs with the software Epanet and obtained good agreement with observed chlorine levels at locations where the hydraulics were well characterized.[6] Other researchers studied chlorine residual propagation in DWDSs and found that residuals varied widely both spatially and temporally; moreover, they observed that a long residence time in storage tanks caused low or nonexistent residual disinfectant concentrations in the DWDS; further, they found that supply system operation has a significant effect on the distribution and concentration of chlorine residuals in the DWDS.[14] Other researchers studied trihalomethane (THM) reaction kinetics and developed a model for predicting their formation in DWDSs. They found that a second-order reaction is a good predictor of THM formation and they observed good agreement between calculated and measured values, confirming that the model is applicable to an actual DWDS.[11]

This work studies the DWDS of Cremona, in the North of Italy, fed by two treatment plants in which ClO_2 disinfection is applied, respectively located to the west and east of the city. The DWDS was monitored for 6 years (2006-2011) analysing residual chlorine, ClO_2^- and ClO_3^- concentrations at 26 points of the network. The software Epanet 2.0 (USEPA) was applied to the DWDS in order to estimate the mixing zone of the water coming from the west and east treatment plants, and in order to simulate the propagation of ClO_2, ClO_2^- and ClO_3^-. The measured and simulated results were compared in order to validate the applicability of the model to the DWDS.

2 METHOD AND RESULTS

2.1 Drinking water supply system of Cremona (Italy)

The drinking water supply system studied is located in the city of Cremona (76,000 inhabitants), in the North of Italy. Groundwater is withdrawn by two well fields, respectively located in the north-west and east of the city; each well field is capable of providing a maximum flow rate of 38.9 ML d^{-1} and each well withdraws water at a depth between 160 and 200 m. Groundwater contains methane, hydrogen sulfide, ammonia, iron, manganese and arsenic (AsIII). The well fields feed two equal DWTPs, respectively located to the east and west of the city, each using the following treatment train: aeration, biofiltration, chemical oxidation with potassium permanganate and ferric chloride, sand filtration and final disinfection with ClO_2. After treatment, water is supplied to the DWDS by means of pumps with inverters. The DWTP in the east of the city also supplies a reservoir located in the center of the city, after which a pumping station delivers water to the city.

The DWDS is 259 km in length, 65% of the pipes are steel, 19% cast iron, 6% HDPE, 10% other materials and the diameters vary from approximately 600 to 800 mm.

2.2 Evaluation of the chlorite and chlorate concentrations in the distribution system

The Cl_2, ClO_2^- and ClO_3^- concentrations were analyzed at 26 points of the DWDS, from 2006 to 2011 (Figures 1 and 2). The Cl_2 residual concentration was determined with a residual chlorine analyzer.[15] The ClO_2^- and ClO_3^- concentrations were determined using Ion Chromatography (Dionex ICS 5000).[16]

The ClO_2^- concentration from 2006 to 2008 often exceeded the WHO GV of 700 μg L^{-1}, in particular at the points of the system more distant from the DWTPs. From 2009 to 2011 the ClO_2^- concentration was below the GV. The data showed significant variability before 2008, that subsequently decreased; this trend was probably due to the reduction of the ClO_2 dosage in the DWTP after a gradual cleaning of the DWDS pipes in 2008. Overall, the ClO_2^- concentration exceeded the WHO GV for 12-16% of results in the first two years, for 48% of results in 2008, and for 1-8% of results from 2009 to 2011.

The ClO_3^- concentration was significantly below the WHO GV of 700 μg L^{-1} throughout the monitoring period.

Figure 1 *Chlorite concentration measured in the distribution system of Cremona (Italy)*

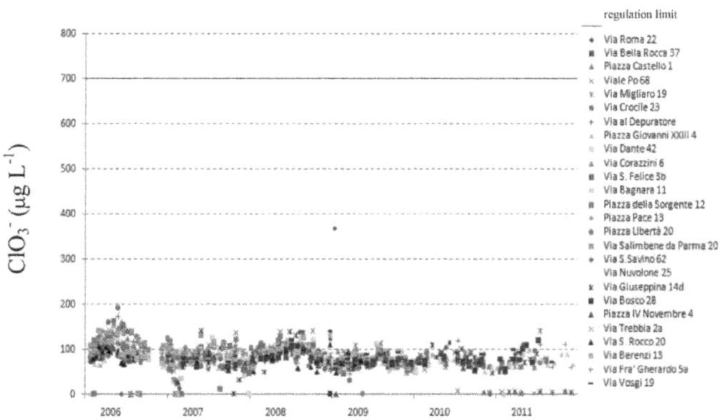

Figure 2 *Chlorate concentration measured in the distribution system of Cremona (Italy)*

2.3 Modelling of the drinking water supply system

The model of the DWDS was developed with the USEPA software, Epanet 2.0. According to the DWDS cartography, the altimetry of each junction and the length, material and roughness coefficient of each pipe were inserted into the software model. The modelled DWDS was simplified in order to simulate only a part of the system, so pipes with a diameter below 150 mm were not considered, since they constituted the peripheral sections of the system. However, the actual flow rates from each removed part of the system were inserted into the appropriate junction in the model. The simplified scheme, characterized by 5 closed paths, 144 junctions and a 23 km total length, was used for all the subsequent simulations.

2.4 Determination of the water mixing zone

In order to determine the mixing zone of the water coming from the two DWTPs, the propagation of a salt (NaCl) in the modelled system was simulated with the software Epanet 2.0. The aim was to evaluate the propagation in the system of a substance without decay or growth. A salt concentration of 70,000 mg L^{-1}, twice that of seawater, was dosed continuously for 24 hours firstly only in the east DWTP and in the central pumping station, and then only in the west DWTP. The propagation of the salt concentration was evaluated in the modelled DWDS after 6 and 7 hours from the start of the salt dosing in the east modelled DWTP. After 6 and 7 hours the salt concentration of 70,000 mg L^{-1} reached the junctions of the network more distant from the DWTP (junctions a, b, c and d of Figure 3). The propagation of the salt was evaluated in the modelled DWDS after 19 and 23 hours from the start of the salt dosing in the west modelled DWTP. After 19 and 23 hours the salt concentration of 70,000 mg L^{-1} reached the junctions of the network more distant from the DWTP (junctions c, e, f, g and h of Figure 3).

The zone of the DWDS where water coming from the east and west DWTP is mixed was determined by comparing the results of the two simulations of salt propagation and it is shown in Figure 3.

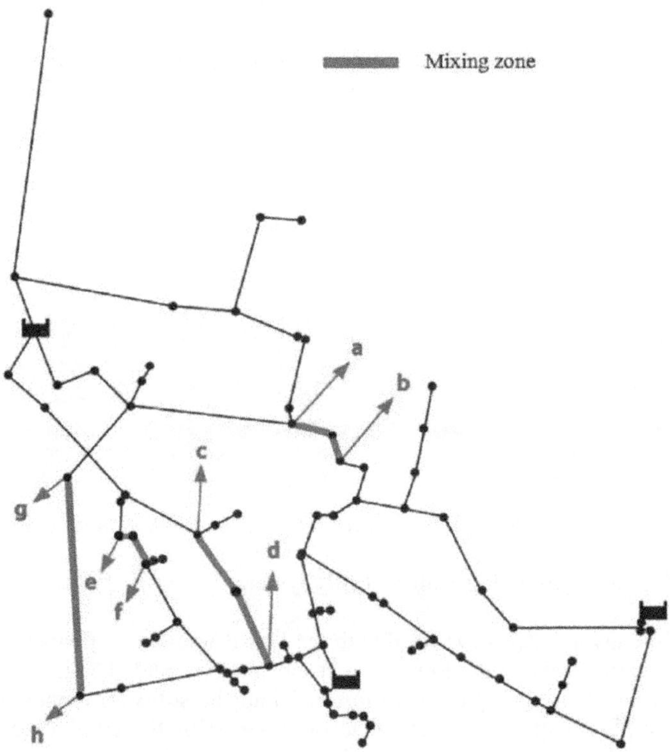

Figure 3 *Water mixing zone of the distribution system of Cremona (Italy) determined with Epanet 2.0*

2.5 Chlorite and chlorate propagation in the distribution system

The ClO_2^- and ClO_3^- reaction kinetics in the DWDS were studied. The wall reaction coefficient (k_w) was not considered. The order of the reaction kinetics and the mass reaction coefficients (k_b) were determined by batch experiments at a laboratory scale on water samples collected at the outlet of the sand filter of the full-scale DWTP. The lab-scale tests were carried out by treating water samples with 5 mg L^{-1} of ClO_2 and analyzing the ClO_2^- and ClO_3^- concentrations after 0, 4, 10, 30, 60, 79, 120, 160, 180, 360, 670, 1440 and 2280 minutes water-disinfectant contact time.

The order of the reaction kinetics and the reaction coefficients (k_b) for the ClO_2^- and ClO_3^- were inserted into the model for each simulation. The average measured initial concentrations of the ClO_2^- and ClO_3^- at the outlet of the real DWTPs were inserted in the model.

The study of the ClO_2^- reaction kinetic showed that the equation that represented the ClO_2^- growth versus time was a zero-order reaction, and the mass reaction coefficient k_b was equal to 0.576 d^{-1}. The ClO_2^- concentration in the modelled DWDS at h 10:00 after 1 month from the start of the disinfectant dosing in both the DWTPs is reported in Figure 4. The results showed that the ClO_2^- concentrations at points near the two DWTPs were close to the concentrations registered at the outlet of the DWTPs (close to 0.70 mg L^{-1}), since the ClO_2^- formation was very slow; with increasing the distance from the DWTPs, a gradual ClO_2^- concentration increase was registered. Moreover, in the peripheral pipes and in the mixing zone of the modelled DWDS, the ClO_2^- concentrations were over the WHO GV of 700 μg L^{-1}. Overall, a good correlation was observed between the measured and modelled ClO_2^- concentrations (Figure 5).

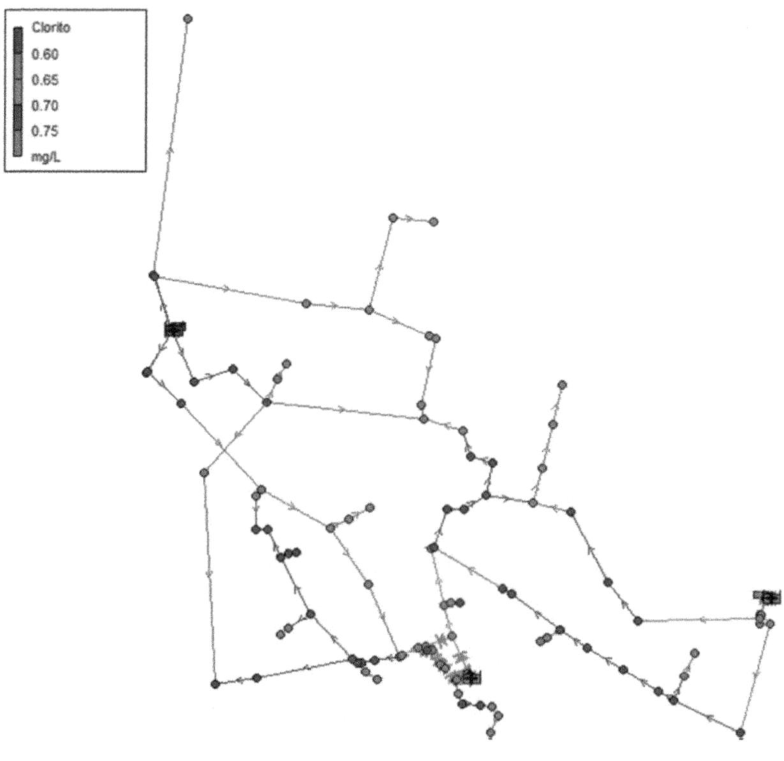

Figure 4 *Simulation of the chlorite propagation in the distribution system of Cremona (Italy) with Epanet 2.0*

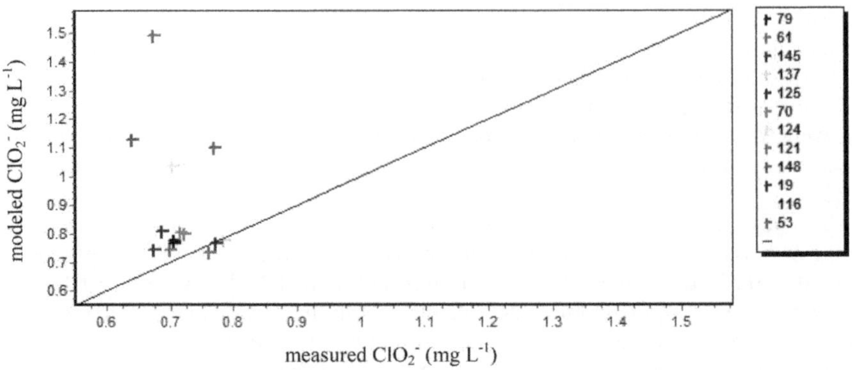

Figure 5 *Comparison between the measured and simulated chlorite concentration in the distribution system of Cremona (Italy)*

The study of the ClO_3^- reaction kinetic showed that the equation that represented the ClO_3^- growth versus time was a zero-order reaction, and the mass reaction coefficient k_b was equal to 0.101 d^{-1}. The ClO_3^- concentration in the modelled DWDS at h 10:00 after 1 month from the disinfectant dosing in both the DWTPs is reported in Figure 6. The results showed a very low ClO_3^- concentration gradient. Near the two DWTPs ClO_3^- formation was negligible, since the concentrations were close to those from the outlet of the DWTPs (close to 0.10 mg L^{-1}); with increasing the distance from the DWTPs, a slow and gradual ClO_3^- concentration increase was observed, with values always below the GV of 700 μg L^{-1}. Overall, a good correlation was observed between the measured and modelled ClO_3^- concentrations (Figure 7).

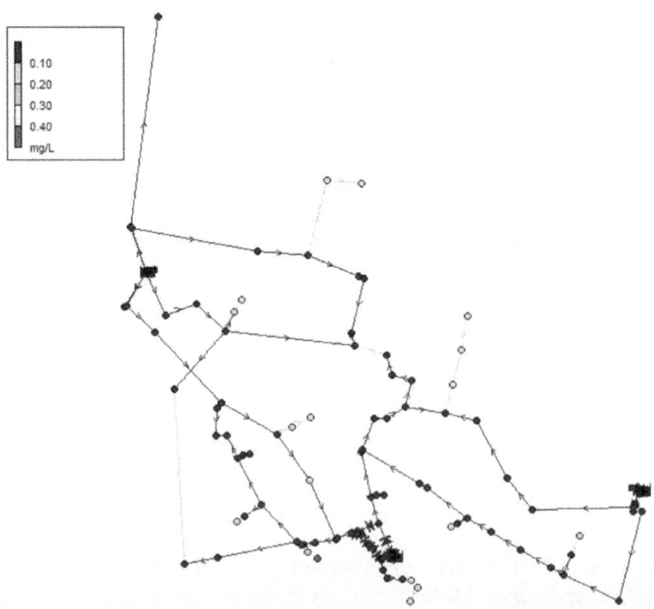

Figure 6 *Simulation of the chlorate propagation in the distribution system of Cremona (Italy) with Epanet 2.0*

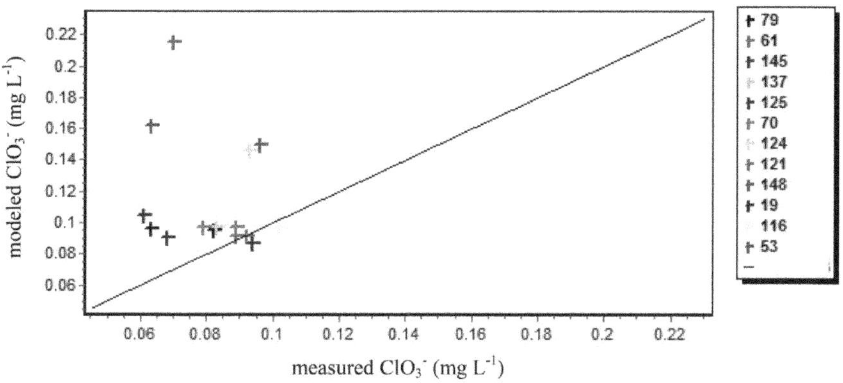

Figure 7 *Comparison between the measured and simulated chlorate concentration in the distribution system of Cremona (Italy)*

3 CONCLUSION

In this work the distribution system of Cremona (North of Italy), was monitored for 6 years analyzing chlorite and chlorate concentrations. The software Epanet 2.0 (USEPA) was applied to the distribution network in order to simulate the propagation of residual chlorine, chlorite and chlorate, and in order to estimate the mixing zone of the water coming from the west and east treatment plants. The measured and simulated concentrations were compared.

The results of the network monitoring showed significant chlorite formation. Predictably this occurred mainly in the points of the network more distant from the treatment plants, where it exceeded the WHO guideline value. Conversely, chlorate formation was much lower and did not exceed the guideline value.

The software Epanet 2.0 allowed the determination of the water mixing zone in the distribution network; the comparison between the concentrations measured in the real system and the values obtained from the simulation showed the usefulness of the model as an instrument to predict DBP concentrations in the distribution system.

References

1 O. Wable, N. Dumoutier, J.P. Duguet, P.A. Jarrige, G. Gelas and J.F. Depierre, *Am Water Works Assoc. Res. Found.*, 1991, 265.
2 G.R. Zhang, L. Kiéné, O. Wable, U.S. Chan and J.P. Duguet, *Environ. Technol.*, 1992, **13**, 937.
3 L. Kiéné, W. Lu, Y. Lévi, *Water Sci. Technol.*, 1998, **38**, 219.
4 A.O. Al-Jasser., *Water Res.*, 2007, **41**, 387.
5 WHO World Health Organization. *Guidelines for drinking water quality. Fourth edition*; 2011.
6 L.A. Rossman, R.M. Clark and W.M. Grayman, *J. Environ. Eng.* 1994, **120**, 803.
7 R. Clark, W. Grayman, J. Goodrich, R. Deininger and K. Skov, *J. Water Resour. Plann. Manage.*, 1994, **120**, 871.

8 M. Islam, M. Chaudhry and R. Clark, *J. Environ. Eng.*, 1997, **123**, 1033.
9 M.J. Rodriguez and J.B. Sérodes, *Environ. Modell. Softw.*, 1998, **14**, 93.
10 M.J. Rodriguez and J.B. Sérodes, *Water Res.*, 2001, **35**, 1572.
11 X. Li and H. Zhao, *Chemosphere*, 2006, **62**, 1028.
12 B.J. Courtis, J.R. West and J. Bridgeman, *Urban Water J.*, 2009, **6**, 407.
13 H.I. Mohamed and G. Abozeid, *Arab J. Sci. Eng.*, 2011, **36**, 173.
14 R. Clark, W.M. Grayman, R.M. Males and A.F. Hess, *J. Environ. Eng.*, 1993, **119**, 349.
15 APAT IRSA CNR 4080, 2003.
16 UNI EN ISO 10304-4, 2001.

Disinfection By-products Prevention and Treatment

NITROSAMINE CONTROL DURING WASTEWATER RECYCLING

W.A. Mitch

Department of Civil and Environmental Engineering, Stanford University, Palo Alto, CA 94305

1 INTRODUCTION

Nitrosamines are among the most potent carcinogenic by-products of disinfection.[1] Although they appeared on the Priority Pollutant List in the U.S., it was not until the late 1990s that their formation as a by-product of chloramine disinfection rose to prominence.[2] Nitrosamines have been a particular concern for wastewater recycling operations due to the higher concentrations of organic nitrogen precursors in municipal wastewaters.[2] Indeed the discovery at Water Factory 21, one of the oldest indirect potable reuse plants in the U.S., that N-nitrosodimethylamine (NDMA) formed during chloramination upstream of microfiltration units, and that NDMA passed through reverse osmosis membranes with only moderate rejection, led to one of the first plant shutdowns until the installation of UV treatment to degrade NDMA downstream of the reverse osmosis units.[2] More recently, potable reuse plants have been experimenting with the installation of ozone upstream of microfiltration units to minimize biofouling. However, pilot studies have indicated significant formation of NDMA by direct ozone reactions.[3] In an evaluation of trace organic chemicals of concern for wastewater reuse scenarios, the National Research Council indicated that disinfection by-products were of far higher potential health concern than pharmaceuticals at the concentrations typically found in wastewaters; among disinfection by-products, NDMA was of highest concern.[4]

To date, research and regulatory focus relevant to nitrosamines has been on NDMA. However, it is important to note that the seven nitrosamines on the U.S. Environmental Protection Agency's (EPA) Integrated Risk Information System (IRIS) database exhibit cancer potencies within an order of magnitude. Table 1 provides the drinking water concentrations of these nitrosamines that are associated with a 10^{-6} lifetime cancer risk. Given the expected error in such toxicity estimates, one might suspect that these compounds exhibit comparable cancer potencies despite the wide range of chemical properties (e.g., hydrophilic N-nitrosodiethanolamine to hydrophobic N-nitrosodi-n-butylamine). Preparatory to considering the regulation of nitrosamines as a group, the U.S. EPA developed an analytical method, Method 521, to measure low ng/L levels of the first six nitrosamines listed in Table 1, along with N-nitrosopiperidine, in surveys of U.S. drinking waters. Generally, NDMA was found in a wide array of chloraminated drinking waters, but the other nitrosamines were rarely found, and when found, tended to occur at lower concentrations than NDMA.[5]

This chapter outlines two approaches for the control of nitrosamines during wastewater recycling. The first approach is to identify key nitrosamines and their precursors. The objective is to focus research and control efforts on those nitrosamines that contribute significantly to the toxicity of a recycled water. Identification of precursors can indicate whether source control efforts may be an effective control strategy. The second approach is to alter the disinfection process to minimize the formation of nitrosamines.

Table 1 *Drinking water concentrations associated with 10^{-6} lifetime cancer risk for nitrosamines*

Nitrosamine	*Concentration (ng/L)*
N-nitrosodimethylamine	0.7
N-nitrosodiethylamine	0.2
N-nitrosodi-n-propylamine	5
N-nitrosodi-n-butylamine	6
N-nitrosoethylmethylamine	2
N-nitrosopyrrolidine	20
N-nitrosodiethanolamine	10

2 METHOD AND RESULTS

2.1 Identification of Key Nitrosamines and Precursors

We pursued three options to identify important nitrosamines and associated precursors. To date, NDMA has been the most frequently detected of the EPA Method 521 nitrosamines in disinfected wastewaters; N-nitrosomorpholine (NMOR) has also been detected, but generally enters wastewater via sewage, rather than forms via disinfection.[6] Although NDMA is the most prevalent of the EPA Method 521 nitrosamines detected in drinking waters and wastewaters, it is unclear whether NDMA is the most prevalent nitrosamine in general, or whether EPA Method 521 targets the wrong nitrosamines.

First, to gauge the relative importance of NDMA relative to the entire pool of nitrosamines, we developed a total nitrosamine (TONO) assay, analogous to the total organic halogen (TOX) assay.[7] Application of the TONO assay in conjunction with EPA Method 521 indicated that NDMA constitutes ~5-10% of the total nitrosamine pool in a range of disinfected drinking waters[8], recreational waters[7], and in preliminary data, recycled wastewaters.

Second, attempts to identify important nitrosamines in drinking water are hindered by the poor characterization of specific, prevalent precursor structures in the humic substance precursor pool in conventional drinking water supplies. However, because the organic precursors in wastewater have had less time to degrade than humic substances, precursors in wastewater effluents are more likely to resemble the compounds originally disposed to sewers. Thus, it should be possible to identify specific nitrosamine DBPs based upon the identification of precursors that are disposed to sewers in large quantities. Previous research has indicated that secondary, tertiary, and quaternary amines can serve as precursors for their analogues secondary nitrosamines during chloramination.[9] Recognizing that the tertiary amine, triethanolamine, is employed for pH control as a major ingredient of a wide array of consumer products, we reasoned that it could serve as a significant precursor for N-nitrosodiethanolamine (NDELA) in municipal wastewaters (Figure 1). Because NDELA is very polar, it is not amenable to analysis by EPA Method 521, which employs gas chromatography. We developed a LC-MS-based analytical method. Preliminary results from recycled wastewater trains indicate that NDELA occurs at concentrations comparable to NDMA.

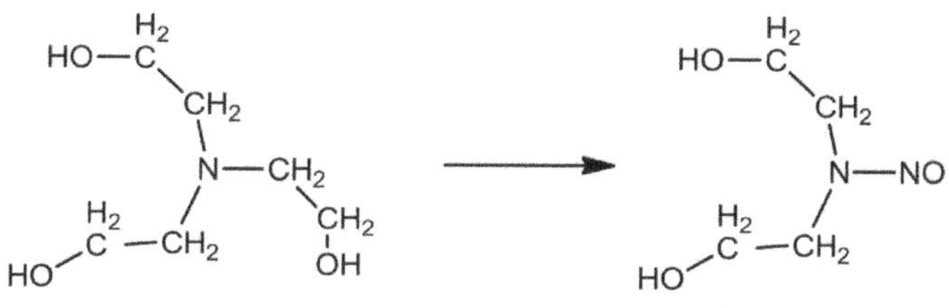

Figure 1 *Triethanolamine as a precursor to N-nitrosodiethanolamine (NDELA)*

Third, it is important to recognize that the wide array of compounds disposed to sewers renders it likely that a range of compounds may contribute to the formation of specific nitrosamines. Rather than attempt to identify each of these compounds, an alternative would be to identify whether specific input streams (industrial, human waste, greywater) are significant. The identification of specific streams may indicate the feasibility of mitigating nitrosamines by source control. For example, if precursor loadings from industrial discharges are important, industrial pretreatment schemes could be identified. If particular domestic streams of precursors are important, this knowledge may help future researchers to identify specific precursors (e.g., constituents of consumer products) in these streams that are important. Preliminary results with a range of greywaters indicate that laundry contributes up to an order of magnitude higher concentrations of NDMA and total nitrosamines compared to other greywater streams.

2.2 Modification of Disinfection Practices

Wastewater recycling utilities frequently practice chloramination by applying chlorine to ammonia-containing (i.e., non-nitrified) municipal wastewater effluents. Chloramines may be applied upstream of microfiltration units to inhibit biofouling in potable reuse treatment trains, or applied to the final effluent to maintain a disinfectant residual in the distribution system for non-potable systems (e.g., irrigation systems). More recently, potable reuse treatment trains have experimented with the application of ozone upstream of microfiltration units instead of chloramines, because ozone oxidation of dissolved organics reduces the fouling potential of these organics on the membrane systems.[10] However, these utilities often observe substantial NDMA formation as a result of direct reactions with ozone.[3]

We evaluated a combination of medium pressure UV (MP UV) and preformed monochloramine as an alternative disinfection scheme. In research with wastewater-impacted drinking waters, we found that application of ~200 mJ/cm^2 germicidal fluence of MP UV was nearly as effective at deactivating NDMA precursors as ~140 mg.min/L free chlorine or 0.2 mg.min/L ozone.[11] Low pressure UV was generally less effective. Concentrations of NDMA and total nitrosamines entering wastewater treatment plants can be substantial.[12] In addition to effectively deactivating precursors to nitrosamine formation during subsequent chloramination, MP UV features the added advantage that it destroys nitrosamines entering into the sewage. In contrast, ozonation forms substantial concentrations of NDMA, and free chlorine exposure cannot be achieved unless the wastewater is fully nitrified.

Wastewater recycling utilities may frequently practice chloramination to maintain a disinfectant residual. In previous research, we determined that the traces of inorganic dichloramine that always co-occur with monochloramine in chloraminated solutions are the predominant chloramine precursor to nitrosamines.[13] During typical chloramination practice, wastewater recycling utilities add free chlorine to ammonia-containing effluents, forming chloramines in situ. The dose of free chlorine generally is such that ammonia is present in molar excess, and thus monochloramine would be the predominant inorganic chloramine. Regardless, there will always be some dichloramine present in equilibrium with monochloramine. However, dichloramine formation can be enhanced beyond that anticipated based upon equilibrium calculations due to a mixing effect. Chloramine formation reactions are rapid. At the site of free chlorine injection, the molar ratio of free chlorine to ammonia can be high prior to complete mixing of the free chlorine into solution. This condition fosters greater dichloramine formation than would be anticipated based upon equilibrium calculations for a completely mixed solution.[14]

An alternative would be to form chloramines in a separate stream via addition of free chlorine followed by ammonia, a condition that would promote monochloramine formation at the expense of dichloramine by the same mixing effect. This "preformed monochloramine" solution could then be applied to the wastewater effluent, minimizing the exposure of the nitrosamine precursors to dichloramine. In a pilot test at an indirect potable reuse treatment train, this strategy effectively reduced NDMA formation during chloramination upstream of microfilters.[15] Preliminary results indicate that the strategy also effectively reduces NDMA formation over longer chloramine contact times, such as found in distribution systems for recycled water applied for irrigation. Combination of MP UV pre-oxidation with post-treatment with preformed monochloramine has the potential to significantly reduce NDMA concentrations in recycled wastewater.

3 CONCLUSIONS

Due to their high cancer potencies, and the high levels of organic nitrogen precursors in wastewater effluents, nitrosamine disinfection by-products have emerged as key contaminants of concern for wastewater recycling operations. To mitigate nitrosamines, it is important to characterize their precursors and identify novel disinfection schemes that can minimize their formation. Despite current interest in NDMA, NDMA is a minor component of total nitrosamines in recycled wastewaters. Noting that triethanolamine is a major component of a range of consumer products, we identified N-nitrosodiethanolamine at concentrations comparable to NDMA in recycled water systems. Greywater laundry effluent contributes significant loads of NDMA and total nitrosamines to sewage. The combination of medium pressure UV followed by application of preformed monochloramine can effectively minimize the concentrations of NDMA in recycled water systems.

References

1. U.S. Environmental Protection Agency, Office of Research and Development (ORD), National Center for Environmental Assessment *Integrated Risk Information System*; U.S. Environmental Protection Agency: Washington, DC, 1993. http://www.epa.gov/iris/ (Accessed August 6, 2014).
2. W.A. Mitch, J.O. Sharp, R.R. Trussell, R.L. Valentine, L. Alvarez-Cohen, and D.L. Sedlak. *Environ. Eng. Sci.* 2003, **20 (5),** 389-404.
3. M. Sgroi, P. Roccaro, G.L. Oelker, and S.A. Snyder. *Environ. Sci Technol.*, 2014, **48**, 10308-10315.
4. National Research Council. *Water Reuse: Potential for Expanding the Nation's Water Supply Through Reuse of Municipal Wastewater.* (Council, N. R., Ed.), **2012**, National Academies Press, Washington, D.C.
5. C.G., Russell, N.K. Blute, S. Via, X. Wu, Z. Chowdhury, and R. More, *Journal AWWA* 2012, **104 (3).** http://dx.doi.org/10.5942/jawwa.2012.104.0037.
6. I.M. Schreiber, and W.A. Mitch *Environ. Sci. Technol.,* 2006, **40 (10),** 3203-3210.
7. P. Kulshrestha, K.D. McKinstry, B.O. Fernandez, M. Feelisch, and W.A. Mitch, *Environ. Sci. Technol.*, 2010, **44,** 3369-3375.
8. N. Dai, and W.A. Mitch, *Environ. Sci. Technol.* 2013, **47**, 3648-3656.
9. A.D. Shah, and W.A. Mitch, *Environ. Sci. Technol.,* 2012, **46,** 119-131.
10. T. Zeng, C.J. Wilson, and W.A. Mitch *Environ. Sci. Technol.* 2014, **48**, 5118-5126.
11. A.D. Shah, S.W. Krasner, T. C.-F. Chen, U. von Gunten, and W.A. Mitch, *Environ. Sci. Technol.* 2012, **46**, 4809-4818.
12. D.L. Sedlak, R. Deeb, E. Hawley, W.A. Mitch, T. Durbin, S. Mowbray, and S. Carr, *Wat. Environ. Res.*, 2005, **77 (1),** 32-39.
13. I.M Schreiber, and W.A. Mitch, *Environ. Sci. Technol.,* 2006, **40 (19),** 6007-6014.
14. I.M Schreiber, and W.A. Mitch, *Environ. Sci. Technol.,* 2005, **39 (10),** 3811-3818.
15. W.A. Mitch, G.L. Oelker, E.L. Hawley, R.A. Deeb, and D.L. Sedlak, D.L. *Environ. Eng. Sci.,* 2005, **22 (6),** 882-890.

ELECTROCHEMICAL DISINFECTION WITH BORON DOPED DIAMOND ELECTRODES AND REDUCTION OF ELECTROLYSIS BY-PRODUCTS

M. Fryda [1], B. Behrendt-Fryda-[1], A. Hampel [1], L. Heesch [1], Th. Matthée [1], M.Fischer[2], W. Schmidt[3]

[1] CONDIAS GmbH, Fraunhofer Straße 1b, 25524 Itzehoe, Germany
[2] MLE, Bernhard-Voß-Straße 27, 01445 Radebeul, Germany
[3] DVGW Technologiezentrum Wasser, Außenstelle Dresden, Wasserwerksstraße 2, 01326 Dresden, Germany

1 INTRODUCTION

Environmental pollution is a growing challenge for water utilities. Even today organic compounds are not always efficiently removed during water treatment. For more than 20 years boron doped diamond electrodes (BDD) have been well known for electrochemical processes with very high oxidative power.[1-5] Electrochemical processes based on BDD electrodes are used in several industrial disinfection applications such as the pharmaceutical industry (ultra-pure water ozonization) or ballast water treatment (marine application). Using such an electrochemical advanced oxidation process makes it possible to degrade organic compounds.[6,7] Due to the high oxidative power of BDD electrodes there is a potential risk for the generation of electrolysis by-products (EBP) such as bromate or chlorate as well as halogenated organic compounds (AOX).[8,9] This investigation presents some applications of BDD for disinfection, recycling and wastewater treatment processes in different water types by modification of BDD composition and structure.

2 METHOD AND RESULTS

2.1 Production of nano-modified BDD electrodes

The Hot-Filament Chemical Vapour Deposition process (HFCVD) of producing BDD electrodes depends on five important process parameters (concentration of hydrocarbon, boron doping, process pressure, power and coating thickness), which were systematically varied using design of experiment (DOE) for the HFCVD conditions of standard BDD electrodes (which were defined as the reference point of the variations). BDD electrode limitations were investigated and several tests (physical, electrochemical and functional) were performed to characterise these new types of nano-modified boron doped diamond electrodes.

Variation of CVD process parameters yielded reproducible different diamond crystal sizes (100nm up to 5000nm), crystal orientations (randomly oriented and <110> or <111> - textured) and phase purities (defined as the ratio of the diamond phase sp^3 and an amorphous carbon phase). DOE statistical analysis identified process power (factor A), process pressure (factor B) and hydrocarbon concentration (factor C) as controlling parameters for these changes in crystallinity. In Figure 1 examples of electrode surface morphology are shown and modifications are clearly visible. In the case of poor phase purity (high amorphous carbon or graphite content) electrochemical stability was inacceptable, due to very quick diamond film decomposition by oxidation of non-diamond carbon phases.

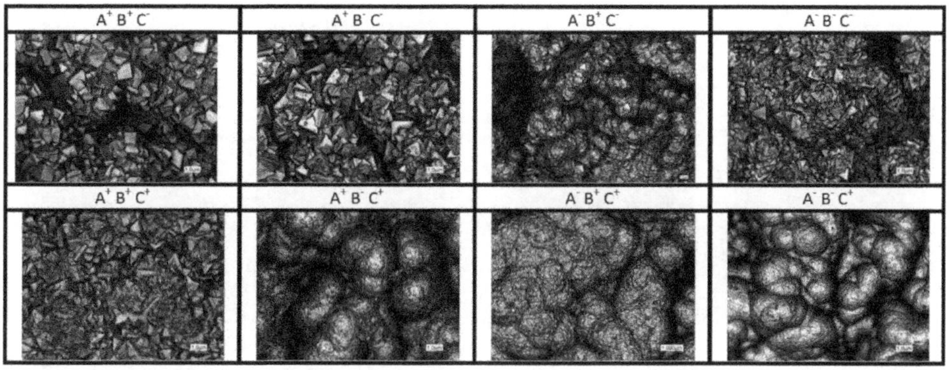

Figure 1 *Morphology of nano-modified BDD electrodes by Laser-Scanning Microscopy (LSM)*
(factor A is process power,
factor B is process pressure and
factor C is process hydrocarbon concentration;
indices + and – stay for high and low levels of the specific factor)

2.2 Electrochemical properties of nano-modified BDD electrodes

For electrochemical characterisation of nano-modified BDD electrodes cyclovoltammography (current-voltage curves; CV) was used with acidic and alkaline electrolytes. In Figure 2 examples of CV curves are shown for 1m sulphuric acid. For statistical validation and as a comparison the reference point factors were used several times and resulting standard BDD electrodes produce consistent and reproducible CV curves (upper left diagram in Figure 2). There is a weak oxidation peak at 2.0V which indicates oxidation of sulphuric acid to peroxodisulphuric acid. Further on there is a high overvoltage for oxygen evolution so that hydroxyl radical formation starts at 2.4V.

Variations of doping level induce a disturbance of the diamond lattice which results in changes of film conductivity but also in morphological changes (see LSM pictures in Figure 2). Therefore the electrochemical properties of modified BDD electrodes are also different. In the case of sulphuric acid electrolyte, oxidation of sulphuric acid at 2.0V is quite different and current response varies from 50mA up to 230mA which is nearly a factor of 5 higher and indicates different selectivity for sulphate oxidation. Also the formation potential of hydroxyl radicals varies in the range 2.25-2.55V.

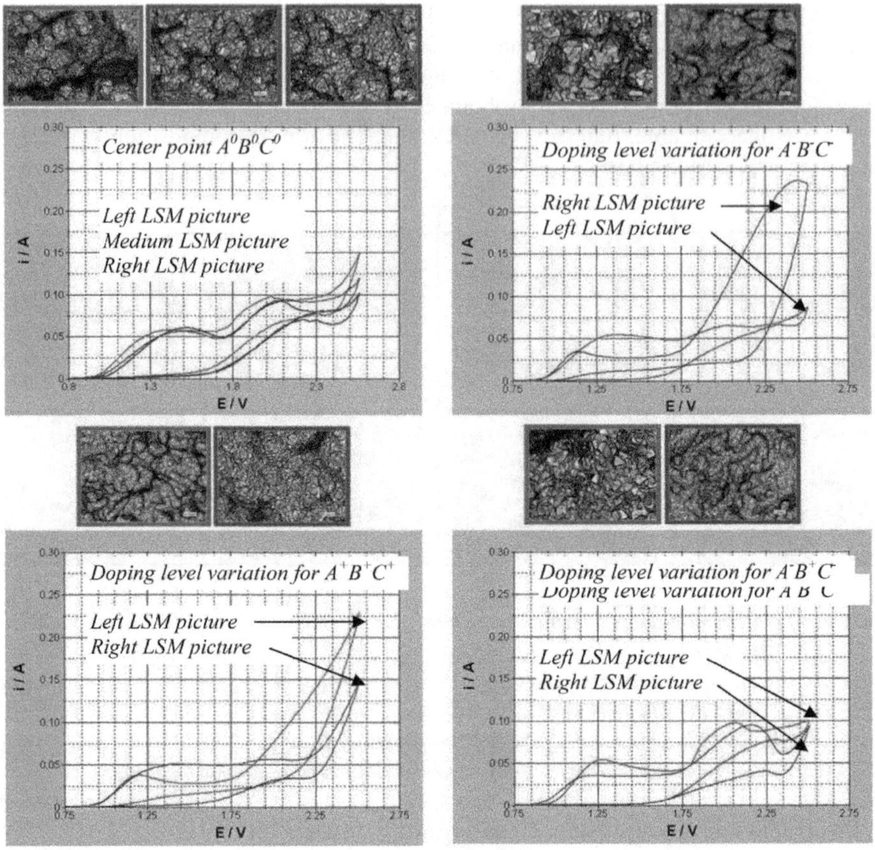

Figure 2 *Cyclovoltammograms of nano-modified BDD electrodes in 1m sulphuric acid (frame colour of morphology picture indicates corresponding CV curve)*

Further investigations of the electrochemical formation of ozone and electrolysis by-products (EBP) by hydroxyl radical formation from water molecule electrolysis were performed in two model water matrices described as "SOFT" and "HARD" water. Compositions of these two matrices are described in Table 1.

Table 1 *Composition of model water matrices "SOFT" and "HARD"*

	Model water SOFT	Model water HARD
Total hardness [°dH]	5	20
Chloride [mg/l]	20	100
Bromide [mg/l]	0,1	1
Sulfate [mg/l]	60	200
Nitrate [mg/l]	10	50

Investigation of the electrochemical formation rate of ozone and EBP was performed in a single pass of different model waters through an optimized electrolysis cell with varying flow rates (60l/h – 560l/h or contact times of 0.05s – 0.5s) and electrical currents (1.5A – 5A); the applied anode area was $3cm^2$. Figure 3 shows concentration results for trihalomethanes (THM). Direct electrochemical formation of THM in such an inline process with BDD seems to be very low because there is nearly no difference between SOFT and HARD water matrices, even if TOC is added. Typically THM concentrations are below 3µg/l if reaction is stopped by addition of sodium sulphite after 30min. Only for HARD water without additional TOC but an ongoing formation reaction with electrochemically produced ozone is increased THM (160µg/l) found, especially bromoform. Formation speed of THM under these conditions is relatively slow. Therefore THM seems not to be an EBP but more likely a disinfection by-product formed by an ongoing reaction with remaining ozone.

Due to the fact that formation of AOX was also found in low concentrations (typically ~10µg/l, in singular cases up to 100µg/l), organic by-products have only minor importance for an inline electrolysis process with nano-modified BDD electrodes.

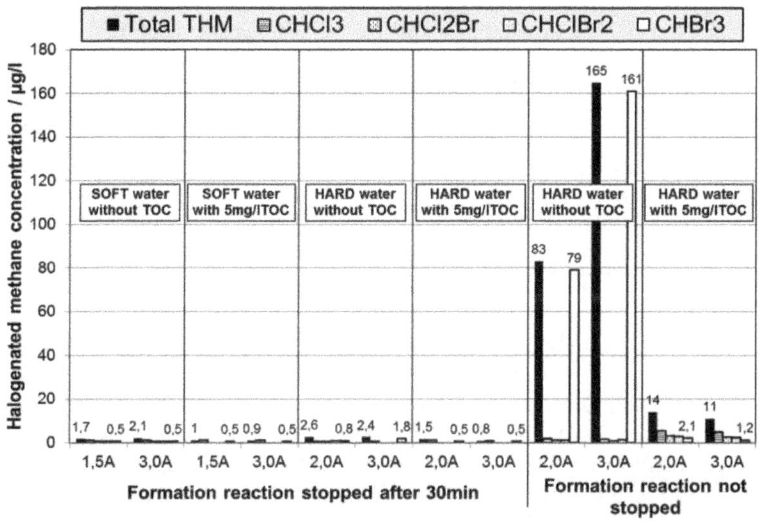

Figure 3 *THM concentrations for a nano-modified BDD electrode disinfection process, for different water matrices, electrical current and a stopped or not stopped oxidation process.*

Formation of inorganic EBP is much more important because chloride and bromide can be oxidized by an electrolysis process with BDD.[8,9] In the case of standard BDD electrodes, formation of bromate (~0.3mg/Ah) and chlorate (~4.5mg/Ah) is proportional to electrical load impact. Whereas ozone formation with stable modified electrodes is comparable or slightly better than standard BDD electrodes (35 to 55mg/Ah). During these investigations nano-modified electrodes were produced, which have a much lower formation rate for EBP. In Figure 4 an overview of concentration ratios of EBP/ozone is presented. These values are an average of flow rate and current variations. High ratios of EBP/ozone result mainly from a low ozone formation with these BDD modifications, due to a higher non-diamond carbon phase which consumes oxidative species such as hydroxyl radicals or ozone. Nevertheless several modifications demonstrate better ratios compared to standard electrodes by a factor of up to 3. Under certain conditions even lower EBP/ozone values (5µg/l bromate and 10µg/l chlorate per 1000µg/l ozone) were measured, which are comparable to typical pure ozonization processes,[10] even though the electrochemical reaction process goes via hydroxyl radicals.

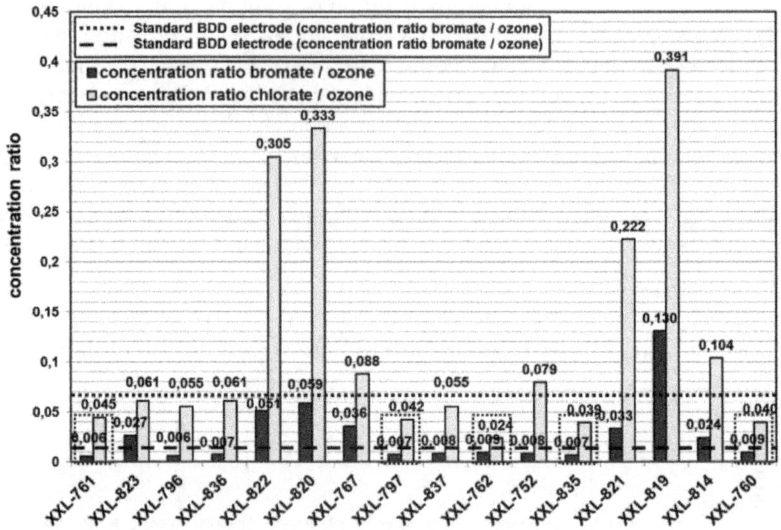

Figure 4 *Ratio of electrolysis by-product / ozone concentration for investigated nano-modified BDD electrodes.*

Statistical analysis of DOE data results in a model description of ozone and EBP formation depending on production parameters of nano-modified BDD electrodes. Figure 5 shows model descriptions of ozone and EBP formation as well as expected electrode life time in 3D block diagrams.

Ozone production rate increases from 0 to 55mg/Ah with increasing CVD process pressure and decreasing hydrocarbon concentration. Ozone production rate is nearly stable at a high level (40-55mg/Ah) for all modifications of high pressure or low hydrocarbon concentration.

The ratio of EBP chlorate / ozone shows a similar behaviour. Again increasing process pressure and decreasing hydrocarbon concentration results in better values, in this case smaller ratios (ratios decrease from 0.5 to 0.05). High pressure or low hydrocarbon concentration show constant low ratios and therefore the most promising CVD conditions for good nano-modified BDD electrodes.

In addition to high ozone production rate and low EBP formation, high stability of BDD electrodes is an important factor for usability. Electrode stability was investigated in an accelerated life time test with extremely high current densities and corrosive electrolytes to get a high electrochemical corrosion rate. The most stable regions of nano-modified BDD electrodes are high process pressure and low hydrocarbon concentration.

Based on these DOE results there seems to be a high possibility in terms of EBP/ozone ratio for even better BDD modifications, which have to be investigated in the future.

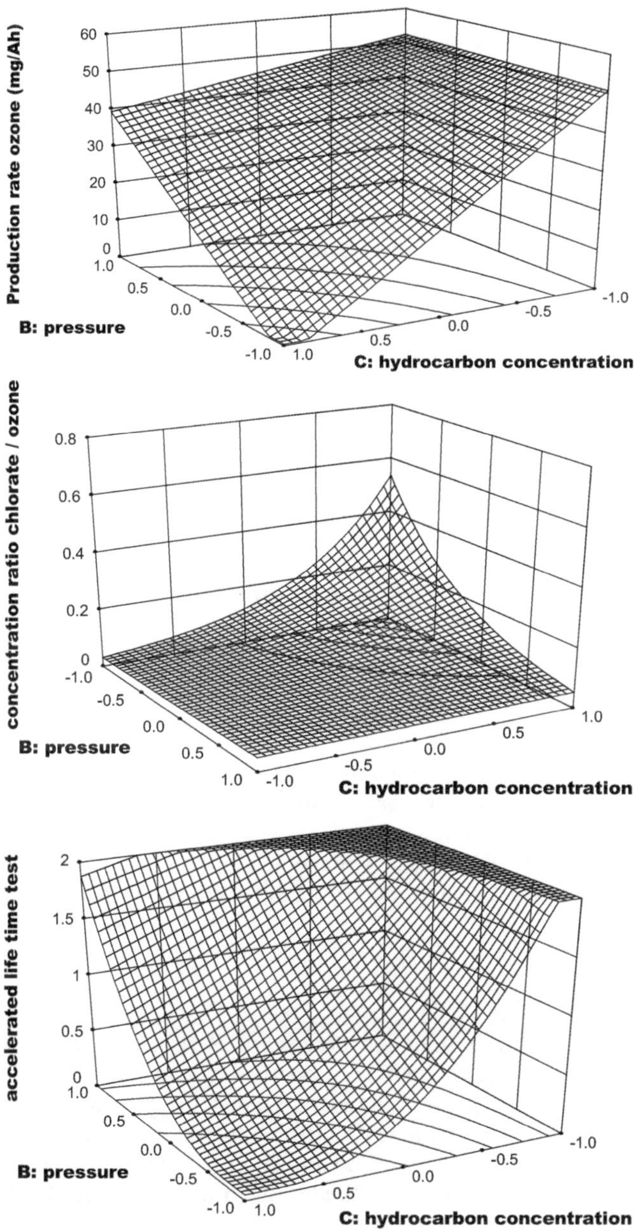

Figure 5 *Design of experiment analysis yields models for dependence of BDD electrode properties on production parameters. Block diagrams show model predictions of ozone and chlorate concentration as well as electrode stability in an accelerated life time test at medium CVD process power.*

With optimized, nano-modified BDD electrode,s bacteria reduction *(E.coli DSM 11250)* investigations were performed using an inline process with SOFT and HARD water matrices and flow rates of 60-560l/h in a single path. In Figure 6, exemplary results for bacteria reduction factor (LOG no. initial bacteria minus LOG no. remaining bacteria) and chlorate formation are shown.

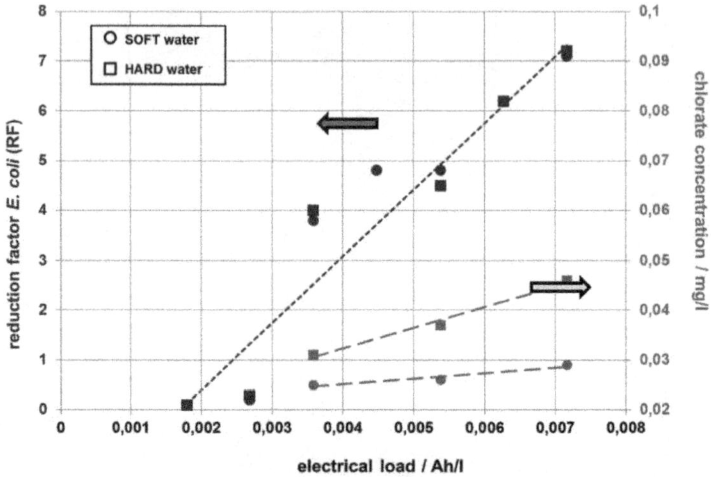

Figure 6 *Reduction factor (RF) of E.coli and chlorate concentration for an electrochemical inline disinfection process based on nano-modified BDD electrodes.*

For both water matrices RF is comparable and with very low energy impact, bacteria counts could be reduced by 7 orders of magnitude. EBP (chlorate) is low, but as expected water matrix HARD increases chlorate concentration. Under these conditions an electrical load of 0.0055Ah/l at a voltage 15V is necessary for a RF of 5, which corresponds to an electrical consumption of 0.0825 kWh/m^3 of treated water.

3 CONCLUSION

Changing parameters in the production process of BDD electrodes result in different BDD films with specific electrochemical properties. Characterization of the production rate of ozone and by-products demonstrates that by-product generation can be significantly suppressed, and parameters for improved modifications are identified. Whereas they are not suitable for disinfection of drinking water due to the concentrations of bromate and perchlorate, nano-modified BDD electrodes show interesting and potent possibilities for other environmental applications. In these investigations nano-modified electrodes showed low energy consumption in disinfection processes, and significantly reduced production of by-products, particularly organic by-products such as AOX, THM and nitrosamine, which are negligible.

Acknowledgement

We thank the Bundesministerium für Bildung und Forschung for funding this project under reference number 03 X 0087 A/B and our colleagues from Fraunhofer IST and TU München for their co-operation in this joint research project and fruitful discussions.

References

1 C. A. Martinez-Huitle, M. A. Quiroz Alfaro, *J. Environ. Eng. Manage.*, 2008, 18(3), 155
2 M. Fryda, L. Schäfer, I. Tröster, *Recent Res. Dev. Electrochem.*, 2001, 4, 85
3 I. Tröster, L. Schäfer, M. Fryda, T. Matthée, *Water Sci. Technol.*, 2004,49(4), 207
4 A. Kraft, M. Stadelmann, M. Wünsche, M. Blaschke, *Electrochem. Comm.*, 2006, 8(5), 883
5 A. Cano, P. Cañizares, C. Barrera-Díaz, C. Sáez, M. A. Rodrigo, M. A., *Chemical Engineering Journal*, 2012, 211-212, 463
6 J. Boudreau, D. Bejan, S Li, N. J. Bunce, *Industrial and Engineering Chemistry Research*, 2010, 49, 2537
7 E. Brillas, S. Garcia-Segura, M. Skoumal, C. Arias, *Chemosphere*, 2010, 79, 605
8 H. Bergmann, *GWF-Wasser/Abwasser*, 2010, 151 (6), 604
9 M. H. Bergmann, J. Rollin, T. Iourtchouk, *Electrochimica Acta*, 2009, 54, 2102
10 Karl Höll, *Wasser*, ed. A. Grohmann, Walter de Gruyter, Berlin New York, 2002, p638

EFFECTS OF WATER TREATMENT TECHNOLOGY AND UV DISINFECTION ON NOM FRACTIONS AND BDOC: EXPERIENCES FROM FULL-SCALE FACILITIES IN NORWAY

Bjornar Eikebrokk

SINTEF, Dept. of Infrastructure – Water and Environment, NO-7465 Trondheim, Norway

1 INTRODUCTION

In Norway, 85 % of the drinking water supply is from surface water sources, mainly lakes. The source waters normally contain natural organic matter (NOM) with TOC levels in the range 1-5 mg/L, although a few sources with levels of more than 15 mg TOC/L are being used. Furthermore, the source waters can be characterized as very soft (< 3 mg Ca/L) and corrosive, with low alkalinity (< 0.1 meqv/L) and low turbidity levels (< 0.5-1 NTU). The microbial quality is normally good, with well protected watersheds and low levels of microbial indicators (e.g. *E.coli*) during most of the season. However, both the NOM (TOC) and the microbial indicator concentration levels normally reach a peak in the autumn during periods of heavy rain. In the southern part of Norway, increasing trends in NOM concentration levels have been observed since the early 1980s, probably due to the effects of climate change, e.g. increasing temperatures, more primary production, increasing precipitation, and less acid rain.

Since 2009, UV-disinfection is the predominant primary disinfection process in Norway, applied at close to 1000 water treatment plants (WTPs) and supplying 70-80 % of the population (2011). The 90-100 utilities that still use chlorine for primary disinfection are operated in compliance with the Norwegian water quality standard, i.e. with a low-dose strategy requiring detectable free chlorine (i.e. 0.05 mg/L) after 30 minutes contact time.

For water treatment prior to disinfection, one out of three primary water treatment processes is normally applied: (i) Enhanced coagulation (EC), (ii) Nanofiltration (NF), or (iii) Ozonation and biofiltration (OBF). EC is applied at some 120 systems including the large ones. NF is applied at some 110 small systems, and/or for source waters with the highest NOM levels. OBF is a relatively new process in Norway, and it is now being applied at some 40 facilities. OBF has gained popularity due to its simplicity and the chemical-free approach.

Enhanced coagulation processes are normally capable of removing most NOM fractions effectively except the neutral hydrophilic fraction.[10,11,12,34,39] NF is normally even more effective than EC for all NOM fractions. However, since this technology is usually selected for source waters with the highest NOM concentration levels, residual NOM and BDOC concentrations may still be significant after NF treatment. Ozone can transform hydrophobic (humic) NOM fractions into more biodegradable fractions, e.g. carboxylic acids, ketoacids, aldehydes, ketones, phenols,[28,29] thereby increasing BDOC and AOC. A majority of the DBPs produced by chlorine or ozone are still unknown, and close to 60 % of the AOC in ozonated drinking water remains unidentified.[28] Thus it is important that the subsequent biofiltration is capable of removing BDOC in treated water to levels that do not sustain regrowth and biofilm formation in the distribution system. This is not always achieved, and a majority of the OBF facilities apply secondary disinfection in the form of low-dose chlorination (typically 0.1-0.2 mg/L) to control regrowth. Microbial growth in distribution systems can affect the drinking water quality in many ways: (i) non-compliance with water quality regulations, (ii) presence of opportunistic pathogens, (iii) aesthetic problems, e.g. discoloured water, taste and odour, invertebrates, (iv) technical problems, e.g. corrosion, pressure loss, leakages, etc.

The potential for biological growth in water distribution systems is determined by several factors including[36]: i) the concentration and nature of biodegradable compounds serving as an energy source for microorganisms (e.g. AOC, BDOC), ii) water temperature, iii) hydraulics and residence time, iv) materials in contact with water, v) presence of inorganic nutrients (e.g. N, P), vi) presence of oxygen or another electron acceptor, vii) sediments, viii) the concentration and nature of disinfectant residuals. Some countries rely on the presence of chlorine-based disinfectant residuals throughout the distribution system as a means of controlling regrowth, while others (e.g. The Netherlands, Denmark, Norway and areas of Germany, Switzerland and Luxembourg) distribute water without chlorine residuals. In the latter case, biostability is an important prerequisite in order to avoid regrowth, biofilm formation and related problems

In UV treatment, biodegradable organics may be formed through adsorption of photons (direct photolysis) or through reaction of hydroxyl radicals with NOM. The latter mechanism is however considered negligible.[7] UV treatment can remove conjugated bonds and thus reduce colour and UV-absorbance, reduce molecular weight, and alter the structure and reactivity of NOM.[3,16,20,23,26,35] UV disinfection does not form THMs and HAAs, but nitrite can be formed from nitrate by adsorption of UV light. The European drinking water standard for nitrite (0.1 mg/L) can however be met for both LP and MP UV lamps with UV doses up to 70 mJ/cm^2. The formation of UV-disinfection by products including nitrite can be significantly reduced if the MP UV lamps are provided with quartz sleeves with a cut-off of the low wavelengths.[3] In Norway, this is a prerequisite for the approval of MP UV lamps used in drinking water disinfection.

Studies in Australia showed that increasing doses of UV, transformed larger amounts of the very hydrophobic (VHA) fractions of NOM into more hydrophilic and more biodegradable fractions[2], similar to the effects of ozone. With respect to biodegradability, studies in the Netherlands showed that AOC formation during UV disinfection was not significant at UV doses up to 90 mJ/cm^2, regardless of the applied type of UV lamps (LP or MP)[17]. Studies in Sweden did not detect any enhanced biofilm growth or pathogen incidence within a municipal distribution system when primary disinfection was shifted from conventional chlorination to UV.[22] Contrary to this, studies in the Netherlands showed that even LP-UV at doses as low as 40 mJ/cm^2 could significantly elevate both the AOC concentration and the biofilm formation in treated water.[38] The studies concluded that the regrowth potential of drinking water depends on the character and concentration of NOM.

2 MAIN STUDY OBJECTIVES

With NOM-laden source waters, UV-disinfection systems and chlorine-free distribution, the need for effective NOM removal and reduction of the numerous impacts of NOM[9,11] on treatment and distribution systems is obvious. An important part of this picture is how to maintain biological stability and control regrowth in distribution systems.

Thus, the main objective of this study was to investigate how the applied water treatment technologies and UV-disinfection systems affect NOM and NOM fraction removal efficiencies, as well as the biological degradability of the remaining NOM (BDOC). This information can be applied for diagnosing source water treatability and optimising treatment performance with respect to NOM, NOM fraction and BDOC removal efficiency. In addition, it can be used to provide better links between treatment, UV-disinfection and water distribution performances.

3 METHODS AND APPROACH

Seasonal water samples were collected from more than 30 full-scale utilities in Norway. Sampling included untreated raw water, effluent water from different treatment steps, sampling prior to and after UV-disinfection, as well as sampling from the distribution system. Due to the prevailing pristine drinking water sources and the good source water quality, the described treatment technologies are normally applied as single-stand processes in combination with corrosion control and disinfection, thus allowing a direct comparison of NOM fraction and BDOC removal performance between the technologies.

Water quality analyses included conventional parameters like pH, turbidity, colour, UV-absorbance, TOC, DOC and metals. Colour and UV-absorbance were analysed by a spectrophotometer (Hitachi U-3000) in accordance with the Norwegian Standard Methods (NS 4787 and NS 9462, respectively).

3.1 NOM Diagnostic Tools

In addition to the analyses of conventional water quality parameters, NOM diagnostic tools like rapid NOM-fractionation and BDOC analysis were used. In addition, biofilm monitors (BFM) were applied to investigate amount and composition of internal deposits formed over time in polyethylene (PE) test pipes.

3.1.1 *NOM Fractionation.* The applied rapid fractionation technique was based on measuring DOC concentrations (Teledyne Tekmar TOC Fusion analyser) in pre filtered (0.45 μm) samples before and after contact with the resins DAX-8, XAD-4 and IRA-958 according to the method (Figure 1) described in detail in the literature.[5,13,39] The rapid characterization technique is specifically designed to study water treatment processes, and is based on a full-scale fractionation scheme.[1,5] Based on subtractions of the DOC concentrations of subsequent resin effluents, the organic carbon concentrations of four fractions of NOM were determined: (i) Very hydrophobic acid (VHA), (ii) Slightly hydrophobic acid (SHA), (iii) Charged hydrophilic matter (CHA), and (iv) Neutral hydrophilic matter (NEU).

VHA and SHA is predominantly composed of higher molecular weight humic and fulvic acids, CHA is ascribed to proteins, amino acids and anionic polysaccharides, and the NEU fraction is ascribed to carbohydrates, aldehydes, ketones and alcohols. The hydrophilic substances are typically biologically derived compounds such as polysaccharides and proteins [19,40] The order for ease of removal by coagulation is generally VHA > CHA=SHA > NEU[5]. The NEU fraction is recalcitrant to removal by metal-based coagulation.[9,12] In fact, the NEU fraction concentration may even increase in periods with sub-optimum coagulant doses, presumably due to the NOM-neutralizing effects of the cationic coagulant species and the formation of poorly filterable/adsorbable flocs at sub-optimum coagulant dose conditions.[11] The removal of particular NOM fractions is dependent upon treatment conditions such as applied coagulant dose, pH, etc. and treatment conditions can be optimized based on the character of the organic matter present in the raw water. The rapid fractionation technique can be applied as a tool for operators of treatment plants to control and monitor the treatment process in the most effective way for NOM removal. The rapid fractionation technique can also be used to identify situations where treatment is not effective for DOC removal, either due to lack of optimisation or problems with routine operation. From analyses of a number of raw and treated water samples (coagulation; nanofiltration; ion exchange) from Norway and Australia, great similarities in NOM fraction distributions and removal characteristics were identified despite the great differences in catchments and climatic conditions between the sampling sites.[13]

3.1.2 *BDOC-columns-in-series*. Biodegradable organic carbon (BDOC) represents the fraction of DOC that can be used as energy and a carbon source by microorganisms, and is calculated as the difference between the initial DOC of a water sample and the minimum DOC observed after inoculation with heterotrophic, environmental bacteria and incubation in batch reactors typically for 28 days at 20 °C[30] or in continuous bioreactors with immobilized biofilm on carriers such as sand or glass beads.[21,27,43] Due to the higher biomass in the continuous bioreactors with carriers, the test period can be significantly reduced, and BDOC determination can usually be performed within hours by this rapid method.[21]

The BDOC set-up used in this study was a modification of the column-based BDOC analysis used by others[15,21] and consisted of multiple bioreactors in series (Figure 1)[9]. Each of the glass columns (H = 29 cm, Ø = 2.5 cm, Chromaflex, USA) was filled with 200 g of glass carrier beads (Ø = 6 mm, surface area = 3.76 cm^2/g). The water samples were continuously pumped upwards through the columns by a peristaltic pump (REGLO Analog tubing pump ISM 828, Ismatec, Switzerland). The empty bed volume of each column is 147 mL, while the real volume after subtracting the volume of the glass beads is about 70 mL. The columns were originally inoculated by pumping a mixture of raw water and effluent water from an ozone-biofiltration pilot plant located at Trondheim Water Works (VIVA) through the columns. Later, the feed water was changed to ozonated water from the same pilot plant. When columns were not in use for BDOC testing, biological activity was maintained by recirculation of ozonated water from a 5 L sample beaker through the columns and then back to the beaker. The water was changed at intervals of a few days. In this mode, the columns were operated with two in series (three parallel lines) to maintain a high biological activity. The columns were operated at room temperature and were covered with black plastic to prevent algal growth.

During the BDOC test, the columns were operated with six in series. A three-way valve configuration allowed effluent water sampling from each column. The water sample flow was adjusted to 1.63 mL/min, i.e. an EBCT of close to 1.5 hrs in each column. This flow was chosen for practical reasons and from experience with the actual DOC degradation rates of the tested NOM water samples. The sample flow can be adjusted to match the DOC degradation rates of the actual sample. The 4-liter sample bottles will normally provide enough sample volume for rapid NOM fractionation as well as BDOC testing. With this approach, the resulting EBCTs (i.e. 9 hours in total) were considered reasonable. Water samples from the investigated water facilities were sent to SINTEF's laboratory in 4-liter fluorinated high-density polyethylene bottles (Nalgene), and pre-filtered through a 0.45 μm membrane filter. Sodium bicarbonate (Na_2CO_3) was used for controlling pH at 7-7.5. If a sample could not be analysed immediately, it was frozen. Feeding of the test water into the columns was started in the afternoon and the pumping through the columns continued overnight before DOC sampling was carried out, thus providing about 17 hrs of acclimation time before sampling. This is in line with the adaptation requirements found within the EU project TECHNEAU.[9,11]

After acclimation, water sampling was started from the last column. As soon as enough sample volume was collected (one hour), the flow was redirected from the previous (i.e. second-last) column into a new sample beaker. This procedure continued until effluent samples from all six columns were collected. In addition, inlet (raw) and outlet samples were analysed. The sample beaker was weighed before and after sampling to verify a correct flow rate. The collected (i.e. seven) samples were filtered through 0.45 μm membrane filters before dissolved organic carbon (DOC) analysis (Teledyne Tekmar TOC Fusion analyser). All values were then compared to the DOC of the inlet (raw) water sample, and the difference in DOC between the inlet sample and the effluent sample from the actual column constitutes the biodegraded DOC according to the EBCT after that specific column. The difference in DOC between the inlet sample and the effluent sample from the last column (i.e. EBCT of 9 hrs) gives the final BDOC, while the results from the intermediate samples (i.e. EBCTs with 1.5 hrs intervals) indicate the biodegradation kinetics of the sample. In order to provide information on kinetics at lower EBCTs, i.e. close to the normal design values of 20-30 minutes, the volume of the first column was reduced to about one third in some of the experiments (as shown in Figure 1).

Parallel testing of water samples in the conventional batch BDOC measurement (28 days) and the bioreactor-based method used here (9 hrs) showed good agreement. In ozonated water samples the measured BDOC levels were 0.46 and 0.42 mg/L, and in ozonated and biofiltered water samples the BDOC concentrations were 0.28 and 0.25 mg/L for the batch method and the six bioreactor-in-series method, respectively. In a separate set of experiments, additional nutrients (N, P, Ca, K; 1 mg/L each) were added to the water samples prior to the BDOC analysis. Only minor influence on DOC degradability was detected (< 0.1 mg BDOC/L), indicating that organic carbon was the limiting substrate.

3.1.3 *Biofilm Monitors* (BFM). In order to study biofilm formation and internal deposition in pipelines, biofilm monitors (Figure 1) were used. The biofilm monitors (BFMs) were a modified version of a Dutch set-up described in detail elsewhere[37]. The up-flow BFM used here was made from pre-used PE with an initial length of 1 m. The BFMs were fed with untreated raw water or with distributed water from different sample points in the distribution system. The upward water flow in the BFM was controlled at 0.2 m/s, and at certain intervals (typically some months), three pieces (30 mm long; 30 cm^2 internal surface area) of the Ø 32 mm PE pipe were cut off (using a special tool as indicated in Figure 1) and sent to SINTEF's laboratory in special boxes along with a few drops of water to maintain humidity. In the laboratory the biofilm and deposits including metals and adsorbed material were removed from the internal pipe walls (ultrasound for

5 minutes) and analysed for TOC, ATP and metals (Thermo Electronics HR-ICP-MS-Element 2; Milestone Ultraclave) after addition of Milli-Q water to 50 mL. The inside area of the piece of pipe was calculated and the area-specific amounts of TOC, adenosine triphosphate (ATP) and metals (Fe, Al, Mn, Ca, etc.) retained in the deposits were calculated as mass per area (cm^2), i.e. as μg of TOC and metals per cm^2, and pmol of ATP per cm^2 of internal pipe surface area.

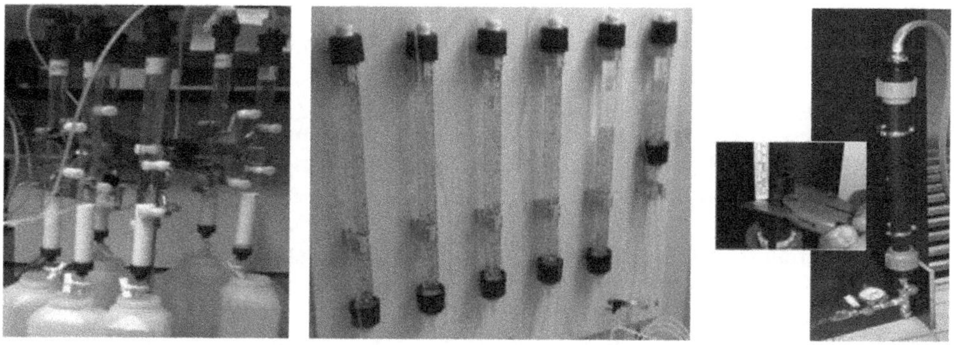

Figure 1 *Pictures of the rapid NOM-fractionation set-up (left), the columns-in-series BDOC (middle), and the biofilm monitor (BFM) including the PE-pipe cutting tool*[8]

4 RESULTS AND DISCUSSION

4.1 Enhanced Coagulation Treatment

Typical results from the enhanced coagulation-contact filtration (EC) facilities are presented in Figure 2 (upper part). It appears from the results that EC is capable of removing DOC effectively, and the low BDOC concentrations in the raw waters are maintained at levels below 0.1 mg/L in treated waters. The NEU fraction concentration is relatively unaffected by enhanced coagulation treatment (6 % removal on average). Both the CHA and NEU concentration levels are however low in most Norwegian source waters, typically in the range of 0.3-0.5 mg/L and 0.4-0.5 mg/L, respectively. The hydrophobic fractions dominate, with average VHA and SHA fraction concentrations of 67 and 15 % of the total average DOC level, respectively. In summary, the VHA and SHA fractions account for 82±5% of total DOC (4.4±2.5 mg/L) in the investigated surface water sources.

4.2 Nanofiltration Treatment

From the results presented in Figure 2 (lower part), is appears that NF treatment is capable of removing DOC even more effectively than EC treatment. NF is typically applied at small to medium-sized facilities with high levels of NOM in the source water. In spite of this, the DOC levels are low in UF-treated water, and the low BDOC concentrations in the raw waters are maintained at levels below 0.1 mg/L in treated waters. In addition, it appears from the results that the NEU fraction is removed more effectively by NF than EC (24 versus 6 % on average for the two treatment methods).

Due to the low concentrations in typical source waters, the remaining hydrophilic and biodegradable NOM fraction concentrations are maintained at relatively low levels in EC and NF treated waters.

4.3 Ozone-biofiltration Treatment

When using ozone-biofiltration (OBF) treatment, the NOM fraction removal pattern is different from that of EC and NF. As shown in Figure 3, ozone is transforming the VHA fraction into the more biodegradable CHA fraction thereby significantly increasing the BDOC level. Part of the ozone-produced CHA and BDOC is reduced in the alkaline filter and in the biofilter. Still a BDOC residual of some 0.2 mg/L is present in treated water, and the BDOC is being further reduced in the distribution system. This pattern is supported by the ATP pattern. The ATP and thus the microbial activity present in the raw waters are reduced to almost zero after ozonation. The ATP values increase during alkaline filtration and is then reduced to zero after biofiltration, UV-disinfection and low dose chlorination (0.1-0.2 mg/L) to the inlet of the clean water tank. In the distribution system the ATP increases significantly, in good agreement with the decline in BDOC substrate concentration. Figure 3 illustrates well the positive effects of the alkaline pre-filter. Besides stabilising the pH at levels close to pH8, several metals (e.g. Fe, Al, Mn) in the ozonated inlet water are precipitating within the alkaline filter bed. Both Al and Fe hydroxides are known as excellent NOM adsorbents.[7,18] Thus the alkaline filter is capable of removing significant amounts of DOC and BDOC at this facility, thereby contributing to the overall DOC and BDOC removal and reducing the BDOC load to the subsequent biofilter.

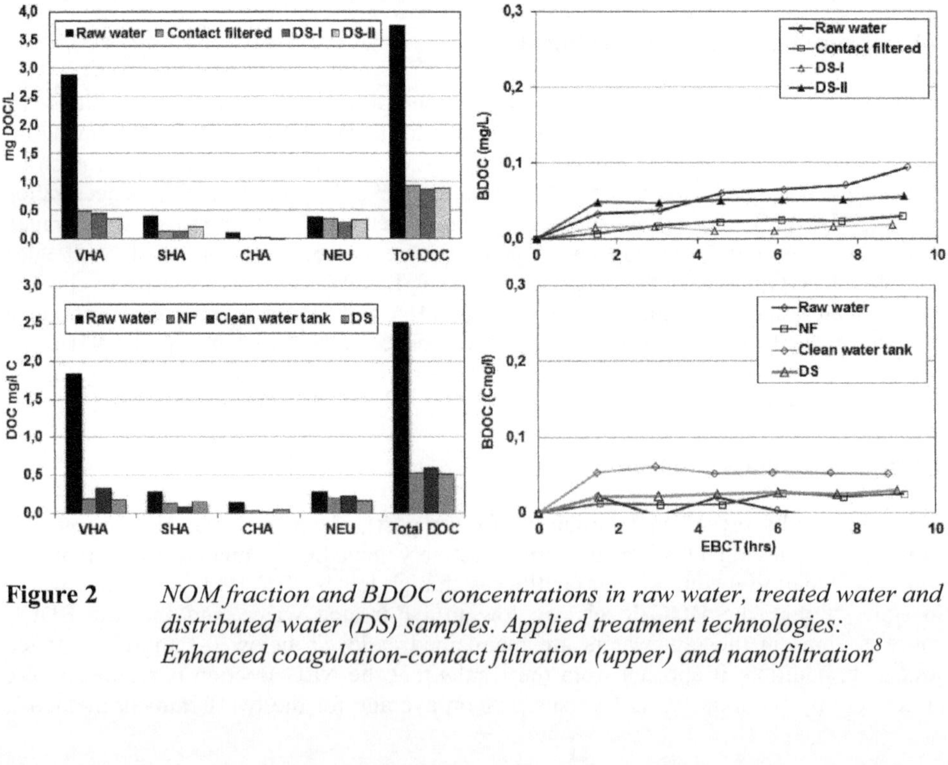

Figure 2 *NOM fraction and BDOC concentrations in raw water, treated water and distributed water (DS) samples. Applied treatment technologies: Enhanced coagulation-contact filtration (upper) and nanofiltration*[8]

For the investigated facilities, Table 1 shows a summary of the treatment technology-specific results in terms of DOC, hydrophilic NOM fractions (CHA+NEU) and BDOC concentrations in raw and clean water samples. The data shown is Avg±Stdev values for 22, 20 and 3 water samples from 10 EC, 12 OBF and 3 NF utilities, respectively.

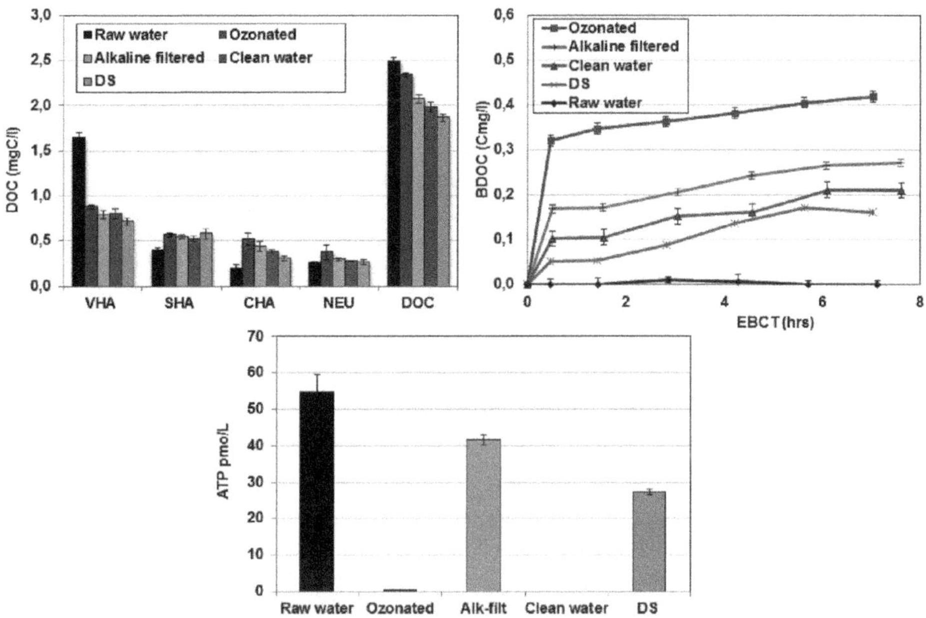

Figure 3 *NOM fraction and BDOC concentrations in raw water, treated water and distributed water (DS) samples. Applied treatment technology: Ozonation, alkaline pre filtration and biofiltration*[8]

Table 1 *Effects of water treatment technology on levels of DOC, hydrophilic NOM fractions (CHA, NEU) and BDOC from 10 EC, 12 OBF and 3 NF utilities (Avg±Stdev values for 22, 20 and 3 water samples, respectively)*

Treatment	**DOC (mg/L)**		**CHA+NEU fraction (% of DOC)**		**BDOC (mg/L)**		**N Utilities/ Samples**
	Raw	Clean	Raw	Clean	Raw	Clean	
EC	4.6±2.3	1.8±0.8	17±4	33±17	0.07±0.07	0.06±0.04	10/22
OBF	3.5±1.7	2.6±1.3	20±6	40±6	0.09±0.08	0.39±0.26	12/20
NF	10.3±7.5	2.2±1.6	12±4	26±14	0.65±0.89	0.06±0.08	3/3

4.4 UV Disinfection

The effects of UV disinfection on different carbon fractions are presented in Table 2. The results include samples taken prior to and after the UV disinfection process at 7 facilities using EC, OBF or NF treatment technologies, and LP- or MP-lamps operated close to or lower than the (biodosimetric) design dose of 40 mJ/cm^2. All MP-UV lamps were equipped with doped quartz sleeves thus cutting off the low wavelengths. The results show only minor effects of UV disinfection on DOC, NOM fraction concentrations and BDOC.

Table 2 *NOM fraction, DOC and BDOC concentrations prior to and after UV treatment measured at 7 Norwegian waterworks (40 mJ/cm^2 design dose; doped UV quartz sleeves)*[8]

Concentration (mg/L)	UV influent	UV effluent
DOC	1.95±1.19	1.92±1.13
VHA	0.75±0.49	0.72±0.51
SHA	0.46±0.28	0.42±0.24
CHA	0.43±0.36	0.41±0.33
NEU	0.34±0.11	0.35±0.13
VHA+SHA	1.18±0.77	1.17±0.72
CHA+NEU	0.78±0.47	0.76±0.44
BDOC	0.24±0.19	0.24±0.20

4.4 NOM Fractions and Biodegradability

Figure 4 shows that BDOC correlates well with the hydrophilic NOM fraction concentrations, i.e. with CHA alone and with CHA+NEU. The data includes raw water, clean water and effluent water samples from different treatment steps at the investigated utilities. The lowest values are typically found in raw waters, and in EC and NF treated waters. The highest values are typically found in water samples taken after the ozone contact tank at the OBF facilities, and the intermediate values are typically found in biofilter effluent samples and in clean and distributed waters samples from the OBF facilities. In order to avoid regrowth in distribution systems a maximum BDOC level of 0.15-0.30 mg/L is proposed.[24,33,41,42] No correlation ($R^2 < 0.1$) was found between BDOC and the hydrophobic NOM fractions (VHA, SHA). Thus the results presented in Figure 4 indicate that the CHA and the sum of the hydrophilic NOM fraction concentrations (CHA+NEU) have to be controlled at relatively low levels, i.e. 0.3 and 0.8 mg/L (as DOC) respectively, in order to provide a biologically stable water (BDOC < 0.15 mg/L).

4.4 Biofilm Monitoring (BFM)

The biofilm monitors (BFMs) accumulate significant amounts of TOC, ATP and metals on the inner surface of the pieces of PE pipes. At irregular time intervals, the deposits are sloughed off and then they start to rebuild again. The examples presented in Figure 5 are taken from two OBF-facilities (A and B) and show results after about 9 (May/Jun) and 12 months (Sep) exposure (BFM flow). Significant amounts of TOC, ATP and metals accumulate on the inner surface of the pipes. For samples taken at the same time, the amounts of TOC are well correlated to the amounts of Fe (R^2 of 0.81-0.96). The

TOC to Fe relationships are however different in the two sampling rounds, with a lower TOC to Fe ratio in September than in May/June.

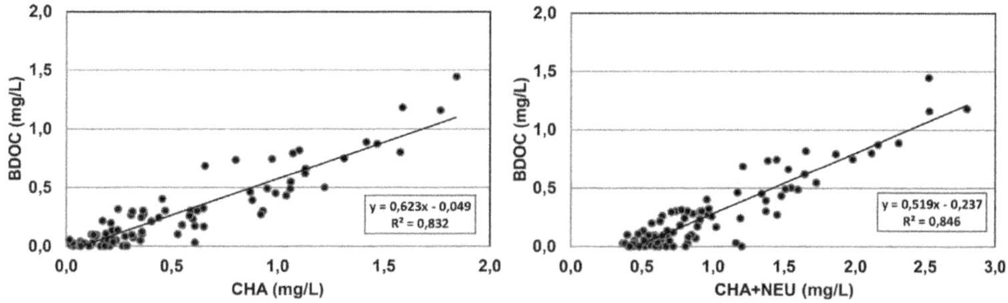

Figure 4 *Correlation between BDOC and hydrophilic NOM fraction concentrations based on raw water samples, samples from various treatment steps, and clean water samples at 10 Norwegian water treatment plants applying EC, OBF or NF*

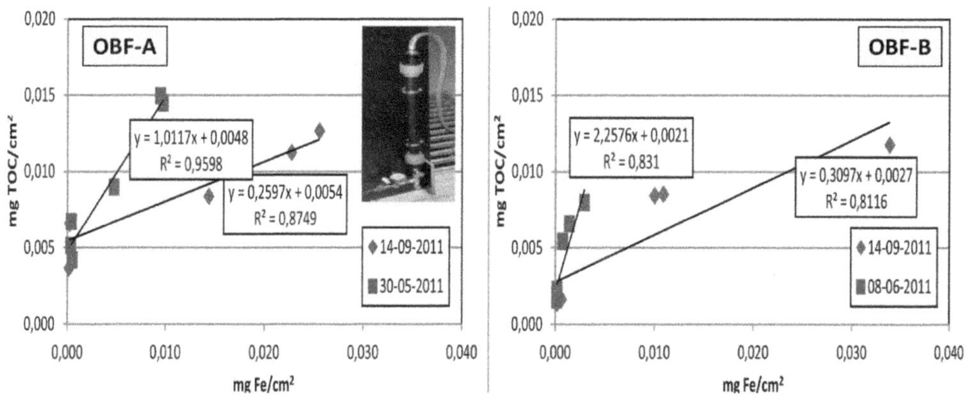

Figure 5 *Amounts of TOC and iron (Fe) in internal PE-pipe piece deposits at two OBF-facilities (A and B), after about 9 (June 2011) and 12 months (Sep 2011) exposure to raw water and clean/distributed waters.*

Long-term data on OBF exposure from the OBF-A facility are presented in Table 3. Here data for the composition of the BFM deposits after 3-4 years of exposure. Among the metals, depositions of iron (Fe) and aluminium (Al) dominate, but manganese (Mn) and calcium (Ca) is also found in significant amounts. The average Fe and Mn concentrations in raw and treated (distributed) water samples were 49 and 14 μg Fe/L, and 5.3 and 0.8 μg Mn/L, respectively for the three BFM sampling rounds. No coagulant is applied at this facility, and the metals in the raw water originate from the source (lake) water. Most of the Fe and Mn are removed in the alkaline prefilters that are located between the ozone contact tank and the biofilters at this facility. For all substances examined, the depositions are significantly reduced from raw to distributed (DS) water.

Figure 6 presents summary results from 8 sampling rounds at the OBF-A and OBF-B facilities during 2011-2015. The amount of TOC in the BFM deposits was found to correlate well with the sum of iron and manganese (R^2 of 0.76) both in raw and treated (distributed) water. Also the amount of cellular ATP (living cells) was found to correlate reasonably well with the amount of iron and manganese in the deposited layers inside of the BFM pipeline (R^2 of 0.56). The TOC to Fe/Mn relationships are different in deposits from raw and treated (distributed) water, with a higher TOC to Fe/Mn ratio in the deposits formed by distributed water. This may be due to differences in NOM fraction distribution between raw and OBF treated waters. The CHA fraction concentrations are significantly higher in OBF treated water. This may explain the higher TOC content in the deposits because the CHA fraction is effectively removed by coagulation, i.e. adsorbs well to metal coagulant hydroxides and thus also to metal precipitates within the BFM.

Table 3 *Composition of BFM deposits after long term exposure, i.e. 1014, 1237 and 1608 days in raw (Raw) and distributed water (DS) at the OBF-A facility*

BFM Location	Exposure (days)	BFM-deposition (μg Me and TOC/cm²; pmol ATP/cm²)					
		Al	Fe	Mn	Ca	TOC	ATP
Raw	1014	22,6	37,7	18,6	6,3	71,3	58,4
DS		1,7	2,1	1,2	7,7	14,2	11,7
Raw	1237	52,8	43,5	10,9	7,5	47,4	17,3
DS		1,9	1,9	1,5	5,1	19,8	10,1
Raw	1608	9,3	14,9	6,8	2,8	21,0	18,8
DS		2,4	2,2	1,4	5,4	10,0	4,5

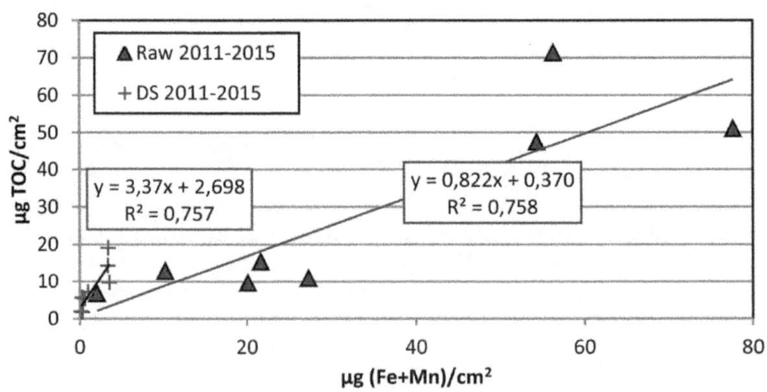

Figure 6 *Data from 8 BFM sampling rounds (2011-2015) from BFMs fed with raw water and treated (distributed) water at the OBF-A and OBF-B utilities (Average values from the 3 pieces of PE pipe cut each time)*

The analysis of the deposits, i.e. the layers formed on the PE pipe walls inside of the BFMs shows a significant accumulation of metals over time. Iron (Fe) and aluminium (Al) seem to dominate, but also Mn and Ca are found in significant amounts on the deposits formed on the internal pipe walls. Some microbiologists tend to interpret the high biofilm formation rates often measured on corroded iron pipes as an effect of NOM adsorption to corrosion products thereby increasing the availability of NOM as a substrate to microorganisms. In our study the BFM-pipe is made from PE, and the adsorption in this case cannot be due to adsorption to iron corrosion products.[4]

An alternative or supplementary explanation is however that metals originating from the raw water and/or from coagulant residuals precipitate at the pipe walls and act as adsorbents for NOM and microorganisms in the water phase. It is well known that metal (Fe and Al) hydroxides are effective adsorbents of NOM. This is well illustrated by the fact that EC processes can be modelled solely as an adsorption to Fe – or Al precipitates with very good results.[7,18] The concentration of Fe in Norwegian water sources is normally within the range of 20-80 μg/L. The precipitation of metals in the pipelines may be triggered by the corrosion control treatment i.e. the elevated pH, alkalinity and Ca content normally found as a final treatment step before the treated water enters the distribution system where time and area (pipe walls) are available for metal precipitation and subsequent adsorption of NOM and microorganisms (ATP).

Therefore, the BFM results presented above indicate that adsorption of NOM (TOC) and microorganisms (ATP) to metal (e.g. Fe) hydroxide precipitates on the pipe wall is an important deposition mechanism. The good correlations found between the deposited amounts of Me and TOC, and between Me and ATP per unit pipe wall area indicate that adsorption is a relevant mechanism.

A relevant question may arise from these results: What is the relative importance of biology versus chemistry in the formation of layers on the BFM pipe wall (i.e. biofilm formation and growth from organic substrate utilization versus adsorption of TOC and microorganisms (ATP) to metal precipitates)? The question is relevant because the curative actions are different: Removal of organic substrate (BDOC) versus removal of metals.

5 SUMMARY AND CONCLUSIONS

Some countries including Norway apply chlorine-free water distribution systems. In addition, the concentration of natural organic matter (NOM) in source waters shows an increasing trend in parts of the Nordic region. The character of NOM may also change due to the effects of climate change and less acid rain. Under these circumstances, the effects on NOM and the formation and control of biodegradable by products from alternative disinfectants (i.e. ozone and UV) are more user-relevant issues than chlorination by- products. In this study, the effects of applied water treatment technology and UV-disinfection on NOM fractions and biostability (BDOC) were investigated at a number of full-scale WTPs in Norway.

From utilities applying enhanced coagulation (EC), nanofiltration (NF), or ozone-biofiltration treatment, samples were collected from untreated raw water, from different treatment steps and from the distribution system. The samples were analysed for typical routine parameters in addition to more unconventional analyses like rapid NOM fractionation, BDOC and ATP.

For the investigated OBF-facilities the results show that the hydrophobic NOM fractions were transformed by ozone into hydrophilic and more biodegradable fractions, thereby increasing BDOC significantly. None of the investigated biofilters were however capable of removing all of the ozone-induced BDOC increase, thereby inducing biological regrowth and elevated HPCs in the distribution system.

Furthermore, the results indicate that with the UV-disinfectant doses normally applied (40 mJ/cm^2), no significant effects of UV on DOC, NOM fractions or BDOC were identified, regardless of the applied water treatment technology.

At the WTPs applying EC and NF the BDOC levels were reduced to levels close to zero after treatment. For all facilities, BDOC levels were found to correlate well ($R^2 > 0.85$) with the hydrophilic NOM content. The analyses of the deposits on the PE pipe walls within the biofilm monitors (BFMs) show that the amounts of TOC and ATP correlate well with the amount of metals (i.e. Fe, Mn), thus indicating that adsorption of NOM and microorganisms to metal precipitates on the pipe wall is an important mechanism.

The need for more knowledge of the relative importance of "true" biofilm formation versus TOC/ATP adsorption to metal precipitates is particularly important when it comes to effective control measures, i.e. removal of substrate (AOC, BDOC) for biofilm control versus removal of adsorbents (metals from raw water and/or from treatment/coagulants) for control of the adsorption of NOM and ATP (microorganisms) to metal precipitates on the pipe walls. With a continuing climate change driven trend towards increasing temperature and NOM levels, it will be increasingly important for the waterworks to provide better links between treatment and distribution and to apply multiple control measures: (i) effective control of NOM and BOM (AOC, BDOC) in order to control the formation of biofilms and deposits in distributions systems, (ii) effective corrosion control, and (iii) effective removal of metals originating from the raw water source and/or residual metal coagulants.

Acknowledgement

Part of this work was undertaken under the TECHNEAU (No. 018320) and PREPARED (No. 244232) R&D projects financed by EU under the FPs 6 and 7, resp. G. Thorvaldsen, SINTEF is acknowledged for the analyses of NOM fractions, BDOC and ATP, and S. Lierhagen, NTNU for the analyses of metals by ICP-MS.

References

1. Bolto, B., Abbt-Braun, G. Dixon, D. Eldridge, R., Frimmel, F. Hesse, S. King, S. and Toifl, M., *Water Sci. and Technology*, 1999, **40** (9), p 71-79
2. Buchanan, W., Roddick, F., Porter, N. and Drikas, M., *Environ. Sci. Technol.,* 2005, **39**, p 4647-4654
3. Buchanan, W., Roddick, F., Porter, N. and Drikas, M., *Water Science and Technology: Water Supply*, 2004, **4** (4), p 103-111
4. Camper, A. *Microbial Growth in Drinking-Water Supplies*, 2014, IWA Publishing, London, UK. ISBN: 9781780400402; 9781780400419, p 73-94
5. Chow, C.W.K., Fabris, R. And Drikas, M., *J. Water Supply: Research and Technology - AQUA*, 2004, **53** (2), p 85-92
6. Croué, J.P., Martin, B., Deguin, A. and Legube, B. *Natural Organic Matter in Drinking Water,* 1994, AWWA, Denver, CO, p 73.
7. Edwards, M., *Jour. AWWA*, 1997 **89** (5), p 78.
8. Eikebrokk, B. *Proc. The 9th Nordic Drinking Water Conference, Helsinki, Finland,* 2014, June
9. Eikebrokk, B. and Juhna, T. *Optimizing water treatment: Enhanced coagulation and ozonation-biofiltration. Summary report on performance assessment tools and optimization.* Techneau-Report D5.3.2B, 2010, www.techneau.eu

10. Eikebrokk, B., Fabris, R., Drikas, M. and Chow, C. in Hahn et al. (Eds), *Chemical Water and Wastewater Treatment IX*, IWA Publishing, London, ISBN: 1843391457, 2007, p 207-219
11. Eikebrokk, B. Juhna, T., Melin, E. and Osterhus, S.W., *Water treatment by enhanced coagulation and ozonation-biofiltration*. Techneau-Report D5.3.2A, 2007, www.techneau.eu
12. Fabris, R., Chow, C.W.K. and Drikas, M. *Journal of Water Supply: Research and Technology – AQUA*, 2012, **61** (4), p 201-219
13. Fabris, R., Chow, C.W.K., Drikas, M. and Eikebrokk, B. *Water Research*, 2008, **42**, p 4188-4196
14. Fabris, R., Chow, C., Tran, T., Gray, S. and Drikas, M. *Report No. 38, CRC for Water Quality and Treatment.* ISBN 1876616636, 2008, p 30
15. Frias, J., Ribas, F. and Lucena, F. Water Research, 1992, **26**, p 255-258
16. Frimmel, F.H. *Envir. Intl*., 1998 **24** (5/6), p 559
17. Ijpelaar, G.F., Harmsen, J.H. and Heringa, M. *UV disinfection and UV/H_2O_2 oxidation: By-product formation and control.* TECHNEAU Deliverable D2.4.1.1, 2007, www.techneau.eu
18. Kastl, G., Sathasivan, A. and Fisher, I., *Research report No. 57, CRC for Water Quality and Treatment, Salisbury, Australia,* ISBN 18766 16830, 2008.
19. Leenheer, J.A., *Environ. Sci. Techn.* 1981 **15** (5), p 578-587
20. Liu, W., Andrews, S.A., Sharpless, C., Stefan, M., and Linden, K.G. and Bolton, J.R. *Water Sci. & Techn.: Water Supply*, 2002, **2** (5/6), p 515
21. Lucena, F. Frias, J. and Ribas, F., *Environmental Technology*, 1990, **12**, p 343-347
22. Långmark, J. Storey, M.V. Ashbolt N.J. and Stenstrøm, T., *Water Research*, 2007, **41**, p 3327-3336
23. Magnuson, M.L., Kelty, C.A., Sharpless, C.M., Linden, K.G., Fromme, W., Metz, D.H. and Kashinkunti, R., *Envir. Sci. & Technol*., 2002, **36**, p 232:252
24. Niquette, P., Servais, P. and Savoir, R., *Water Research*, 2001, **35**, p 675-682
25. Piriou, P., Dukan, S. and Kiene, L., *Water Science and Technology*, 1998, **38** (8-9), p 343-347
26. Reckhow, D.A., Linden, K.G., Kim, J., Shemer, H. and Makdissy, G. *Jour. AWWA*, 2010, **102** (6), p 101-113
27. Ribas, F., Frias, J. and Lucena, F., *Journal of Applied Bacteriology*, 1991, **71**, p 371-378
28. Richardson, S. D. *Trends in Analytical Chemistry*, 2003, **22** (10), p 666-684
29. Sadiq, R. and Rodriguez, J., *Science of the Total Environment*, (2004), **321**, p 21-46
30. Servais, P., Billen, G. and Hascoet, M.C., Determination of the biodegradable fraction of dissolved organic matter in waters. *Water Research*, 1987, **21**, p 445-450
31. Servais, P., Laurent, P. and Randon, G., *Water Supply: Res. and Techn. – Aqua*, 1995, **4**, p 10-17.
32. Servais, P., Laurent, G. and Randon, G., *Proc. AWWA-AWQTC Conference, Miami, FL,* 1993, Nov. 7-10
33. Servais, P., Laurent, P. and Randon, G. *Water Supply: Res and Techn – Aqua*, 1995, **4**, p 10-17
34. Sharp, E., Parsons, S.A., and Jefferson, B., *Water Science and Technology: Water Supply*, 2004, **53** (7), p 67-76
35. Tranvik, L. J., *Aquatic Humic Substances – Ecology and Biogeochemistry*, Springer Verlag, Berlin Heidelberg, 1998, p. 268-283

36. van der Kooij, D. and van der Wielen, P.W.J.J. *General introduction. In: Microbial Growth in Drinking-Water Supplies.* IWA Publishing, London, UK. ISBN: 9781780400402; 9781780400419, 2014, p 1-32
37. van der Kooij, D. and Veenendaal, H.R. *Proc. AWWA Water Quality Technology Conference*, 1992, Nov 15-19, Denver, Co, p 1099-1110
38. van der Maas, P., van der Woerdt, D. Wübbels, G. and van der Kooij, D. *Proc. IWA World Water Congress & Exhibition, Montréal, Canada*, 2010, Sep 19-24
39. van Leeuwen, J., Chow, C., Fabris, R., Withers, N., Page, D. and Drikas, M., *Wat. Sci. Tech.: Water Supply*, 2002, **2**(5), p 427–433
40. van Leeuwen, J., Page, D., Spark, K., Fabris, R., and Sledz, L. *Proc. NOM Innovations and Applications for Drinking Water Treatment. Victor Harbor, Australia*, 2004, March 2-5.
41. Volk, C. and Joret, J.C., *Rev. Sci. de l'Eau*, 1994, **7**, p 131-152
42. Volk, C. and LeChevallier, M.W. *J.AWWA*, 2000, **92**, p 64-76
43. Volk, C.J., Volk, C.B. and Kaplan, L.A. *Limnology and Oceanography*, 1997, **42**, p 39-45

IMPACTS OF FERRATE TREATMENT ON NATURAL ORGANIC MATTER, DISINFECTION BY-PRODUCTS AND BROMIDE

Yanjun Jiang, Joseph E. Goodwill, John E. Tobiason, and David A. Reckhow*

Department of Civil and Environmental Engineering, University of Massachusetts, Amherst, Massachusetts 01003, United States. E-mail: reckhow@umass.edu.

1 INTRODUCTION

Ferrate (or Fe(VI)), i.e., iron in the hexavalent form, has received increased interest among drinking water treatment researchers because it can contribute to multiple treatment objectives. In addition to acting as an effective disinfectant against microorganisms,[1–5] Fe(VI) can oxidize a wide range of aquatic trace pollutants while producing little or no hazardous byproducts.[6,7] In addition, the main Fe(VI) decomposition product, ferric iron, might provide benefits as an in situ coagulant. Jiang and Wang in 2003 found Fe(VI) was more effective than ferric sulfate at lower doses for removing UV_{254} absorbing substances and dissolved organic carbon (DOC), and thus its use helped decrease the trihalomethane formation potential (THMFP).[8] Other studies also demonstrated the effectiveness of Fe(VI) to remove humic substances.[9,10]

Pre-oxidation of natural organic matter (NOM) by strong oxidants (e.g., ozone) could alter the structure and characteristics of NOM as indicated by decreased UV_{254} absorbance and lowered molecular weight distribution, thereby influencing subsequent coagulation and disinfection byproduct (DBP) precursor removal.[9] Conflicting effects have been observed for pre-ozonation on subsequent NOM and DBP precursor removal by coagulation.[9,11–14] The deleterious effect of pre-oxidation on coagulation was attributed to the oxidized NOM being more hydrophilic and fragmented (i.e., lower molecular weight), which rendered it more difficult to remove by coagulation.[15,16] As a result, intermediate-oxidation processes (e.g., intermediate ozonation, which places ozonation after coagulation/settling, but before filtration/chlorination) have been used by many utilities.

Under acidic conditions, the oxidation-reduction potential of Fe(VI) (2.2 V) is higher than any other oxidant commonly used in water treatment.[17,18] Therefore, Fe(VI) may also alter the nature of NOM and impact the subsequent coagulation for NOM and DBP precursor removal. Despite extensive research on the ability of Fe(VI) to oxidize specific contaminants, little is known about the effectiveness of Fe(VI) to remove NOM and DBP precursors in a conventional treatment train. In addition, it is believed that one major advantage of Fe(VI) over ozone is that Fe(VI) does not oxidize bromide and thus no bromate or brominated DBPs would form as a result of Fe(VI) oxidation. This conclusion is based on the observation that Fe(VI) decay rate was the same with and without the presence of bromide in phosphate-buffered pure water.[6] However, the oxidation of bromide by Fe(VI) has not been studied in waters with other buffers or in natural waters.

The principal objectives of this research are to investigate the impacts of direct Fe(VI) oxidation, pre-ferrate (Fe(VI) added prior to coagulation) and intermediate-ferrate (Fe(VI) added after coagulation) on NOM and DBP precursors, and to determine the oxidation of bromide by Fe(VI) under various conditions (e.g., bromide dose, buffer type and strength, and deionized versus natural water).

2 MATERIALS AND METHODS

2.1 Chemicals and Reagents

Potassium ferrate (K_2FeO_4, of 87% purity), reagent-grade 2,2'-azino-bis(3-ethylbenzothiazoline-6-sulfonate) (ABTS), potassium bromide, sodium phosphate dibasic, and sodium borate decahydrate were purchased from Sigma-Aldrich (St. Louis, MO, US) or Fisher-Scientific (Fair Lawn, NJ, US). All aqueous solutions were prepared with ultrapure water produced by a Milli-Q system (Advantage A10, Millipore, Billerica, MA).

2.2 Experimental Methods

Fe(VI) oxidation (25 or 50 μM) was initiated by adding a weighed amount of K_2FeO_4 solids to the buffered waters (10 mM borate or phosphate) under rapid mixing at 20 °C at pH 6.2 or 7.5. Samples were withdrawn at appropriate time intervals with the Fe(VI) concentrations determined immediately using the ABTS method, monitoring absorbance at 415 nm.[19] After reaction for about 1 h, after which no Fe(VI) residual can be detected, samples were collected to determine concentrations of hypobromous acid (HOBr) and bromate.

Raw water samples were collected from the source waters of drinking water treatment plants at three locations in MA and CT (see Table 1 below). Table 1 shows the major chemical characteristics of the waters. Each water was treated under four scenarios: 1) Direct Fe(VI) oxidation followed only by chlorination; 2) Ferric chloride coagulation followed by settling/filtration/chlorination; 3) Pre-ferrate treatment: Fe(VI) addition followed by coagulation/settling/filtration/chlorination; and 4) Intermediate-ferrate treatment: Fe(VI) addition after coagulation/settling and before final filtration and chlorination. The effectiveness of each scenario regarding control of DBPs was assessed, and the impacts of varied Fe(VI) oxidation conditions (e.g., Fe(VI) dose and pH) were also studied. Four groups of DBPs, including trihalomethanes (THMs), dihaloacetic acids (DHAAs), trihaloacetic acids (THAAs), and dihaloacetonitriles (DHANs) were investigated in this study.

2.2.1 Scenario I (Direct Fe(VI) Oxidation Followed Only by Chlorination). The process of direct Fe(VI) oxidation was the same as described above. After reaction for about 1 h when no Fe(VI) residual could be detected, samples were collected for chlorination.

Table 1 *Raw water characteristics for all bench scale testing*

Sample Location	TOC (mg/L)	DOC (mg/L)	UV_{254} (cm^{-1})	SUVA (L/mg/m)	pH
Gloucester, MA	5.8	5.7	0.294	5.2	5.5
Norwalk, CT (epi)	3.3	3.2	0.138	4.4	7.2
Norwalk, CT (meso)	4.4	4.3	0.161	3.7	7.2
Norwalk, CT (hypo)	2.8	2.8	0.106	3.8	6.5
South Deerfield, MA	2.1	2.1	0.075	3.6	7.0

2.2.2 Scenario II (Ferric Chloride Coagulation Followed by Settling/Filtration/Chlorination). Coagulation was conducted using standard jar test procedures. Samples were divided into six 1 L-aliquots and placed into jars with the stirring paddle occupying the center. Different doses of ferric chloride were added into each jar and rapidly mixed at a stirring speed of 200 rpm for 1 min followed by a 10-min period of slow-mixing at a speed of 20 rpm. The pH was monitored and kept at 5.5 by dropwise addition of sodium hydroxide or sulfuric acid solutions. Thereafter, the samples were allowed to settle for 30 min, and a 20-mL aliquot was decanted from each jar for UV_{254} absorbance measurement. The sample with the lowest UV_{254} absorbance was selected as the one with the optimal ferric chloride dose. This sample was filtered (glass fiber filter, GF/F, effective size cutoff of 0.7 μm, Whatman, Clifton, NJ), and stored for subsequent chlorination. The optimal ferric chloride dose was denoted as $C_{opt,FeCl3}$.

2.2.3 Scenario III (Pre-ferrate Treatment, i.e., Fe(VI) Addition Followed by Coagulation/Settling/Filtration/Chlorination)). The process of Fe(VI) pre-oxidation is the same as described in scenario I, with the oxidized samples collected for subsequent coagulation. The coagulation/settling process was the same as stated in scenario II, and the sample with the lowest settled UV_{254} absorbance was filtered and collected for subsequent chlorination, Note that the sample coagulated with the same ferric chloride dose as the optimal ferric chloride dose ($C_{opt,FeCl3}$) determined in scenario II was also usually the one with the lowest UV_{254} absorbance among the samples in scenario III.

2.2.4 Scenario IV (Intermediate-ferrate Treatment, i.e., Fe(VI) Addition After Coagulation/Settling and Before Final Filtration and Chlorination). Samples were first coagulated with the optimal ferric chloride dose as decided in scenario II. After settling and filtration of the supernatant (with GF/F), Fe(VI) (20, 35, or 50 μM) was added for intermediate oxidation at pH 6.2, and then final filtration and chlorination were conducted.

2.2.5 Chlorination and Analytical Methods. Chlorination was conducted on both raw and treated samples using 300 mL chlorine demand free, glass-stopped bottles. The chlorine doses were determined according to preliminary chlorine demand tests on raw waters, and the target chlorine residual was 3.0–5.0 mg/L as Cl_2 after being incubated headspace-free at 20 °C in the dark for 72 h. All samples were buffered with 10 mM phosphate at pH 7.0. The DBPs analyzed for all samples included THMs, HAAs, and DHANs. Four chlorine- and bromine-containing THMs ($CHCl_3$, $CHBrCl_2$, $CHBr_2Cl$, and $CHBr_3$), and three DHANs (dichloro-, bromochloro-, and dibromoacetonitrile (DCAN, BCAN, and DBAN)) were quantified by liquid/liquid extraction with pentane followed by gas chromatography with electron capture detection (GC/ECD) according to USEPA Method 551.1. Nine HAAs (monochloro-, monobromo-, dichloro-, bromochloro-, dibromo-, bromodichloro-, dibromochloro-, trichloro- and tribromoacetic acid (MCAA, MBAA, DCAA, BCAA, DBAA, BDCAA, DBCAA, TCAA, and TBAA, respectively)) were analyzed by liquid/liquid extraction with methyl-tertiary-butyl-ether (MtBE) followed by derivatization with acidic methanol and by GC/ECD according to USEPA Method 552.2. Ultraviolet absorbance at 254 nm (UV_{254}) was determined using a Genesys 10S UV-Vis spectrophotometer (Thermo Scientific, Waltham, MA). Total organic carbon (TOC) and dissolved organic carbon (DOC) concentrations were measured with a TOC-V_{CPH} total organic carbon analyzer (Shimadzu Corp., Kyoto, Japan). Samples were filtered with GF/F for UV_{254} and DOC measurements. Bromine in the +1 oxidation state (i.e., HOBr/OBr^-) and bromate were determined by HPLC/MS/MS and UPLC/MS/MS, respectively.[20]

3 RESULTS AND DISCUSSION

3.1 Ferrate Decomposition in Aqueous Solutions

Bench-scale testing showed that Fe(VI) decomposition reactions are more complicated than previously recognized. The kinetic models proposed by Carr in 2008 and Lee et al. in 2014 did an excellent job of describing the decomposition kinetics we observed in phosphate buffered pure waters.[21,22] However, natural waters and even pure waters buffered with borate or carbonate showed much different decomposition rates (see Figure 1). We attribute this to surface reactions that accelerate Fe(VI) decomposition but are partly quenched by different ligands. Note that the initial Fe(VI) decomposition rates for buffer concentrations below 2 mM were speculated results (dotted lines), and the lines for the three buffers should converge at zero buffer concentration (see Figure 1). The constituents of alkaline and colored waters (bicarbonate and NOM) inhibited the catalytic effects of Fe(VI) decomposition products and stabilized Fe(VI) in natural waters.[23] This is good news for possible applications of Fe(VI) in water treatment as it means that the Fe(VI) species will persist longer than previously predicted, resulting in higher CT values (area under the concentration versus time curve) for disinfection. The observed CT values are high enough to meet criteria for 2 log removal of viruses and *Giardia* based on the current EPA assessment of Fe(VI). Higher pHs result in greater Fe(VI) stability and therefore higher CT values.

3.2 Effects of Ferrate Oxidation on DBP Precursor Removal

Direct Fe(VI) oxidation had little effect on total organic carbon (TOC) concentrations (data not shown), indicating that Fe(VI) oxidation did not mineralize NOM. Figure 2 shows the effect of direct Fe(VI) oxidation on THM, DHAA, THAA, and DHAN formation potentials (FPs) for each water under varied Fe(VI) dose and pH conditions. TTHM is the sum of the four chlorine- and bromine-containing THMs, DHAA is the sum of DCAA, BCAA, and DBAA, THAA is the sum of TCAA, BDCAA, DBCAA, and TBAA, whereas DHANs include

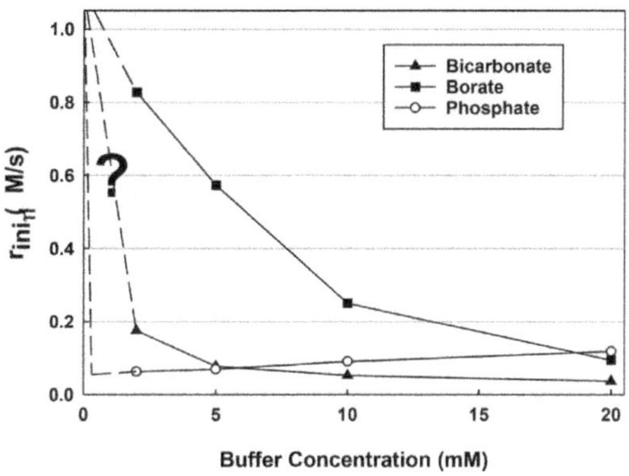

Figure 1 *Initial Fe(VI) decomposition rate in the presence of different solutes (based on data in Jiang et al., 2015).*[23] *The symbols represent the calculated rate and the lines connect sets of data points to show trends. Experimental conditions: pH = 7.5, T = 20 ℃, 50 μM of Fe(VI).*

DCAN, BCAN, and DBAN. The relative DBP formation for each water was calculated as the DBP formation potential (DBPFP) following Fe(VI) oxidation divided by that of raw water.

Substantial decreases in THM, THAA, and DHAN yields were observed for all waters, and the decreases in DBPFP increased with Fe(VI) dose. With 25 and 50 μM of Fe(VI), THM precursors were destroyed by 16–27% and 23–39%, respectively, DHAN precursors showed slightly higher levels of destruction (15–40% and 22–46%, respectively), THAA precursors were destroyed by -8–28% and 8–46%, respetively, and DHAA precursors were less affected (-2–10% and 0–17% decrease, respectively). Precursor oxidation by ferrate is similar to that observed for ozone at equivalent doses. Figure 3 compares the effects of ozone and Fe(VI) on THM precursor removal at equivalent doses, and the results are comparable, given typical values of bicarbonate. Ferrate directly destroyed THM precursors by 16–39% at doses of 0.3–1.4 mg-Fe/mg-C. Like ferrate, ozone also showed a greater impact on THM and THAA precursors than on DHAA precursors.[24] Hua and Reckhow investigated DBP formation from various NOM fractions. Hydrophobic NOM generally constituted the major pool of precursors for THMs and THAAs whereas the hydrophilic

NOM formed more DHAAs.[25] It is likely that similar to ozone, Fe(VI) can transform some hydrophobic NOM into more hydrophilic form, thus a greater decrease in THM and THAA formation potential was obtained than for DHAA. In addition, precursor destruction does not seem to diminish at low pHs where Fe(VI) decay is much faster (lower Fe(VI) exposure), thus precursor destruction is either faster when the reactants are more protonated, or ferrate decomposition products play an important role.

Figure 4 shows the effect of Fe(VI) pre-oxidation on coagulation for THM, DHAA, THAA, and DHAN precursor removal. The relative DBP formation here was defined as the DBPFP of the sample treated by ferrate/coagulation/filtration/chlorination process (scenario III) divided by that of the sample treated by coagulation/filtration/chlorination process (scenario II). Fe(VI) pre-oxidation generally resulted in decreased THM and THAA formation potential compared to samples without ferrate pre-oxidation. For waters pre-oxidized by Fe(VI) at doses of 25 and 50 μM, THMFP decreased by 1–16% and 3–19%,

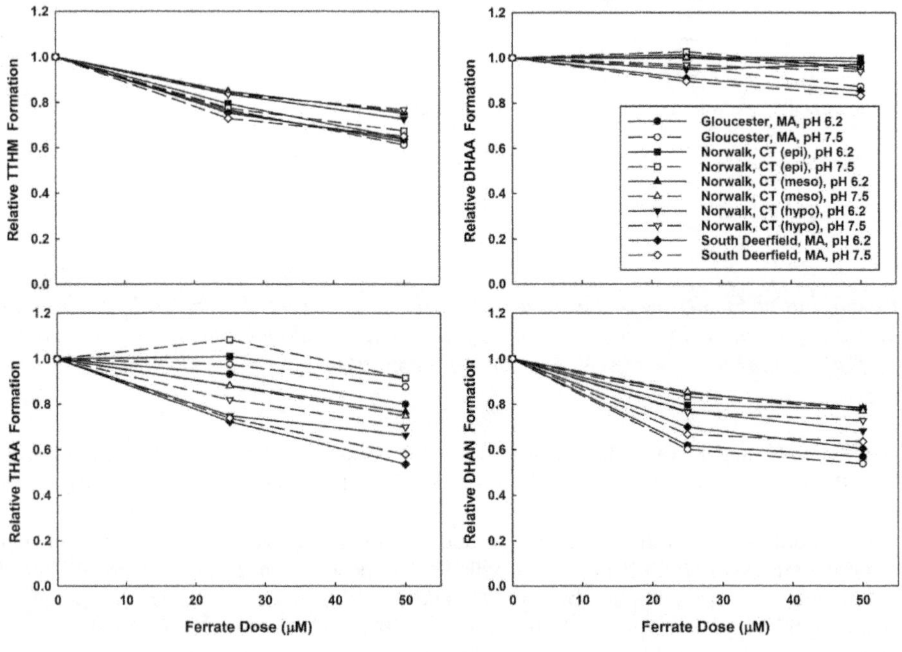

Figure 2 *Effect of direct Fe(VI) oxidation on relative THM, DHAA, THAA, and DHAN formation under varied Fe(VI) dose and pH conditions.*

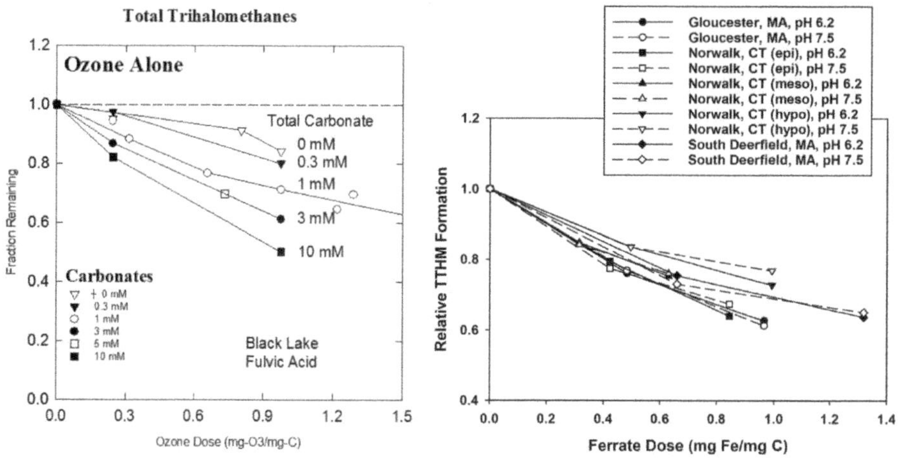

Figure 3 *Comparison with ozone (ozone figure from Reckhow, 1986, unpublished manuscript)*

respectively, and THAA precursors were destroyed by 2–30% and 10–42%, respectively, except for the Norwalk water (epi) pre-oxidized at pH 7.5 which showed some increase. Fe(VI) pre-oxidation slightly increased DHAAFP compared to samples without Fe(VI) pre-oxidation. Fe(VI) might transform hydrophobic NOM into more hydrophilic form, which is less readily removed by coagulation, and thus there might be some increase in DHAAFP due to Fe(VI) pre-oxidation. Fe(VI) pre-oxidation also caused some increase in DHANFP in certain waters compared to coagulation treatment alone. Similar adverse effects have been observed for preozonation on TOC and DBP precursor removal by coagulation.[12,27] In addition, the pH of Fe(VI) pre-oxidation impacts the effect of Fe(VI) on coagulation for DBP precursor removal. Pre-oxidation at pH 6.2 led to lower DBPFP than pre-oxidation at pH 7.5 for all waters. This might be because Fe(VI) has higher redox potential at lower pH, and thus could react with DBP precursors to a greater extent.

These results showed that the overall impact of Fe(VI) pre-oxidation on DBP precursor removal is somewhat diminished when it is followed by conventional coagulation. The direct precursor destruction noted in scenario I is partly mitigated by the fact that many of the oxidized precursors would have been removed by coagulation anyway. However, pre-ferrate treatment may allow the use of lower coagulant doses due to the Fe(III) introduced by Fe(VI) reduction. Ozone, by comparison, can result in elevated levels of bromate and brominated DBPs.

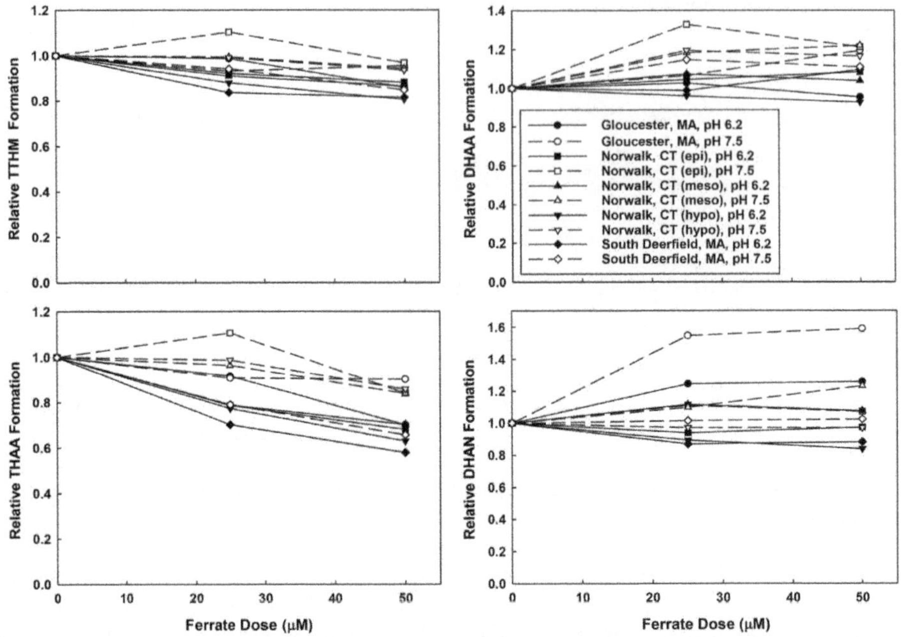

Figure 4 *Effect of Fe(VI) pre-oxidation on coagulation for THM, DHAA, THAA, and DHAN precursor removal under varied Fe(VI) pre-oxidation dose and pH conditions.*

Compared to ferric chloride coagulation, intermediate-ferrate (20, 35, and 50 μM) further decreased THM, DHAA, THAA, and DHAN formation potential by 39–55%, 19–25%, 59–77%, 16–46%, respectively.[28] This is different from intermediate ozonation which does not usually show much better performance than preozonation for DBP precursor removal.[11,16] In addition, size characterization of the iron solids showed that the Fe(III) resulting from Fe(VI) decay during the intermediate-ferrate process existed primarily as particulate iron and could be easily removed by filtration. It is clear that intermediate-ferrate treatment was the most effective method for DBP precursor removal among the four scenarios. The reason for this high effectiveness might be that after coagulation, the remaining DBP precursors are more readily oxidized by Fe(VI) and the oxidized forms can adsorb to iron particles that are easily removed by filtration.[28]

3.3 Ferrate Oxidation of Bromide

Contrary to previous reports, our experiments with a model and real waters indicate that ferrate does react slowly with bromide. Low levels of active bromine ($HOBr/OBr^-$) and bromate have been measured in model and raw waters following Fe(VI) treatment. Some organic bromide (as measured by adsorption/combustion/ICPMS; a TOBr protocol) has also been measured in Fe(VI) treated natural waters.[20]

Tests with buffered, bromide-fortified deionized water showed that the highest levels of bromate were formed at lower pH and in the absence of phosphate. Similar experiments with a real natural water indicated that the presence of the raw water matrix suppressed bromate formation as compared to the borate-buffered DI water. Nevertheless, it seems unlikely that bromate levels could ever exceed the US Maximum Contaminant Level (MCL = 10 μg/L) due to Fe(VI) oxidation for any but the most extreme cases. Bromide to bromate yields were in the range of 0.1–5% for the model waters and lower for the natural water.[20] This is in contrast to ozonation for which the bromide to bromate conversion yields are from 10–30% under typical conditions.

4 CONCLUSIONS

Ferrate was found to result in substantial oxidation of DBP precursors and its use could help with overall DBP control. Precursor removal by the four scenarios considered in this research followed the order of: intermediate-ferrate treatment > ferric chloride coagulation ≈ pre-ferrate treatment >simple Fe(VI) oxidation. At typical Fe(VI) doses, direct Fe(VI) oxidation could significantly decrease DBP formation potentials (DBPFPs), i.e., THMs 16–39%, THAAs -8–46%, DHAAs -2–17%, and DHANs 15–46%. After subsequent coagulation and filtration, the overall impact of Fe(VI) pre-oxidation on DBP precursors is somewhat diminished. The direct precursor destruction noted above is partly mitigated by the fact that many of these oxidized precursors would have been removed by coagulation anyway. The net impact is a small benefit in the ultimate formation of regulated DBPs (i.e., THMs and HAAs). Therefore, the greatest benefit of pre-oxidation with Fe(VI) may be that oxidation and disinfection can be achieved without formation of regulated DBPs and with modest decreases in these DBPs after final chlorination. Waters pre-oxidized by Fe(VI) at pH 6.2 showed lower DBPFPs than those pre-oxidized at pH 7.5. Intermediate-ferrate proved to be the best approach to control DBPs. Fe(VI) can slowly oxide bromide, forming low levels of active bromine and bromate. Nevertheless, this is not expected to cause a serious problem for Fe(VI) application as the bromide to bromate yields are very low.

References

1 J. Q. Jiang, S. Wang and A. Panagoulopoulos, *Chemosphere*, 2006, **63**, 212.
2 J. Q. Jiang, S. Wang and A. Panagoulopoulos, *Desalination*, 2007, **210**, 266.
3 F. Kazama, *FEMS Microbiol. Lett.*, 1994, **118**, 345.
4 F. Kazama, in *Water Science and Technology*, 1995, vol. 31, pp. 165.
5 T. Schink and T. D. Waite, *Water Res.*, 1980, **14**, 1705.
6 V. K. Sharma, *Environ. Sci. Technol.*, 2010, **44**, 5148.
7 V. K. Sharma, *Coord. Chem. Rev.*, 2013, **257**, 495.
8 J. Q. Jiang and S. Wang, *Environ. Eng. Sci.*, 2003, **20**, 627.
9 N. J. D. Graham, T. T. Khoi and J. Q. Jiang, *Water Sci. Technol.*, 2010, **62**, 929.
10 J.-H. Qu, H.-J. Liu, S.-X. Liu and P. J. Lei, *J. Environ. Eng.*, 2003, **129**, 17.
11 P. C. Chiang, E. Chang and C. H. Liang, *Chemosphere*, 2002, **46**, 929.
12 C. R. O'Melia, W. C. Becker and K. K. Au, *Water Sci. Technol.*, 1999, **40**, 47.

13 E. C. Wert and F. L. Rosario-Ortiz, *Ozone Sci. Eng.*, 2011, **33**, 14.
14 P. Bose and D. A. Reckhow, *Water Res.*, 2007, **41**, 1516.
15 W. C. Becker and C. R. O'Melia, in *Water Science and Technology: Water Supply*, 2001, vol. 1, pp. 81.
16 P. C. Singer, C. Arlotta, N. Snider-Sajdak and R. Miltner, *Ozone Sci. Eng.*, 2003, **25**, 453.
17 V. K. Sharma, *Adv. Environ. Res.*, 2002, **6**, 143.
18 R. H. Wood, *J. Am. Chem. Soc.*, 1958, **80**, 2038.
19 Y. Lee, J. Yoon and U. von Gunten, *Water Res.*, 2005, **39**, 1946.
20 Y. Jiang, J. E. Goodwill, J. E. Tobiason and D. A. Reckhow, *Environ. Sci. Technol.* (Bromide Oxidation by Ferrate (VI). To be submitted April 2015)
21 J. D. Carr, *Abstr. Pap. Am. Chem. Soc.*, 2006, **232**, 495.
22 Y. Lee, R. Kissner and U. von Gunten, *Environ. Sci. Technol.*, 2014, **48**, 5154.
23 Y. Jiang, J. E. Goodwill, J. E. Tobiason and D. A. Reckhow, *Environ. Sci. Technol.*, 2015. **49**, 2841.
24 G. Hua and D. A. Reckhow, *Water Res.*, 2013, **47**, 4322.
25 G. Hua and D. A. Reckhow, *Environ. Sci. Technol.*, 2007, **41**, 3309.
26 D. A. Reckhow, B. Legube and P.C. Singer, *Water Res.*, 1986, **20**, 987.
27 D. A. Reckhow and P. C. Singer, *J. Am. Water Work. Assoc.*, 1984, **76**, 151.
28 Y. Jiang, J. E. Goodwill, J. E. Tobiason and D. A. Reckhow, *Water Res.* (Impact of Ferrate Treatment on Disinfection Byproduct Formation. To be submitted September 2015)

THE REMOVAL EFFECT OF VOLATILE HALOCARBON COMPOUNDS IN DRINKING WATER USING HOUSEHOLD WATER PURIFIER

Masayuki Goto, Reiko Takanashi, Takehiko Kaneko, Yumi Manaka, Koji Kishida and Hadjime Nakajima

Department of Health and Nutrition, Wayo Women's University, 2-3-1 Konodai, Ichikawa-shi, Chiba, JAPAN

1 INTRODUCTION

High concentrations of volatile halocarbon compounds (e.g. trihalomethane) in drinking water may harm human health. We examined the effectiveness of filtration using a water purifying cartridge to remove volatile halocarbon compounds (VHCs) in drinking water. The cartridge filters we selected were widely sold for family use in Japan, which consisted of active carbon and hollow fiber. It is well known that the boiling is a kind of useful method to remove VHCs. The purpose of this study is to evaluate the usefulness of commercial filters in comparison with boiling to remove VHCs.

2 MATERIALS AND METHODS

2.1 Water adulteration and drinking water samples

The standard pure water were adulterated with eight VHCs of four levels of concentration such as $CHCl_3$ 0.1, 1, 10, 100μg/L, $CHCl_2Br$ 0.025, 0.25, 2.5, 25μg/L, $CHClBr_2$ 0.04, 0.4, 4, 40μg/L, $CHBr_3$ 0.2, 2, 20, 200μg/L, C_2HCl_3 0.15, 1.5, 15, 150μg/L, C_2Cl_4 0.04, 0.4, 4, 40μg/L, $C_2H_3Cl_3$ 0.04, 0.4, 4, 40μg/L and CCl_4 0.01, 0.1, 1, 10μg/L for the experiments of heating/boiling or filtering test. Five drinking water samples, which were characterized by each VHCs were examined to evaluate water purifier.

2.2 Heating and boiling

Heating was started at 80 °C followed by boiling for 3 or 10 minutes to confirm the removal effect on VHCs.

2.3 Evaluation of cartridge filter

The total 200 L of adulterated water samples were filtered using a commercial cartridge filter (Ube Industries, Ltd, UT-60, Tokyo, Japan) followed by quantitative analysis by gas-chromatography (Figure 1).

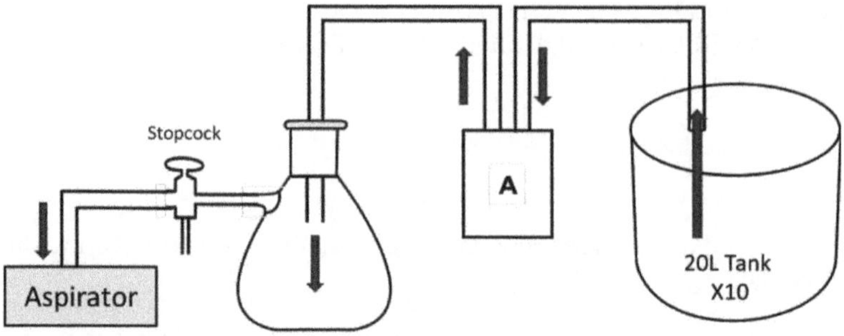

Figure 1 *Schematic diagram of the test for volatile halocarbon compounds using a water purifying cartridge filter.*
A : water purifying cartridge filter.

2.4 Quantitative analysis of VHCs

All samples were analyzed by gas-chromatography (ECD) using 2μL of each extracted sample.

3 RESULTS AND DISCUSSION

In almost all boiled samples except $CHBr_3$, the concentration levels of VHCs were decreased to undetectable levels (Figure 2: $CHCl_3$). However, with the highest level of $CHBr_3$, the level of 200 μg/L was reduced to 1 instead of undetectable levels after 10-minutes boiling (Figure 2: $CHBr_3$).

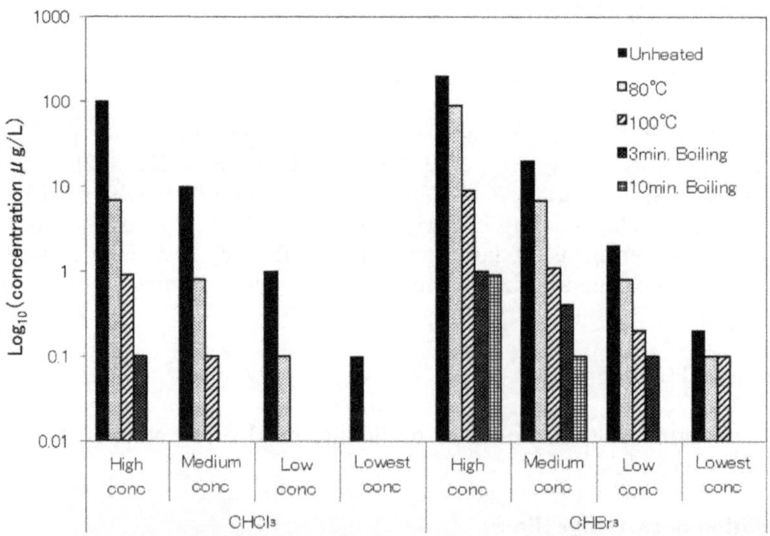

Figure 2 *The effect of heat treatment and initial concentration level on volatile halocarbon compounds' residual. (not detected means <0.01 μg/L.)*

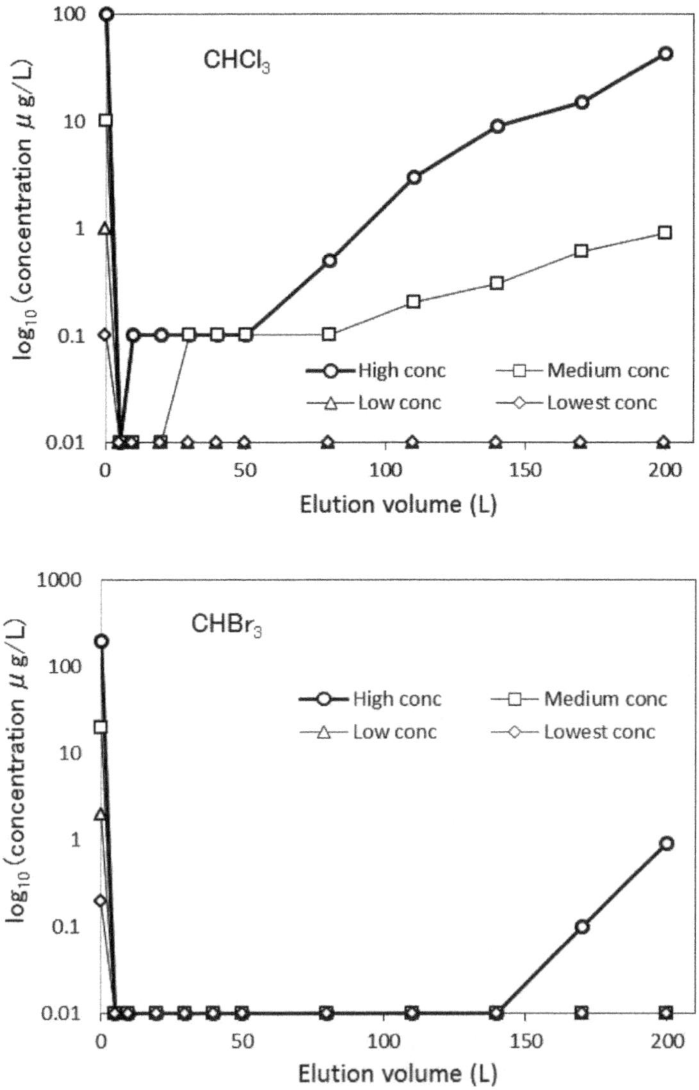

Figure 3 *The concentration levels of volatile halocarbon compounds after usage of commercial cartridge filter.*

The total 200L of each adulterated sample was filtered using a commercial cartridge filter and the VHCs levels analyzed by gas-chromatography.
The highest concentration of $CHCl_3$ was decreased to 43% of the original level. However, in the series of $CHCl_2Br$, $CHClBr_2$, $CHBr_3$, C_2HCl_3, C_2Cl_4, and $C_2H_3Cl_3$, only 0.5% of these VHCs were detected after filtering, and the amount of CCl_4 was undetectable (Figure 3).

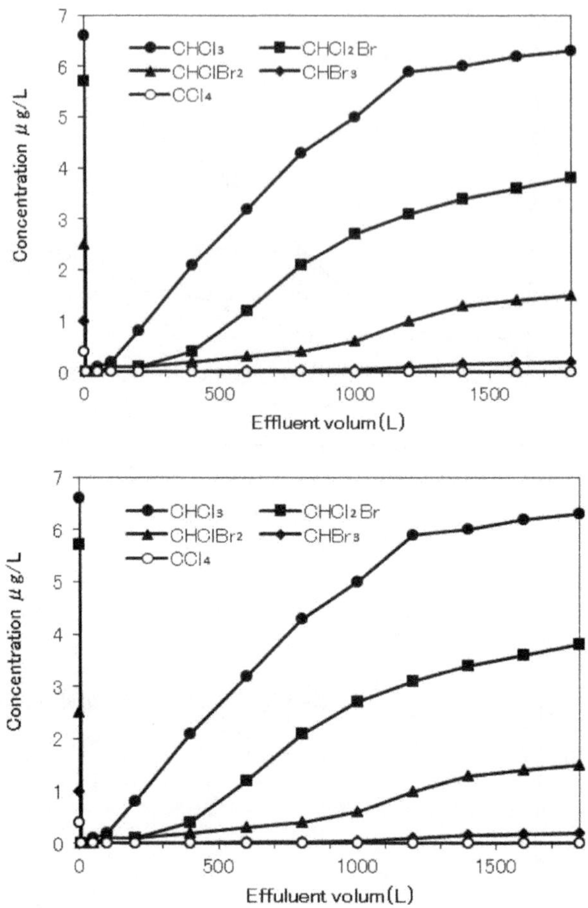

Figure 4 *The concentration levels of volatile halocarbon compounds after usage of commercial cartridge filter in drinkable tap water.*

We also evaluated the VHCs contamination level such as $CHCl_3$, $CHCl_2Br$, $CHClBr_2$, $CHBr_3$ and CCl_4 in samples of drinkable tap water, where the VHCs concentration was 6.6, 5.7, 2.5, 1.0 and 0.4μg/L, respectively. By filtering of 1,800L of contaminated water, the removal capability of VHCs fell (Figure 4).

The levels of $CHCl_3$, $CHCl_2Br$, $CHClBr_2$, $CHBr_3$ and CCl_4 remained at 95, 67, 60, 20 and 0% of original amount, respectively (Figure 4).

4 CONCLUSION

These data suggested that a household water purifier was not sufficiently effective for removal of some kinds of VHCs after repeated usage for a long period of time.

References

1 JV Bruckner, BD Davis and JN Blancato, *Critic. Rev. Toxicol.*, 1989, **20**, 31.
2 S.Rajan, J.Azarah and U. Bauer, *Zbi. Hyg.*, 1990, **189**, 312.

EVALUATION OF CHLORITE AND CHLORATE REMOVAL BY ACTIVATED CARBON IN A PILOT SCALE DRINKING WATER TREATMENT PLANT

S. Sorlini[1], F. Gialdini[1], M. Biasibetti[1], and M.C. Collivignarelli[2]

[1]Department of Civil, Environmental, Architectural Engineering, and Mathematics, University of Brescia, Via Branze 43, 25123 Brescia, Italy
[2]Department of Civil Engineering and Architecture, University of Pavia, Via Ferrata 1, 27100 Pavia, Italy

1 INTRODUCTION

Chlorine dioxide (ClO_2) applied in the disinfection of drinking water can produce chlorite (ClO_2^-) and chlorate (ClO_3^-). It is noted that the ClO_2^- and ClO_3^- formation is about 68% and 9% of the ClO_2 consumed.[1,2] The World Health Organization recommends a guideline value (GV) of 700 μg L^{-1} for both of these compounds in drinking water, since they can cause oxidative damage to human red blood cells.[3] Specific treatment processes can be applied to control the production of ClO_2^-, while no treatment is known for ClO_3^- removal. The ClO_2^- can be removed by adding reduced-sulphur compounds such as sulphur dioxide and sodium sulphite, or by adding iron-based salts, such as ferrous chloride and ferrous sulphate, or applying powdered (PAC) or granular (GAC) activated carbon.[4,5,6] Few studies are available in the literature regarding the ClO_2^- and ClO_3^- removal by GAC. Some researchers found that regenerated GAC can completely remove ClO_2^-, while exhausted GAC reduces the ClO_2^- concentration by 50%.[5] Other researchers found an initial 93.5% ClO_2^- removal using virgin GAC with a 1.65 minute empty bed contact time (EBCT), while using a pre-loaded GAC with a 1.78 minute EBCT they observed a 13-21% ClO_2^- removal. In both cases, ClO_3^- was poorly or not removed.[7]

This work was carried out at a drinking water treatment plant that uses a high ClO_2 dose for disinfection, resulting in a high ClO_2^- concentration in the treated water, exceeding the WHO GV of 700 μg L^{-1}. The plant was monitored for ClO_2 demand, UV absorbance at 254 nm (UV_{254}) and the permanganate index (PI) at the inlet and at the outlet of each treatment stage. A pilot scale study was carried out to evaluate ClO_2^- and ClO_3^- removal with activated carbon using two columns filled with two different mineral activated carbons installed at the outlet of the sand filter of the plant.

2 METHOD AND RESULTS

2.1 Drinking water supply system of Cremona (Italy)

The full-scale drinking water treatment plant of Cremona, located in the North of Italy, treats a maximum water flow of 38.9 ML d^{-1}. The plant is composed of the following processes: aeration for methane and hydrogen sulphide stripping and iron oxidation; biofiltration for manganese, iron and arsenic oxidation and ammonia nitrification; flash mixing with potassium permanganate addition for the oxidation of manganese and arsenic and the precipitation of arsenate with ferric chloride; sand filtration for the removal of iron/manganese hydroxides/oxides and the insoluble arsenic-containing compounds and final disinfection with ClO_2.

2.2 Activities

Monitoring was carried out on samples collected at the inlet and at the outlet of each treatment stage. UV_{254} and PI were analyzed and the water samples were treated at laboratory scale with 5 mg L^{-1} of ClO_2 solution and the ClO_2 demand was determined.

A pilot scale study was carried out to evaluate ClO_2^- and ClO_3^- removal with activated carbon. Two columns were installed at the outlet of the sand filter of the plant and were filled with two different mineral activated carbons: Norit GAC 1240 (mesoporous) and Poractiv M21 (macro-microporous). Each column (0.25 m diameter, 2 m height) was filled from the bottom upwards with 10 cm of coarse gravel (9-18 mm), 20 cm of fine gravel (0.95 mm) and 100 cm of GAC. Each column operated with downflow and was tested under two conditions: 249 L h^{-1} flow with a 10 minute EBCT and 124.5 L h^{-1} flow with a 20 minute EBCT. The columns were backwashed in upflow condition once a week for 30 minutes using water coming from the outlet of the sand filter, with a 500 L h^{-1} flow and a 12 m h^{-1} hydraulic load. Water samples were collected at the inlet and at the outlet of each column during the column operating period. UV_{254} and PI were analyzed and the water samples were treated at laboratory scale with 1 mg L^{-1} of ClO_2 solution and the ClO_2 demand and ClO_2^- and ClO_3^- concentrations were determined.

2.3 Methods

The ClO_2 demand was evaluated after treating water samples collected at the outlet of each treatment stage of the plant with 5 mg L^{-1} of ClO_2 solution, and treating water samples collected at the inlet and at the outlet of the carbon columns with 1 mg L^{-1} of ClO_2 solution. After rapid mixing, aimed at homogenizing the solution, each disinfected water sample was put in a 250 mL dark glass bottle with a cap. After a certain water-disinfectant contact time, 10 mL of the solution were taken from the bottle and the residual ClO_2 was analyzed using a residual chlorine analyzer, according to the DPD method.[8] The ClO_2 demand was calculated as the difference between the initial ClO_2 dose and the residual ClO_2. The ClO_2 demand on water samples collected at the inlet and outlet of each treatment stage was evaluated after 5, 10, 30, 60, 120 and 180 minutes water disinfectant contact time; the ClO_2 demand on water samples collected at the inlet and at the outlet of each pilot column was evaluated after 30, 60 and 120 minutes water-disinfectant contact time.

The UV_{254} was determined with a spectrophotometer (Perkin Elmer Lambda 25), using a 1 cm glass cuvette. PI was evaluated according to the standard method.[9] Chlorite and chlorate were determined with an Ion Chromatograph (Dionex ICS 5000)[10] on water samples collected at the inlet and at the outlet of each pilot column and disinfected at lab-scale with 1 mg L^{-1} of ClO_2 solution, after 120 minutes water-disinfectant contact time.

2.4 Results

The results of the plant monitoring showed that the ClO_2 demand, UV_{254} and PI progressively decreased after each treatment, except for the aeration stage, for which no variations were observed compared to the raw water (Figure 1). After a water-disinfectant contact time below 180 minutes, the ClO_2 demand at the outlet of the sand filter was higher than the demand at the outlet of the flash mixing stage, due to the release of compounds by the sand filter that increased the ClO_2 consumption. Predictably, with increasing contact time the ClO_2 demand also increased; however, this trend progressively decreased after each treatment. The ClO_2 demand at the outlet of the disinfection stage was below 1 mg L^{-1}. UV_{254} increased at the outlet of the flash mixing stage, since the addition of chemicals (ferric chloride and potassium permanganate) in this reactor increased water colour.

The results of the pilot experimentation with GAC columns operating with a 10 minute EBCT showed a significant reduction in ClO_2 demand (70-80%), regardless of the water-disinfectant contact time, mainly in the initial operating phase of each column (Figure 2). After 35 days of operation, with a 30 minute water-disinfectant contact time, the ClO_2 demand reduction was 20% and then it was constant (15-20%). After 60 and 120 minute water-disinfectant contact times the ClO_2 demand reduction after 60 days of operation of each column was 10% and 0% respectively. The macro-microporous carbon Poractiv M21 showed a slightly higher reduction in ClO_2 demand than the mesoporous GAC 1240 throughout the operating period, regardless of the water-disinfectant contact time. With increasing water disinfectant contact time the reduction in ClO_2 demand decreased, most significantly after 40 days of operation of each column.

UV_{254} was significantly reduced with both carbons, particularly in the first 40 days (75-35%) (Figure 3). PI was initially reduced then progressively increased at the outlet of the pilot plant, until it exceeded the inlet (Figure 4).

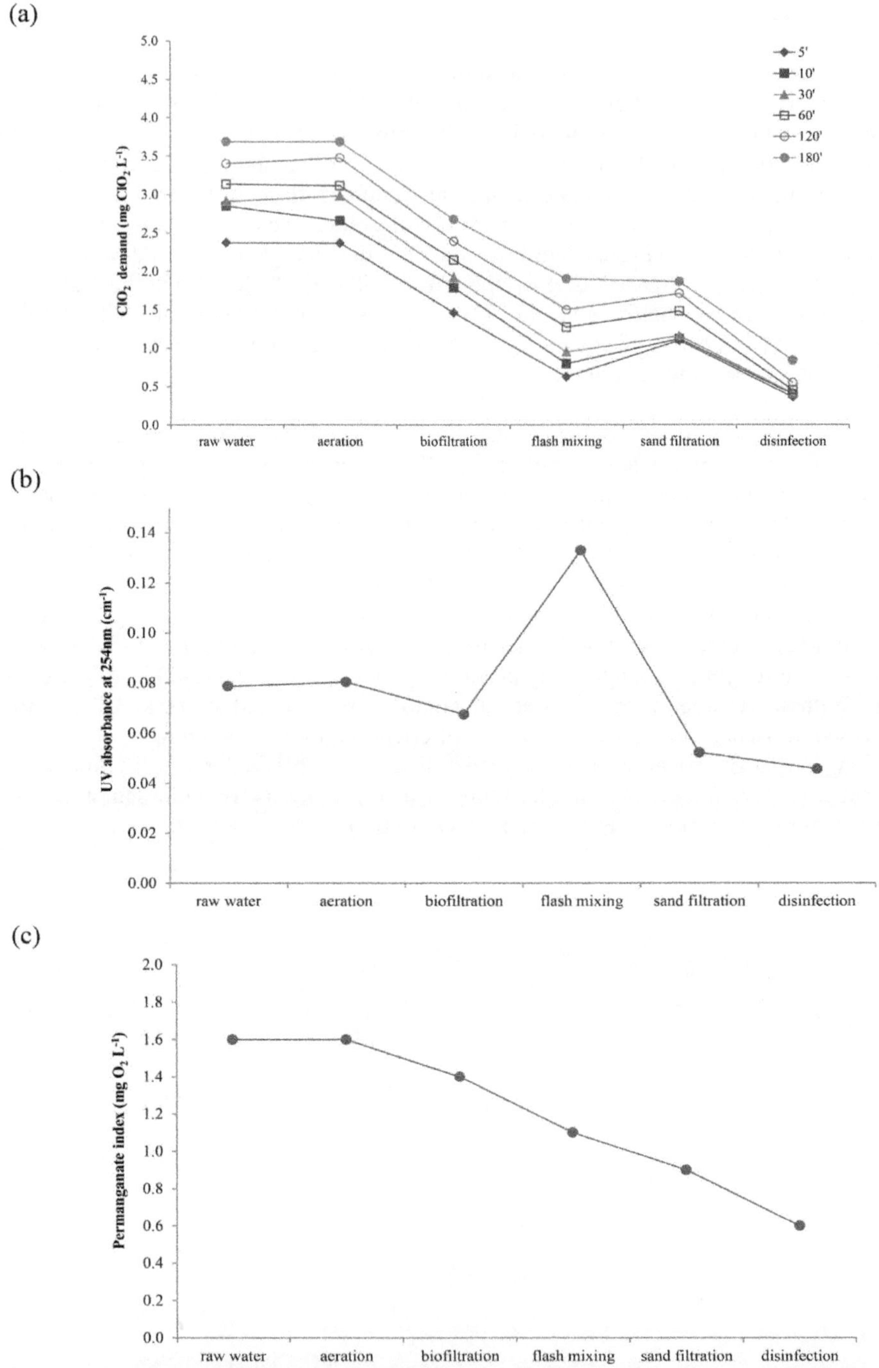

Figure 1 *ClO_2 demand (a), UV absorbance at 254 nm (b) and permanganate index (c) measured on water samples collected at the outlet of each treatment stage of the drinking water treatment plant of Cremona (Italy)*

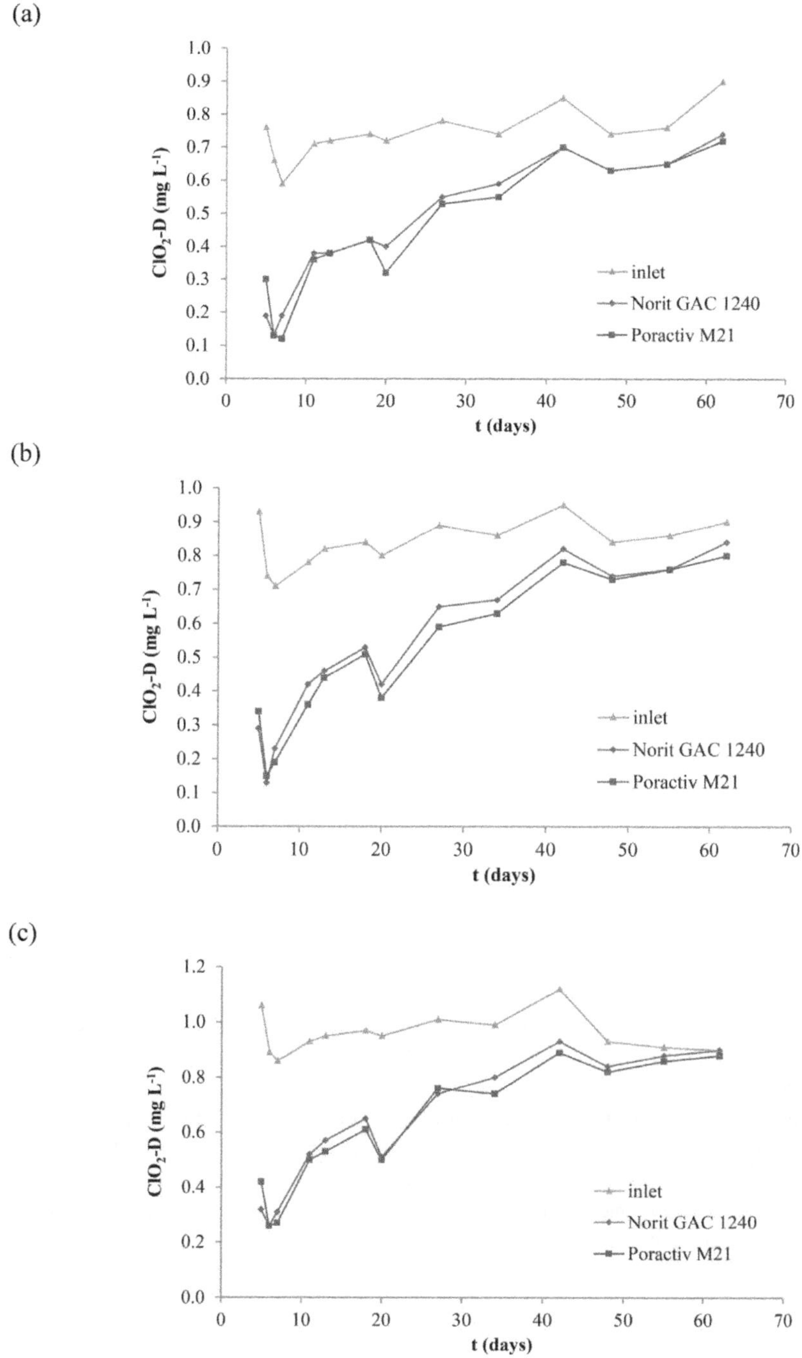

Figure 2 *ClO_2 demand measured on water samples collected at the inlet and outlet of the carbon columns operating with a 10 minute EBCT, and disinfected with 1 mg ClO_2 L^{-1} at a 30 (a), 60(b) and 120 (c) minutes water-disinfectant contact time*

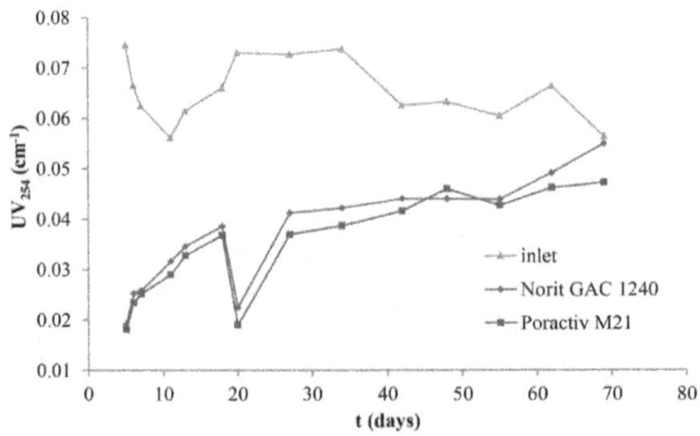

Figure 3 *UV_{254} measured on water samples collected at the inlet and outlet of the carbon columns operating with a 10 minute EBCT*

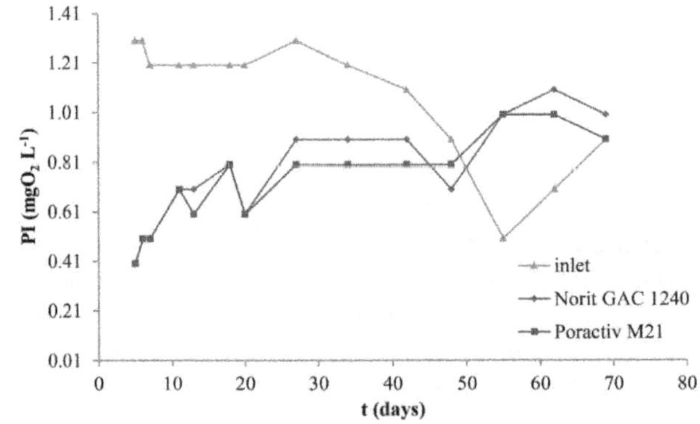

Figure 4 *PI measured on water samples collected at the inlet and outlet of the carbon columns operating with a 10 minute EBCT*

With both carbons, after a 120 minute water-disinfectant contact time, ClO_2^- was reduced by 20-40% during the first two weeks of operation of the columns; from the 3rd to the 6th week the reduction decreased with values of 11-17%, and then below 10% (Figure 5a).

The ClO_3^- concentration at the outlet of both columns, after a 120 minute water-disinfectant contact time, was quite variable during the operating period (Figure 5b), and the reduction in ClO_3^- does not appear to be statistically significant, particularly for the mineral GAC Poractiv M21 where sometimes outlet ClO_3^- concentration is higher than the inlet concentration.

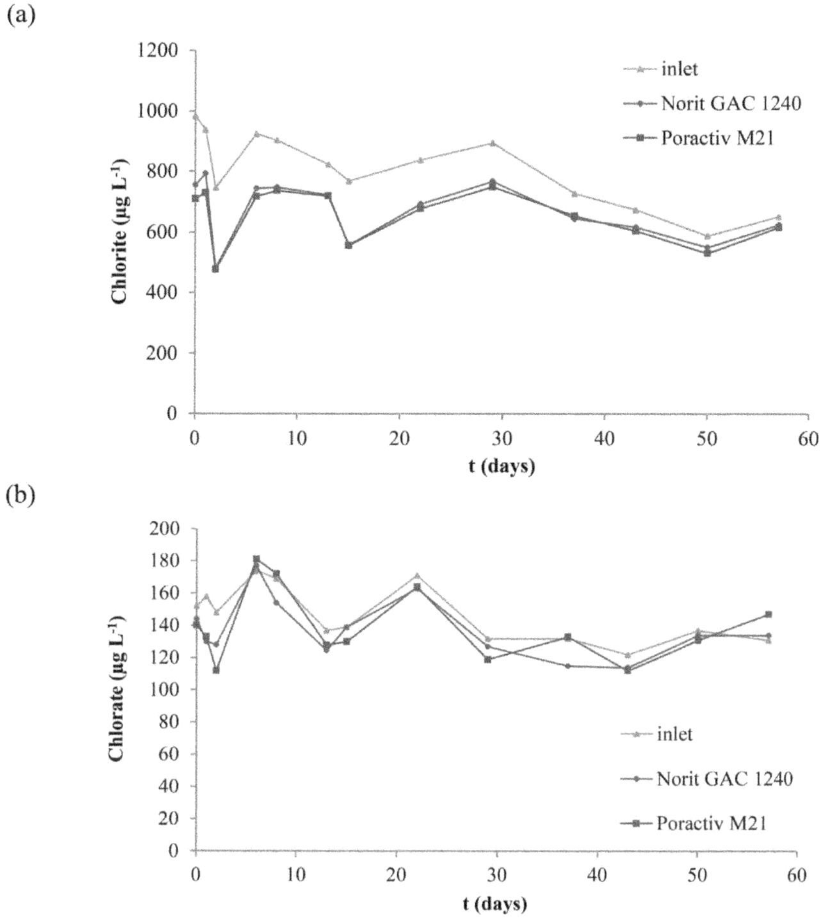

Figure 5 *Chlorite (a) and chlorate (b) measured on water samples collected at the outlet of the carbon columns operating with a 10 minute EBCT, and disinfected with 1 mg ClO_2 L^{-1} at a 120 minute water-disinfectant contact time*

The results of the pilot experimentation with GAC columns operating with a 20 minute EBCT showed a high reduction in ClO_2 demand, UV_{254} and PI (65-70%), regardless of the water-disinfectant contact time, in the initial operating phase of each column; after 20 days the reduction was 40-45% (Figure 6, 7 and 8).

(a)

(b)

(c)

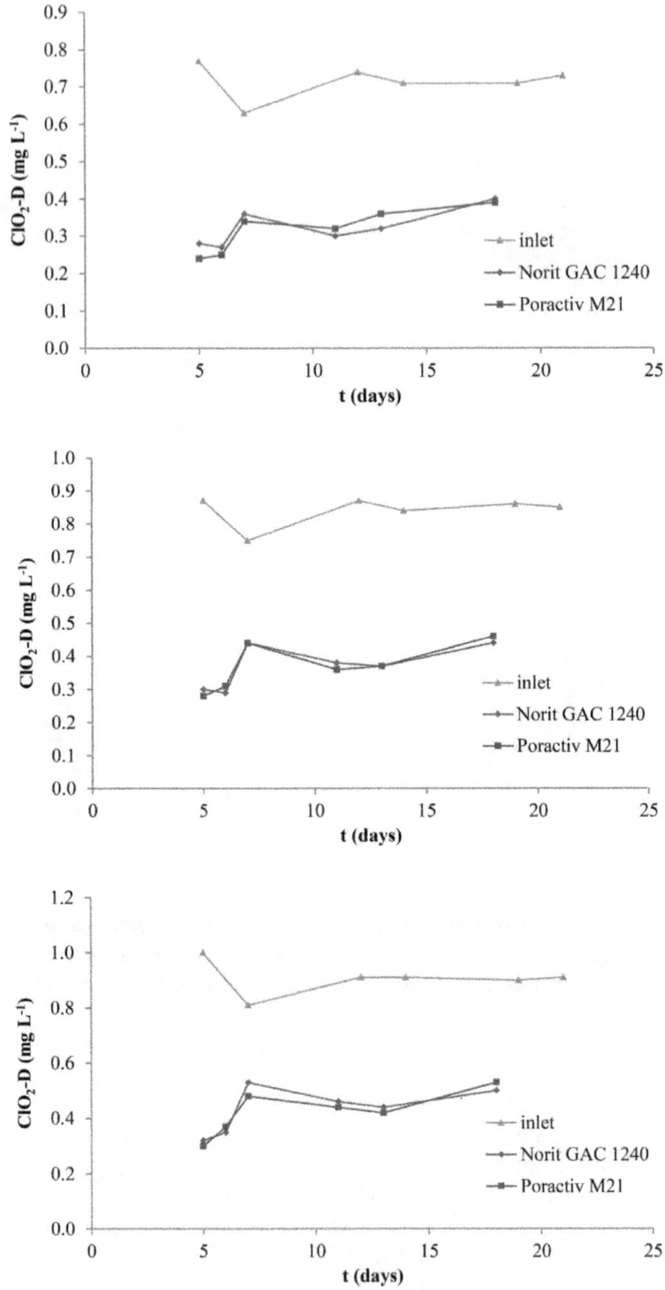

Figure 6 *ClO_2 demand measured on water samples collected at the inlet and outlet of the carbon columns operating with a 20 minute EBCT, and disinfected with 1 mg ClO_2 L^{-1} at a 30 (a), 60(b) and 120 (c) minutes water-disinfectant contact time*

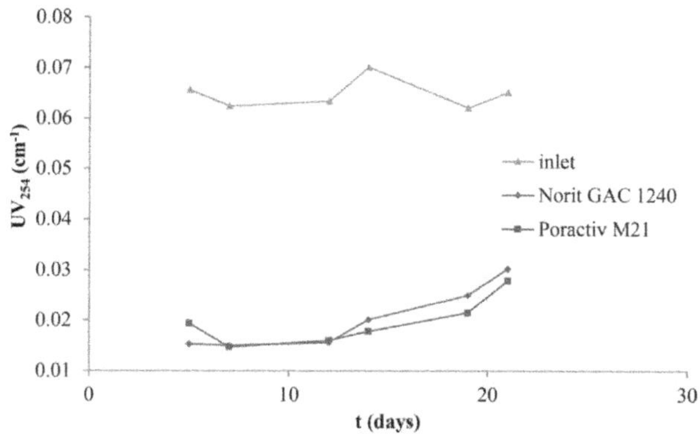

Figure 7 *UV_{254} measured on water samples collected at the inlet and outlet of the carbon columns operating with a 20 minute EBCT*

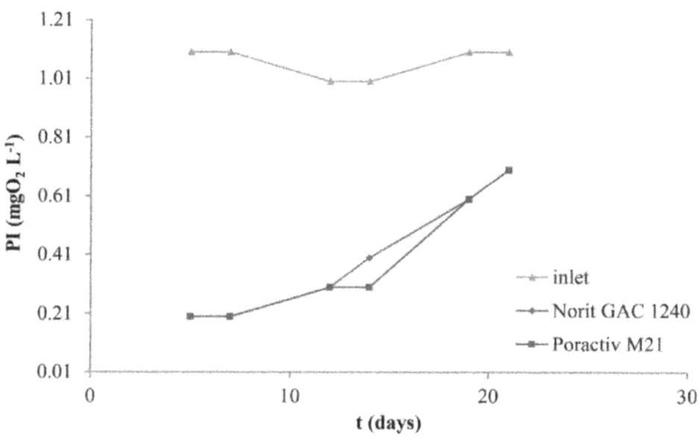

Figure 8 *PI measured on water samples collected at the inlet and outlet of the carbon columns operating with a 20 minute EBCT*

The chlorite and chlorate concentration, after a 120 minute water-disinfectant contact time, was reduced respectively by 31-39% and 5-20% with both carbons (Figure 9).

Comparing the two operating conditions, in the first 10 days a 10 minute EBCT gave a ClO_2 demand reduction higher than with a 20 minute EBCT; after 20 days, the reduction yields were similar. The reduction in UV_{254} was higher with 20 than 10 minute EBCT in the first 15 days of operation with both carbons; after 20 days a significant decrease in the reduction yields was observed for both carbons. The reduction in PI was higher with 20 than 10 minute EBCT in the first 15 days, after which the opposite behavior was observed.

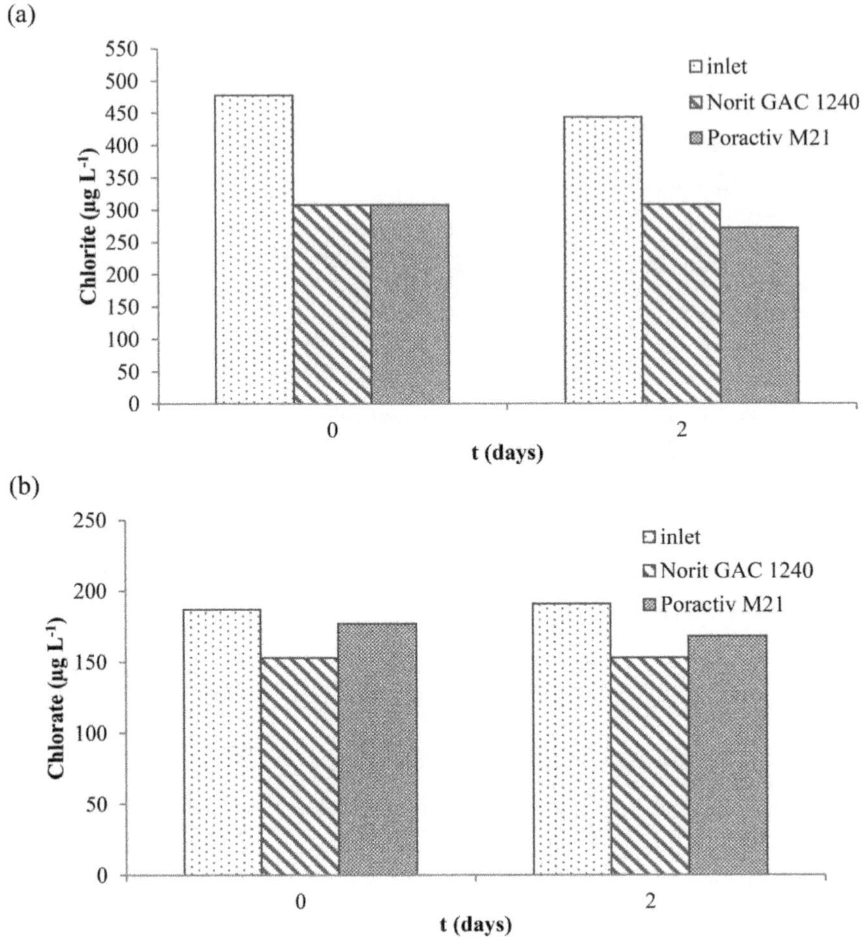

Figure 9 *Chlorite (a) and chlorate (b) measured on water samples collected at the outlet of the carbon columns operating with a 20 minute EBCT, and disinfected with 1 mg ClO_2 L^{-1} at a 120 minute water-disinfectant contact time*

3 CONCLUSION

In this work the drinking water treatment plant of Cremona, in the North of Italy, was monitored analysing the ClO_2 demand, UV absorbance at 254 nm and permanganate index at the inlet and at the outlet of each treatment stage. The results showed a progressive decrease in the concentrations of these parameters after each treatment.

A pilot scale study was carried out to evaluate the ClO_2^- and ClO_3^- removal with activated carbon. Two columns, filled with different mineral activated carbons, were installed at the outlet of the sand filter of the plant. The results showed that with both carbons ClO_2^- was reduced by 20-40% during the first two weeks of operation of the columns; from the 3rd to the 6th week the reduction decreased with values of 11-17%, and then below 10%. The ClO_3^- concentration at the outlet of both columns was quite variable during the operating period, and ClO_3^- with both carbons and the reduction in ClO_3 does not appear to be statistically significant.

References

1 C. Korn, R.C. Andrews and M.D. Escobar, *Water Res.*, 2002, **36**, 330.
2 C. Collivignarelli and S. Sorlini, *J. Water Supply: Res. Technol.-Aqua*, 2004, **53**, 159.
3 WHO World Health Organization. *Guidelines for drinking water quality. Fourth edition*; 2011.
4 A. Katz and N. Narkis, *Water Res.*, 2001, **35**, 101.
5 S. Sorlini and C. Collivignarelli, *Desalination*, 2005a, **176**, 267.
6 S. Sorlini and C. Collivignarelli, *Desalination*, 2005b, **176**, 255.
7 K. L. Dixon and R. G. Lee, *J. Am. Water Works Ass.*, 1991, **48**, 48.
8 ISO, International standard 7393/2, Ist edition, 1985.
9 ISO, Standard method, 1997.
10 UNI EN ISO 10304-4, 2001

APPLICATION OF CHLORINE-RESISTANT NANOFILTRATION MEMBRANE IN SWIMMING POOL WATER TREATMENT TO MINIMIZE THE DISINFECTION BY-PRODUCTS

D. Peng, F. Saravia, G. Abbt-Braun, H. Horn

Chair of Water Chemistry and Water Technology, Engler-Bunte-Institute, Karlsruhe Institute of Technology, Engler-Bunte-Ring 1, 76131 Karlsruhe, Germany. E-mail: di.peng@kit.edu

1 INTRODUCTION

Swimming pools have a great recreational value all over the world. To guarantee the health of swimming pool visitors, disinfection is a necessary step. Chlorine, the most common disinfectant in swimming pools, reacts with the organic matter in swimming pool water and produce undesirable and potentially harmful disinfection by-products (DBPs), such as trihalomethanes (THMs). Disinfection by-products in swimming pool water have been linked to human health effects such as bladder cancer.[1] Understanding the behavior of DBP-formation in pool water is complex due to the irregular new input of contaminants and the recirculation of swimming pool water over long periods of time.[2] Inadequate pool water treatment can adversely affect water quality as the contaminants which cannot be removed by treatment will accumulate.[3, 4]

Nanofiltration (NF) is an attractive alternative to conventional pool water treatment, based on the effective elimination of dissolved organic substances and thus disinfection by-product precursors. The work to be presented here addresses the application potential of nanofiltration for swimming pool water treatment, which minimizes the formation of DBPs and their precursors at relatively low operating pressure.

2 MATERIAL AND METHODS

In order to understand DBP formation and its possible correlation with visitor numbers, water samples were collected in a public swimming pool over short time intervals. Dissolved organic carbon (DOC), organically bound halogens adsorbable on activated carbon (AOX), and THMs were analysed and correlated with the number of visitors. Here DOC was taken as a surrogate for precursors of DBP.

Due to the fact that the bather and contaminant load, chlorine dosage, DBP concentration and fresh filling water can vary considerably both over short and longer time scales, swimming pool water cannot easily be simulated in the laboratory. Therefore, on-site experiments in this public swimming pool with a laboratory-scale cross-flow membrane set-up were carried out. The plant has six flat-sheet modules and each module enables a membrane area of 0.008 m^2 (4 * 20 cm). Three commercial chlorine-resistant NF membranes (Table 1) were parallel tested in duplicate. Although no data was provided, NP030 was expected to be chlorine resistant due to the membrane material polyethersulfone. All filtration experiments were conducted with a cross-flow velocity (V_{CF}) of 0.22 m/s.

Table 1 *NF membrane properties (manufacturer's instructions)*

Membrane	*Manufacturer*	*Material*	*Chlorine-resistance* (ppm)	*MWCO*[a] (Da)
SB90	Trisep	Cellulose acetate	1	183
NTR-7470pHT	Nitto-Denko	Sulfonated polyethersulfone	10	463
NP030	Microdyn-Nadir	Polyethersulfone	n.d.[b]	1343

[a]Molecular weight cut-off (MWCO) was determined at 25°C, 10 bar, V_{CF} = 0.22 m/s.
[b]n.d.: no data provided by the manufacturer

The NF-set-up operated with UF pre-treated water (combination of flocculation, adsorption on powdered activated carbon and ultrafiltration). This is a standardized treatment process for swimming pool water included in the German DIN 19643 (2012). During the experiments DOC, AOX and THM were determined both in feed and permeate.

3 RESULTS AND DISCUSSION

Intensive sampling of pool water showed a time delay of about 2 days between the increase of DOC and formation of AOX (organically bound halogens adsorbable on activated carbon), which indicates the possibility of minimizing DBP formation by quickly removing the organic matter (and thus the DBP-precursors) from the pool water. Similar results were observed also in outdoor swimming pools.[5] But unlike some previous research,[5, 6] the number of visitors did not correlate with the DOC development. This indicates that introduction of anthropogenic pollutants into swimming pool water as well as the DBP formation cannot be predicted simply from the number of visitors.

The characteristics of the feed water to the NF-set-up are shown in Table 2. All three applied membranes showed stable rejection performance during the experiments at the transmembrane pressures of 6, 8, and 10 bar for a total of approx. 6 weeks. NTR-7470pHT and SB90 indicated high removal rates of DOC and AOX (> 80%). Figures 1 and 2 present the results of DOC- and AOX-rejection at 10 bar. NP030 showed a slight increase in the rejection performance, probably because membrane fouling improved the size-exclusion effect during the filtration process. In contrast to some previous research,[7,8] no THM rejection was observed even for the tightest membrane with the lowest molecular weight cut-off (SB90, ca. 200 Da). A possible reason is the small molecular weight of chloroform, which constitutes >90% of the THMs in the swimming pool water.

Table 2 *Feed water properties for the NF experiment at 10 bar ($n \geq 4$)*

Temperature (°C)	*pH*	*DOC* (mg/L)	*AOX* (µg/L)	*THM* (µg/L)
28.9 ± 0.4	7.2 – 7.4	3.2 ± 0.1	387 ± 27	16 ± 2

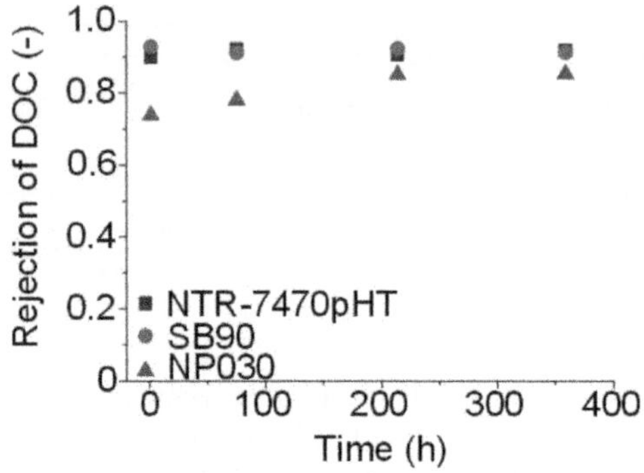

Figure 1 *Rejection performance of DOC in flat-sheet module experiments (Feed: UF-treated pool water, 10 bar, V_{CF} = 0.22 m/s)*

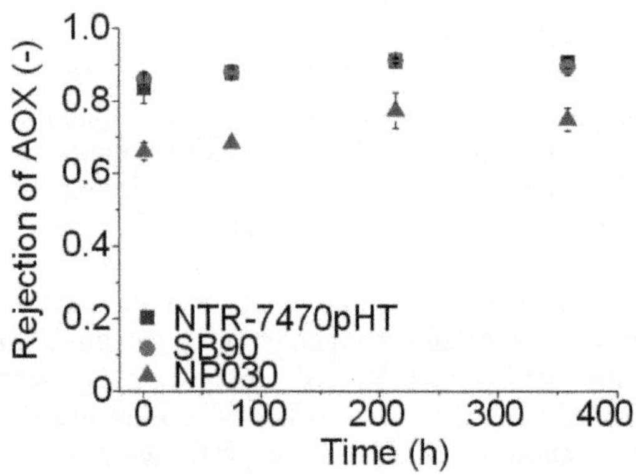

Figure 2 *Rejection performance of AOX in flat-sheet module experiments (Feed: UF-treated pool water, 10 bar, V_{CF} = 0.22 m/s)*

In further filtration experiments with the selected NF-membranes a 70% rejection of THM-FP (THM-formation potential) and 80% of AOX-FP (AOX-formation potential) was observed, again indicating that most of the precursors should also be removed by NF. Based on extensive flat-sheet module experiments, SB90 (Trisep) was selected for investigations with pilot scale spiral wound modules. The results of the pilot plant experiments confirmed the rejection performance of the flat-sheet-modules and showed stable permeability for at least 4 months.

4 CONCLUSION

High rejection rates of DOC and AOX indicate that most of the precursors are removed by nanofiltration. It implies the possibility of minimizing DBP formation by quick removal of dissolved organic matter (and thus DBP-precursors) from the pool water. The results achieved with the pilot scale NF module allowed calculation of the possible minimization of DBPs and precursor-formation, if a branch current is treated with NF-membranes. Such an efficient treatment system not only provides a better pool water quality but also leads to a reduction of fresh water and energy consumption.

Acknowledgements

The authors thank the German Federal Environmental Foundation (DBU) for their financial support (Az: 28707-23), the W.E.T. GmbH for their collaboration, Hydranautics© and Microdyn-Nadir© for providing membrane samples used in this study, and A. Heidt, U. Reichert, and M. Weber for the analytical work.

References

1. C. M. Villanueva, K. P. Cantor, J. O. Grimalt, N. Malats, D. Silverman, A. Tardon, R. Garcia-Closas, C. Serra, A. Carrato, G. Castano-Vinyals, R. Marcos, N. Rothman, F. X. Real, M. Dosemeci and M. Kogevinas, *Am J Epidemiol*, 2007, **165**, 148-156.
2. S. Chowdhury, K. Alhooshani and T. Karanfil, *Water Res.*, 2014, **53**, 68-109.
3. C. Zwiener, S. D. Richardson, D. M. De Marini, T. Grummt, T. Glauner and F. H. Frimmel, *Environmental Science & Technology*, 2007, **41**, 363-372.
4. S. Müller and W. Uhl, in *Fourth International Conference Swimming Pool and Spa*, Porto, Portugal, Editon edn., 2011.
5. T. Glauner, F. H. Frimmel and C. Zwiener, *GWF-Wasser/Abwasser* 2004, **145**, 706-713
6. H. Chu and M. J. Nieuwenhuijsen, *Occupational and Environmental Medicine*, 2002, **59**, 243-247.
7. N. Ates, L. Yilmaz, M. Kitis and U. Yetis, *Journal of Membrane Science*, 2009, **328**, 104-112.
8. A. M. Klüpfel, T. Glauner, C. Zwiener and F. H. Frimmel, *Water Science & Technology*, 2011, **63**, 1716.

EFFICIENCY OF UV-OXIDATION IN REMOVAL OF PHARMACEUTICALS FROM WASTER WATER SAMPLES AND TOXICOLOGICAL EVALUATION BEFORE AND AFTER THE OXIDATIVE TREATMENT

H. Bielak[1], A. Boergers[2], J. Raab, J. Tuerk[2], E. Dopp[1]

[1]IWW Rheinisch-Westfälisches Institut für Wasserforschung gGmbH, Mülheim, Germany
[2]Institut für Energie- und Umwelttechnik e. V., IUTA, (Institute of Energy and Environmental Technology e. V.) Duisburg, Germany

1 INTRODUCTION

Conventional waste water treatment plants are not able to remove micropollutants completely, resulting in the influent of those substances into surface waters. Especially pharmaceuticals are very often resistant against the biological degradation because of their polarity and persistence [1] [2]. As components in waste water treatment plant (WWTP) effluents these substances can influence the aquatic environment (surface- and ground water) and cause adverse effects in aquatic organisms [3]. Low concentrations of pharmaceuticals can also be detected in drinking water which is of particular importance for human exposure [4]. Moreover, according to the WHO it is expected that additive toxicological effects of the different micropollutants may occur [5].
According to the German Federal Environment Agency [6], the yearly consumption of pharmaceuticals, measured as active pharmaceutical ingredients (API), exceeds 25000 tons. Concentration ranges of API in surface water in German rivers are up to the $\mu g\ L^{-1}$ range [6]. The highest influent of API into waste water is due to private households, hospitals and retirement homes [7]. Hospital effluents are a main source for the discharge of very polar and biological stable iodated contrast media (ICM) into the sewage system [8] [9]. Because of their persistence and low sorption ability ICM can be found in ground- and drinking water in concentrations up to 1 $\mu g\ L^{-1}$ [10].
Concentrations of pharmaceuticals in waste water are related to the pharmacokinetic properties of these substances [11]. Ternes et al. [12] have shown that concentrations of pharmaceuticals in waste water from municipal WWTP effluents correlate with the pharmaceutical concentrations that were detected in the receiving water from those WWTP.

In the present study five substances were investigated: (1) amidotrizoic acid with a mean concentration of 2.6 $\mu g\ L^{-1}$ in WWTP effluents [13], (2) the anti-epileptic agent carbamazepine which is persistent in the environment and is verified as one of the most frequent pharmaceuticals in surface water with a maximum concentration of 1.8 $\mu g\ L^{-1}$ [13] [14], (3) the analgetic diclofenac which is used as non-steroidal anti-rheumatic agent [15], (4) the beta-blocker metoprolol which was detected in WWTP effluents at a concentration of 0.48 $\mu g\ L^{-1}$ and in surface water at a concentration of 0.017 $\mu g\ L^{-1}$ [13]

and (5) sulfamethoxazole, a bacteriostatic acting antibiotic that occurs in WWTP effluents and surface waters at a mean concentration of 0.33 and 0.013 μg L^{-1}, respectively [13]. For example at conventional WWTPs the maximum elimination rate for carbamazepine is 7% [14].

Further removal of these substances from waste water is possible e. g. by using advanced oxidation processes (AOP) [16] [17]. However, several studies have shown that formation of oxidation by-products is possible. Information about these transformation products are rare, especially their toxicological properties are often unknown [1] [18] [19].
AOP methods are able to reduce the concentration of pharmaceutical residues in municipal WWTP effluents to the protective goal of 1 ng L^{-1} which is given by the European Medicines Agency (EMA) [2]. The most applied processes are UV/H_2O_2, O3/H_2O_2 and O3/UV [17].

Chemical analyses are able to detect transformation products but cannot give any information about the toxicity of oxidation by-products. Therefore, toxicological testing systems (bioassays) are a necessary supplement to the chemical analysis [13].

The aim of the present study was the investigation of the efficiency of an UV-oxidation system using different matrices and the toxicological evaluation of the detected substances.

2 METHODS

2.1 Sample preparation

The degradation of amidotrizoic acid, carbamazepine, diclofenac, metoprolol and sulfamethoxazole (Sigma-Aldrich, Germany) by oxidation was measured for the individual substances and for a substance mixture consisting of 100 mg L^{-1} of each substance. In the present study toxicological evaluations were performed for the substance mixture of amidotrizoic acid, carbamazepine and ciclofenac.
For the chemical measurement the samples were transferred to autosampler-vials using a syringe filter. The solid phase extraction (SPE) of the samples with ENV+ (Biotage, Germany) and Strata XL (Phenomenex, Germany) cartridges was performed by a Gilson-System (Gilson International BV, Germany).
Before the toxicological examinations all samples were sterile filtrate (pore size 0.2 μm) to remove microbial contaminations.

2.2 Determination of the degradation rate during UV-oxidation

In the laboratory system a low-pressure lamp with an emission wavelength of 254 nm and a maximum UV output of 15 W (TNN 15/32, Heraeus, Germany) was used. The sample volume was 1 L with a flow rate of 0.03 m^3 h^{-1}. The low-pressure UV lamp needed 5 minutes to reach its maximum output. For the UV oxidation experiments a single substance or the substance mixture was added to HPLC water or the WWTP effluents. After 15 minutes of mixing, the UV lamp was activated. After 2-10 minutes the maximal efficiency was reached. Samples were taken before oxidation and at regular time intervals during the oxidation process. The oxidation by an UV/H_2O_2 system was also evaluated. For these experiments 0.3 g L^{-1} H_2O_2 was added after the warm-up phase of the UV lamp of 5 minutes. The residual peroxides in the samples after oxidation were removed by the enzyme catalase from *Aspergillus niger*.

The degradation of each substance was analyzed using LC-MS/MS with an API 3000 and a Q TRAP 3200 (AB Sciex, Germany).
To calculate the degradation rate for the reactions a kinetic of first order was assumed. The half-life period is then calculated by means of the reaction speed, i.e. the positive slope of the correlation curve.

2.3 Cytotoxicity and genotoxicity

Toxicological investigations were performed for samples from the laboratory scale UV-system. Samples with the substance mixture were tested up to 60 min of oxidative treatment time, while samples with a single substance (amidotrizoic acid, carbamazepine and diclofenac) were tested up to 30 min after the oxidation.
All samples and controls were sterile-filtrated (hydrophilic, pore size 0.2 μm) before they were applied in cell culture tests. Samples treated with H_2O_2 were checked for residual peroxides with peroxide test sticks (semi-quantitative, detection limit 0.5 – 25 mg L^{-1}) to exclude false positive results in the cyto- and genotoxicity testing.

Chinese Hamster Ovary cells (CHO-9) (provided by ECACC, UK) were exposed to the test samples in a 1:10 dilution (c = 10 μg L^{-1}) for 24 h. The culturing of the cells was performed in accordance with standard procedures. CHO 9-cells were grown in HAM's F12 medium at 37 °C, 5 % CO2 and 95 % RH.
Cytotoxicity was tested using the MTT (3-(4,5-Dimethylthiazol-2-yl)-2,5-diphenyltetrazoliumbromid) assay. This test detects cytotoxic effects depending on the mitochondrial activity of the cell. A total of 104 cells were seeded in 200 μL growth medium per well in a 96-well plate. After one day of cell growth and additional 24 h of exposure (samples were tested in triplicate), MTT solution was added to the wells for two hours. MTT is a soluble, yellow tetrazolium salt that can be transformed to an insoluble, purple formazan in the mitochondria of intact cells. After lysis of the exposed cells the viability can be determined by a photometrical measurement. Therefore the medium was removed after the incubation time, cells were lysed with a lysis solution for 15 minutes and the absorption at 595 nm was then measured immediately. For the determination of the cell viability relative values were calculated by correlating the absorption values of the samples to the absorption value of the negative control (unexposed cells, 100 % viability). MMA III (monomethylarsenic acid (III), 50 μM) was used as positive control.
To test the genotoxicity of the samples the Alkaline Comet Assay ("single cell gel electrophoresis") was performed according to the protocols of Ostling and Johanson [20] and Singh et al. [21]. In this test, DNA single and double strand breaks of eukaryotic cells can be detected through the migration of DNA in an electrical field during gel electrophoresis, whereby smaller (broken) fragments move faster leading to the formation of a so called comet tail that can be analyzed under the microscope after staining of the DNA. Briefly, 104 cells were seeded in 2 mL growth medium per well in a 24-well plate. After one day of cell growth and additional 24 h of exposure (samples tested in duplicate), 8000 cells were re-suspended in 20 μL phosphate buffered saline, mixed with low melting agarose (0.75 %, 37 °C) and transferred to previously prepared agarose mini gels on ice. After incubation of the gels in lysis solution at 4 °C overnight, the gels were incubated in cooled alkaline electrophoresis solution to unwind the DNA

with subsequent electrophoresis for 20 min (300 mA/ 25 V) followed by neutralization and fixation of the gels. The dried gels were stained with SYBR green solution (0.001 %), fixed on microscopic slides and manually evaluated under a fluorescence microscope (Leitz GmbH & Co. KG, Germany). The positive control used in the Alkaline Comet Assay was N-Ethyl-N-Nitrosourea (ENU, 100 µg L^{-1}, 30 min).

3 RESULTS

3.1 Degradation of the substances

All individual substances in HPLC-water were completely eliminated during the oxidation experiments. The degradation rates of amidotrizoic acid and sulfamethoxazole could not be determined because the degradation of these substances was too fast. The degradation rates differed depending on the use of H_2O_2. The degradation rate for carbamazepine was 8.6 min L^{-1} without H_2O_2 and 5.0 min L^{-1} with H_2O_2, showing an enhancement of the oxidation by adding H_2O_2. For diclofenac no enhancement of the oxidation was observed after the addition of H_2O_2.
The treatment of the substance mixture in HPLC-water showed similar trends in the degradation (Figure 1). The fastest degradation was observable for amidotrizoic acid, diclofenac and sulfamethoxazole. The degradation rate for amidotrizoic acid without H_2O_2 was 0.8 min L^{-1} and 0.7 min L^{-1} with H_2O_2. For sulfamethoxazole the rates without and with H_2O_2 are 1.1 and 1.02 min L^{-1}, respectively. Carbamazepine and metoprolol were more stable and the elimination was slower. Here the degradation rate for metoprolol was enhanced from 30.8 min L^{-1} without H_2O_2 to 3 min L^{-1} with H_2O_2 and for carbamazepine from 34 min L^{-1} to 3.5 min L^{-1} (Figure 1).

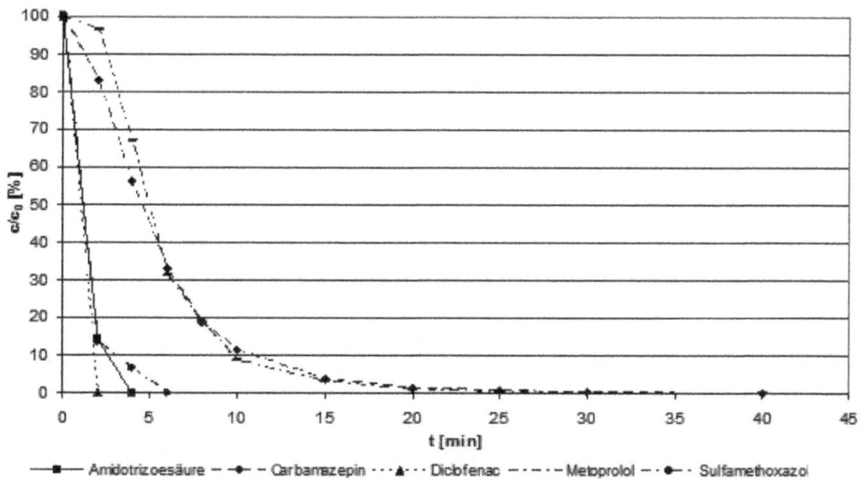

Figure 1 *Degradation curves of amidotrizoic acid, carbamazepine, metoprolol, sulfamethoxazole and diclofenac as substance mixture (c0 = 100 µg L^{-1}) in 1 L HPLC-water with 0.3 g L^{-1} H_2O_2.*

In WWTP effluents diclofenac and sulfamethoxazole (2.5 min L^{-1}) were degraded faster than the other pharmaceuticals. Only 47 % metoprolol and 20 % carbamazepine were eliminated from the WWTP effluent. Half of the concentration of metoprolol was eliminated after 81.5 min, for carbamazepine after 315 min. The additional use of H_2O_2 decreased the degradation of sulfamethoxazole to 1.9 min L^{-1}, of metoprolol to 5.8 min L^{-1} and of carbamazepine to 5.9 min L^{-1}. Diclofenac was eliminated immediately, thus it was not possible to calculate the degradation rate. The same results were obtained in measurements with WWTP effluent samples from another treatment plant (data not shown).

Matrix effects were observed for the degradation of the substance mixture in waste water effluent. The degradation rates were higher than those of the substance mixture in HPLC-water. For amidotrizoic acid the degradation rate was increased to 1.5 min L^{-1}, for carbamazepine to 10.6 min L^{-1}, for metoprolol to 4.7 min L^{-1} and for sulfamethoxazole to 1.7 min L^{-1}.

3.2 Cytotoxicity

To evaluate the cytotoxicity by using the MTT assay, the viability of the cells after exposure was calculated in relation to the viability of the negative control which was set to 100 %. According to DIN ISO 1993-5 there are three grades of cytotoxicity: not cytotoxic (100-81 %, grade 0), weakly to moderately cytotoxic (80-61 %, grade 1-2) and highly cytotoxic (60-0 %, grade 3). This DIN-ISO norm was used for the evaluation of the results in the present study.

In the water effluent samples from both WWTPs enriched with substance mixture (c_0 = 100 μg L^{-1}) no cytotoxic effects were detected after treatment with UV-light. In one WWTP only the sample which was treated with UV/ H_2O_2 showed very weak cytotoxic effects (79.4 ± 7.5 % cell viability) after 24 h of exposure. No cytotoxicity was observed before the oxidative treatment (data not shown).

Amidotrizoic acid, carbamazepine and diclofenac were tested as single compounds in HPLC-water (c_0 = 100 μg L^{-1}). For all the substances the viability was above 90 %, which means that no cytotoxicity occurred before or after the oxidative treatment. However, HPLC-water with the substance mixture treated for 60 min with UV/ H_2O_2 showed weak cytotoxicity effects (81 ± 6.5 % cell viability) after 24 h of exposure (Figure 2).

3.3 Genotoxicity

To detect the genotoxic potential of the tested samples the alkaline comet assay was performed. The prepared gels were stained with SYBR-green staining solution and analyzed manually using the program "Comet Assay IV" (Perspective Instruments, UK) under a fluorescence microscope at a 400x magnification. For the analysis 50 randomly distributed cells were counted and the olive tail moment (OTM) was analyzed [23] [24]. Results were statistically evaluated by using the Mann-Whitney-Test for two independent groups. Then the results of the treated cells were compared to the results of the negative control (cell culture medium). To express the significance, p-values were calculated: $p > 0.05$ not significant, $p \leq 0.05$ significant (*), $p \leq 0.01$ highly significant (**), $p \leq 0.001$ extremely significant (***).

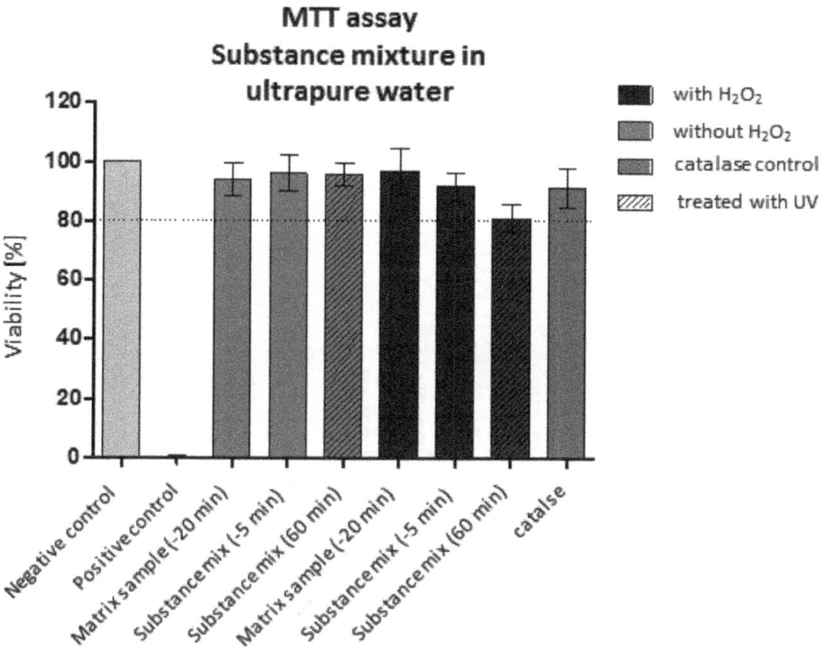

Figure 2 *Viability of CHO-9 cells after 24 hours of exposure to a pharmaceutical mix (amidotrizoic acid, carbamazepine, diclofenac, metoprolol and sulfamethoxazole) dissolved in HPLC-water (100 µg L^{-1}) before (-20, -5 min) and after (60 min) oxidative treatment with UV or UV/H2O.*

In the effluents with additional substance mixture of both tested WWTPs treated with UV no genotoxicity was detected. The OTMs of the samples were not significantly different from the OTM of the negative control, neither before nor after the oxidative treatment (data not shown). Also the single substances (amidotrizoic acid, carbamazepine, diclofenac) in HPLC-water treated with UV showed no significant change in the OTM compared to the negative control in the cells. So, no genotoxicity occurred in the cells after the exposure to the substances before and after oxidative treatment.

In contrast, the substance mixture treated with UV-light in HPLC-water for 60 minutes showed elevated genotoxic effects (OTM = 0.23) after 24 h of exposure, while the untreated sample was not genotoxic (Figure 3).

The genotoxicity of the samples treated with additional H_2O_2 could not be assessed because of high cell damage (cytotoxicity 80 %). To exclude genotoxic effects of catalase in samples treated with additional H_2O_2, catalase in HPLC-water was also tested and showed no genotoxic potential.

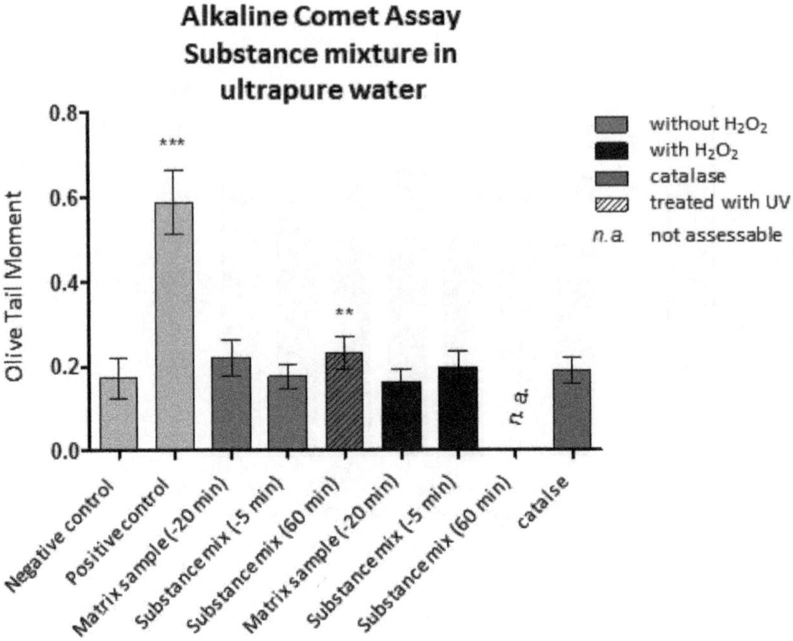

Figure 3 *Genotoxic potential of a pharmaceutical mix (amidotrizoic acid, carbamazepine, diclofenac, metoprolol and sulfamethoxazole) dissolved in HPLC-water (100 µg L^{-1}) before (-20, -5 min) and after (60 min) oxidative treatment with UV or UV/H2O in CHO cells after 24 hours of exposure. ** = $p \leq 0.01$, *** = $p \leq 0.001$, n.a. not assessable because of high cytotoxicity.*

4 CONCLUSIONS

In the present study, the efficiency of an UV-oxidation system using different matrices was investigated and toxicologically evaluated.

It was shown that the application of the substances as a mixture has only a slight effect on the degradation of the individual compounds. Furthermore, the photo-oxidative impact of the Hg-low pressure source was very low which is reflected in long degradation times.

The addition of H_2O_2 accelerated the degradation by the formation of hydroxyl-radicals. Those radicals are very reactive and facilitate the degradation of all substances in the aqueous solution. Also, other matrix substances are degraded by UV-radiation and OH-radicals which leads to competition reactions and to a slower degradation of the substances of interest [22]. The additional radical source leads to a higher amount of hydroxyl-radicals, thereby lowering the competition reaction by the matrix and accelerating the degradation. The matrix effects are also shown in the comparison of HPLC-water with effluent water. In HPLC-water no other substances are present, so that the leading substances amidotrizoic acid, carbamazepine, diclofenac, metoprolol and sulfamethoxazole were degraded faster than in the complex waste water effluent which contains suspended matter and particles leading to turbidity and absorption of some of the UV-light.

In the present study, the toxic effect of oxidation by-products after the UV-treatment of waste water effluents and enriched ultrapure water with and without additional H_2O_2 was compared. Samples treated only with UV showed no cyto- or genotoxic effects. Only in one case the UV treated substance mixture was significantly genotoxic. Slightly cytotoxic results (61-80% cell viability) were obtained for samples with the substance mixture treated with UV/H_2O_2, indicating that combination effects have led to the formation of some toxic by-products. No difference in toxicity was observed between enriched and original WWTP effluent samples and enriched HPLC-water samples. No cyto- or genotoxic effects were detected during treatment of single substances.
The results indicate that the treatment of a mixture of the test substances leads to oxidation by-products which have more toxic potential than those of the individual substances. Comparing the degradation of single substances with the elimination of the substances from the substance mixture in HPLC-water, no significant difference was observed; in both cases the substances were degraded within 60 min. So, the assumption is supported that different by-products are formed during the treatment process. However, toxic results occurred only for samples in ultrapure water, in waste water effluents no genotoxicity was observed after treatment with UV. In this case, matrix effects may lead to the formation of different by-products and also to a slower degradation of the trace substances. These results show that toxicological tests are necessary not only for single substances but especially for substance mixtures.
In summary, the results of the present study show that UV-oxidation can be used for the elimination of pharmaceuticals from WWTP effluents. The additional use of H_2O_2 resulted in higher elimination rates. However, the use of H_2O_2 also led to weak toxicological reactions in the cell culture tests.

References

1. IGF: Industrielle Gemeinschaftsforschung und –entwicklung, *AiF-FV* , 2006, **15852.**
2. O. A. H. Jones, N. Voulvoulis and J. N. Lester, *Water Research*, 2002, **36**, 5013.
3. M. Pedrouzo, F. Borrull, E. Pocurull and R. M. Marcé, *Water Air Soil Pollut*, 2010, **217,** 267.
4. J. Fick, H. Söderström, R. H. Lindberg, C. Phan, M. Tsklind and D. G. J. Larsson, *Environmental Toxicology and Chemistry*, 2009, **28**, 2522.
5. WHO: World Health Organization, *Guidelines for Drinking-Water Quality*, 1996, **2**.
6. H. H. Dieter, K. Götz, K. Kümmerer, B. Rechenberg and F. Keil, *Umweltbundesamt,* Berlin, 2010.
7. A. Wick, G. Fink, A. Joss, H. Siegrist and T. A. Ternes, *Water Research*, 2009, **43**, 1074.
8. A. Haiss, P. Hubner, J. Zipfl and K. Kümmerer, *vom Wasser*, 1998, **91**, 315.
9. H. Schröder, S. Osterhorn and V. Flöser, *gwf Wasser/Abwasser*, 1999, **140**, 20.
10. R. Sattelberger, *Umweltbundesamt GmbH,* Wien, 1999.
11. T. A. Ternes, *Water Research*, 1998, **32**, 3245.
12. T. A. Ternes, H. Siegrist and G. Berthold in Umweltbundesamt Berlin, 2005*,* **29**, 101.
13. BLAC: Bund/Länderausschuss für Chemikaliensicherheit, *Arzneimittel in der Umwelt – Bewertung der Untersuchungsergebnisse*, 2003.
14. B. Ferrari, R. Mons, B. Vollat, B. Fraysse, N. Paxeus, R. L. Giudice, A. Pollio and J. Garric, *Environmental Toxicology and Chemistry*, 2004, **23**, 1344.
15. K. Aktories, U. Förstermann, F. Hofmann and K. Starke, *Elsevier GmbH, Urban-Fischer-Verlag*, 2009.

16. T. Oppenländer, *WILEY-VCH Verlag*, 2003.
17. C. Gottschalk, J. A. Libra, A. Saupe, *WILEY-VCH Verlag*, 2010.
18. J. Tuerk, B. Sayder, A. Boergers, H. Vitz, T. K. Kiffmeyer and S. Kabasci, *Water Science & Technology – WST*, 2010, **61**, 985.
19. M. B. Heringa, D. J. H. Harmsen, E. F. Beerendonk, A. A. Reus, C. A. M. Krul, D. H. Metz and G. F. IJpelaar, *Water Research*, 2011, **45**, 366.
20. O. Ostling and K. J. Johanson, *Biochemical and Biophysical Research Communications*, 1984, **123**, 291.
21. N. P. Singh, M. T. McCoy, R. R. Tice and E. L. Schneider, *Experimental Cell Res.*, 1988, **175**, 184.
22. K. Höll, *Wasser*; de Gruyter, 2002, 161.
23. P. L. Olive, J. P. Banath and R. E. Durand, *Radiation Research*, 1990, **122**, 86.
24. M. de Boeck, N. Touil, G. de Visscher, P. A. Vande and M. Kirsch-Volders, *Mutat. Res.*, 2000, **469**, 181.

DISINFECTION BY-PRODUCT PRECURSOR REMOVAL AND FORMATION POST REVERSE OSMOSIS MEMBRANE FILTRATION

K. Doederer[1], M.J. Farre[1,2], H.S. Weinberg[3], E. Filloux[1] and W. Gernjak[1,2]

[1] The University of Queensland, Advanced Water Management Centre, Level 4 Gehrmann Building (60), 4072 St Lucia, Brisbane, QLD, Australia
[2] ICRA, Catalan Institute for Water Research, Scientific and Technological Park of the University of Girona, H2O Building, Emili Grahit 101, 17003 Girona, Spain
[3] University of North Carolina, Gillings School of Global Public Health, Department of Environmental Sciences and Engineering, Michael Hooker Research Center, Chapel Hill, NC 27599-7431, USA

1 INTRODUCTION

Due to increased water demand and decreased availability of traditional water sources, potable reuse of water from secondary effluent has become an important treatment alternative. Integrated membrane systems using low pressure membranes such as ultra/micro filtration followed by reverse osmosis (RO) membranes have developed to an industrial standard for potable reuse applications[1]. They have a high treatment efficiency for the removal of salts, metals, endocrine disrupting compounds, pharmaceuticals, personal care products and other emerging contaminants[2]. But a major limitation for RO membrane performance is membrane fouling, which can occur as inorganic (scaling), particulate, organic and biological fouling. It has been shown that fouling has adverse effects on membrane operation such as an increase in pressure drop, decrease in salt rejection and flux decline[3]. In order to specifically limit biofouling, the water is generally disinfected with chemical agents, such as chlorine or chloramines[3, 4]. Chloramines are preferably used as they chemically age membranes at a much slower rate than free chlorine. As an unintentional consequence of this treatment for biofouling prevention, disinfection by-products (DBPs) are formed by the reaction between organic and inorganic matter and disinfectants. DBPs are of concern as they have been found to pose potential public health risks in drinking water through routes of ingestion, inhalation and dermal adsorption[5, 6]. Secondary effluents as source water generally contain a high concentration of potential DBP precursors with a total organic carbon (TOC) ranging typically from 5 to 15 mg-C/L. Previous studies demonstrated DBP formation in disinfected secondary effluent before[7, 8] as well as after their passage through RO membranes[7, 9, 10]. Depending on their physico-chemical properties DBPs are able to pass RO membranes to a different extent[11]. Hence, there is the potential of DBP precursor passage by RO membranes. Studies have shown a good TOC removal by RO membranes; for example, a 97% removal with a 3 stage pilot-plant has been observed[12]. Dissolved organic nitrogen (DON) removal by RO membranes ranges between 78% to

99% with DON concentrations measured in the treated water as high as 0.12 mg-N/L[13]. DON may contain DBP precursors such as amino acids which can result in chloral hydrate (CH) and dichloroacetonitrile (DCAN) formation upon chlorination[14].

During advanced water treatment for potable reuse, membrane filtration is not the last treatment step. It is generally followed either by UV treatment to remove potentially formed nitrosamines and inactivated microorganisms, in particular viruses, that may have passed to the permeate, or by a UV/H_2O_2 advanced oxidation process (AOP) in order to remove trace organic chemicals and unknowns in addition to microorganisms. This treatment step is then typically followed by remineralisation and a final chlorination of the treated water to make the water suitable for transport.

UV and UV/H_2O_2 treatment of water with subsequent chlorination promoted the formation of CH and halonitromethanes (HNMs)[15-17]. Therefore, since DBP precursors are present in RO permeate, these processes may result in further DBP formation in the RO-treated water. In Queensland (Australia) strict guidelines for the production of high quality recycled water make the control of DBPs a major challenge. For example, DCAN and bromochloroacetonitrile (BCAN) are two nitrogen-containing DBPs (N-DBPs) regulated at 2 μg/L and 0.7 μg/L, respectively, in recycled water destined for augmentation of the State's drinking water supplies[18]. N-DBPs are of concern in drinking water as there is some evidence that they are more geno- and cytotoxic than the regulated DBPs without nitrogen such as haloacetic acids[19].

Currently, DBP precursor removal by full-scale RO treatment has not been extensively studied and the impact of UV/H_2O_2 on DBP formation upon final chlorination in RO permeate remains mostly unknown. This chapter presents the findings of a study to investigate the fate of DBPs after RO membrane filtration at full-scale advanced water treatment plants (AWTPs). Additionally, DBP precursor removal by full-scale RO membrane filtration is evaluated and the impact of UV/H_2O_2 on DBP formation in RO permeate upon final chlorination studied at the bench scale in the laboratory under controlled conditions is discussed.

2 METHODS AND RESULTS

2.1 AWTP Sampling Campaigns

Two advanced water treatment plants (AWTP 1 and AWTP 2) located in South East Queensland (Australia) with treatment trains shown in Figure 1 were sampled three times each. Samples were collected from AWTP 1 once in October 2010 and twice in November 2010, and from AWTP 2 twice in February 2011 and once in March 2011. RO permeate was collected from AWTP 2 for bench-scale experiments in July, August, and October, 2014. Analytical duplicate grab samples were collected in clear, acid-prewashed sample vials containing quenching agent and sealed with screw caps lined with PTFE/silicone septa. Due to the volatility of the DBPs, samples were collected headspace free, kept on ice for transport and stored at 4°C for extraction within 24 hours. The quenching agent for removal of residual disinfectant was ascorbic acid which had been added at twice the required stoichiometric concentration to the vials.

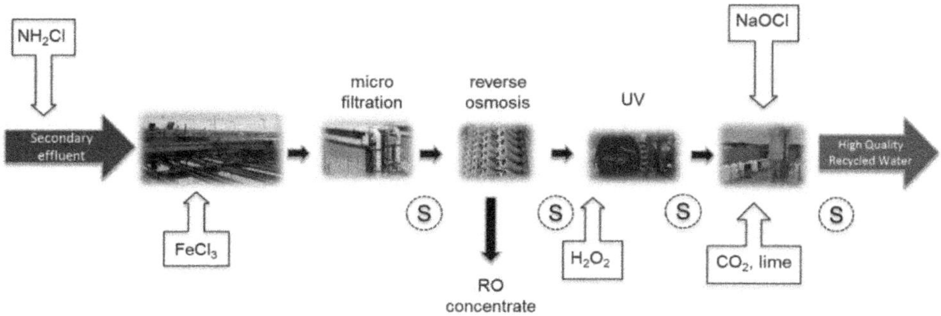

Figure 1 *Simplified scheme of the AWTPs studied. S corresponds to sampling points*

2.2 Analysis of Water Quality Parameters

Conductivity and pH were measured with a Mettler Toledo SevenEasy. UVA_{254} was measured using a Varian-Cary 50 bio spectrophotometer (Varian, Australia) with 1 cm path length. For the determination of the concentration in the NaOCl stock solution and during the preparation of monochloramine, free and total chlorine concentrations were measured in duplicate using the N,N-diethyl-p-phenylenediamine (DPD) colorimetric method following Standard Method 4500-ClG[20]. Molecular weight distribution of natural organic matter (NOM) fractions was evaluated using a Shimadzu high pressure liquid chromatograph equipped with SIL-20A HT autosampler and size exclusion column (Toyopearl HW-50S) with a 250 mm x 20 mm packing. The unit was connected to a SPD-M20A diode array detector and a Sievers 900 portable online TOC analyzer with inorganic carbon remover. The analyses used a 25 mM phosphate mobile phase (pH 6.85), 1 mL/ min flow rate, 1000 µL injection volume, and 100 min analysis time. Dissolved organic carbon (DOC) and total nitrogen (TN) was measured using a Shimadzu TOC-L total organic carbon analyser with a TNM-L total 132 nitrogen analyser and ASI-L autosampler. Ammonia, nitrite, total NOx and phosphate were measured on a Lachat flow injection analyzer as per the Lachat QuickChem method 31-107-06-1-A. DON was calculated as the difference between TN and the sum of NH_4–N, NO_3–N and NO_2-N. For each sample acceptable precision was a relative standard deviation of less than 3% from the analysis of three separate aliquots of sample. The detection limit for the DOC and TN was 0.1 mg/L while for NH_4-N, NO_3-N, and NO_2-N it was 1 µg/L.

2.3 Analysis of DBPs

After sample collection, duplicate 30 mL aliquots were pH-adjusted to 3.5 using 0.2 M sulphuric acid and liquid–liquid extracted with 3 mL methyl tert-butyl ether (MtBE) (containing 100 µg/L internal standard) in the presence of 10 g of pre-baked (at 500°C) sodium sulphate. Samples were vortexed for 1 minute. After settling for 5 minutes, 1.5 mL of the MtBE layer was transferred to gas chromatography (GC) vials for separate injection onto two different columns (DB1 and DB5) and for parallel double pulsed splitless injection (200ºC) on an Agilent 7890A gas chromatograph with ^{63}Ni electron capture detector (GC-ECD). Chromatographic separation of the DBPs was performed using the DB-5 column (30 m x 0.25 mm i.d., 1.0 µm film thickness, Agilent), while the DB-1 column (30 m x 0.25 mm i.d., 1.0 µm film thickness, Agilent) was used for analyte confirmation. The oven temperature program was 35°C for 25 min, ramped to 100°C at 2°C/min and held for 2 minutes, then ramped to 200°C at 5°C/min, and final ramp at 50°C/min to 280°C leading to a total run time of 81.1 minutes. The extraction and analysis procedures for DBPs were adapted from Weinberg *et al.*[21]. This long run allows for analysis of all the targeted DBPs from a single injection. The reporting limit for the haloacetamides was 0.5 μg/L and 0.1 μg/L for all other DBPs.

2.4 Experimental Methodology

2.4.1 UV irradiation. The UV/H_2O_2 experiments were carried out with a commercial collimated beam unit supplied by CalgonCarbon (Coraopolis, PA, USA) equipped with one 40 W low pressure (LP) UV lamp emitting mainly at 253.7 nm. All sample solutions were placed in a crystallisation dish and stirred during irradiation with a sample depth of 5 cm. UV irradiance was measured at the water surface using a F600-UV-VIS-SR detector (StellarNet, Tampa, FL, USA) with a W diffuser connected to a BLACK-Comet C50 radiometer (StellarNet, Tampa, FL, USA). The UV dose was calculated by multiplying the irradiance at the 253.7 nm by the Petri factor (0.97), water factor, radiometer sensor factor and reflection factor[22]. RO permeate was irradiated with either LP UV or LP UV in combination with H_2O_2.

2.4.2 DBP formation. The DBP formation potential (FP) was determined after UV or UV/H_2O_2 treatment applying chlorine to achieve a residual of 1.5±0.6 mg-Cl_2/L after 24 hours of contact time. In order to simulate anticipated full-scale WTP conditions, sample solutions were not buffered for the DBP FP tests. The chlorine residual of the chlorinated samples was measured using the DPD method with powder sachets obtained from the Hach corporation, Loveland, CO, USA). After UV/ H_2O_2 treatment, sufficient chlorine was added to the sample to quench residual H_2O_2 stoichiometrically (1 mg-H_2O_2/L requires 2.09 mg-Cl_2/L) and to achieve a residual of 1.5±0.6 mg-Cl_2/L[17, 23].

2.5 Fate of DBPs Post-Membrane Filtration at Full-Scale AWTPs

Six sampling campaigns were conducted at two AWTPs where samples were taken at the RO feed, the combined permeate, post UV/H_2O_2 treatment and the finished water after final chlorination. DBP concentrations measured at AWTP 1, where values correspond to the average of three sampling campaigns, are shown in Figure 2. As similar trends were observed for both AWTPs, data is only shown for AWTP 1. The sum of the concentrations of 4 THM species (THM4) namely, trichloromethane (TCM), bromodichloromethane (BDCM), dibromochloromethane (DBCM) and tribromomethane (TBM), ranged from 2.7 to 10.1 μg/L. The maximum individual THM species' values measured in this study (Figure 2) were significantly lower than the 100 to 200 μg/L Public Health Regulation Standard for TCM, TBM and DBCM for recycled water to augment a supply of drinking water in Queensland[18]. However, the maximum level of BDCM (3.8 μg/L) is close to the stricter limit of 6 μg/L[18]. A non-regulated THM detected was dichloroiodomethane (DCIM) at 0.7±0.2 μg/L which was poorly removed by the RO membranes (<40%). Since regulation limits are set at very low maximum levels for DCAN (2 μg/L) and bromochloroacetonitrile (0.7 μg/L) , the dihaloacetonitriles (DHANs) are of particular concern[18]. Average concentrations measured in the RO permeate for DCAN were 0.8±0.2 μg/L and 0.6±0.2 μg/L for BCAN. Another group of N-DBPs are the HNMs. Unfortunately, only trichloronitromethane (TCNM) but not tribromonitromethane (TBNM) could be analysed in the plant samples but from the ratio of TCM to TBM it can be deduced that TBNM would have substantially lower levels than TCNM assuming a similar mechanistic pathway and kinetics of formation. TCNM was found in the RO feed at a concentration of 0.3±0.1 µg/L, but was below the limit of detection in the RO permeate. The over 95% removal of CH by RO membranes was also seen for 1,1,1-trichloropropanone (1,1,1-TCP), which is in agreement with previous studies[11].

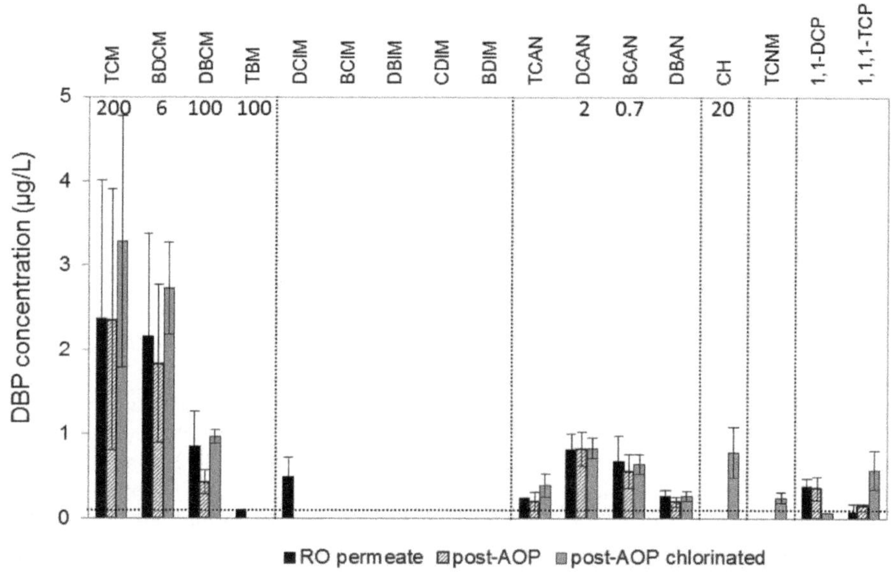

Figure 2 *DBP occurrence in the combined reverse osmosis permeate, post advanced oxidation and after final chlorination at AWTP 1. Error bars indicate the standard deviation between three sampling campaigns. Numbers indicate the regulation limit of the Queensland regulation for recycled water to augment a supply of drinking water[18] and the horizontal line the limit of detection.*
TCM = trichloromethane, BDCM = bromodichlormethane,
DBCM = dibromochloromethane, TBM = tribromomethane,
DCIM = dichloroiodomethane, BCIM = bromochloroiodomethane,
DBIM = dibromoiodomethane, CDIM = chlorodiiodomethane,
TCAN = trichloroacetonitrile, DCAN = dichloroacetonitrile,
BCAN = bromochloroacetonitrile, DBAN = dibromoacetonitrile,
CH = chloral hydrate, TCNM = trichloronitromethane,
1,1-DCP = 1,1-dichloropropanone, 1,1,1-TCP = 1,1,1-trichloropropanone

Figure 2 also plots DBP concentrations after UV/H_2O_2 treatment. While HANs and haloketones (HKs) do not appear to be affected by this treatment, it resulted in degradation of some brominated DBPs and DCIM. This is in accordance with findings of Chen and co-authors[24] who showed that the bromine- or iodine-substituted species compared to their chlorinated analogues are more susceptible to photolysis. While the RO membrane only poorly rejected DCIM, it could be removed by the UV/H_2O_2 treatment at the AWTP. The auxiliary effect for DBP removal during the UV/H_2O_2 oxidation was also observed for DBCM, TBM and DBAN, which showed a low rejection by the RO membranes.

The final treatment step in the advanced treatment train involves chlorination of the UV/H_2O_2 treated water. Concentrations of some of the DBPs increased after the final disinfection with chlorine. DBCM doubled in concentration while negligible increases were observed for the DHANs. CH and TCNM on the other hand, both of which were completely removed by the RO membranes, were formed again after the final chlorination step at concentrations of 0.8±0.3 μg/L and 0.2±0.1 μg/L, respectively. It should be noted that all the DBPs measured in the final treated water were below the regulated maximum levels, although concentrations for BDCM, DCAN and BCAN were close to the regulation limits for augmentation of drinking water supplies in Queensland (Australia).

2.6 DBP Precursor Occurrence Post-Membrane Treatment

RO membranes are expected to retain the bulk of DBP precursors in the influent water; for example a 97%[12] and 98%[25] removal of DOC has been reported. However, the additional DBP formation upon final chlorination observed at the AWTPs may be the result of low precursor concentrations remaining in the permeate. DOC, organic and inorganic nitrogen levels in RO permeate and their removals at both AWTPs are listed in Table 1.

A total nitrogen concentration of 0.5±0.1 mg-N/L was measured, which is a similar magnitude to the median level of 0.9 mg-N/L for raw waters of 28 drinking water treatment plants surveyed in the US[26]. The full-scale three stage RO treatment AWTP train was able to remove 95% of the DON with less than 0.1 mg-N/L remaining in the permeate. An important part of the DON content in water are amino acids[27], which are small molecules with molecular weights between 75 and 204 Da potentially able to pass RO membranes at least partially. The chlorination of amino acids was shown to form nitriles that with subsequent chlorine substitution can lead to dichloroacetonitrile (DCAN) formation[28]. Inorganic nitrogen species showed a lower removal of <84% (Table 1).

Table 1 *Water quality data for RO permeate of AWTP 1. Errors indicate the standard deviation between three measurements. NA = not available as samples for TN could not be analysed*

	AWTP 1		AWTP 2	
	concentration (mg/L)	removal (%)	concentration (mg/L)	removal (%)
DOC	0.07 ± 0.00	99	0.09 ± 0.00	98
DON	0.08± 0.04	95	NA	NA
NH_4-N	0.26 ± 0.05	79	0.34 ± 0.08	69
NO_3-N	0.23 ± 0.03	84	0.11 ± 0.05	78

DOC on the other hand was very well removed and concentrations were reduced by 99% to μg-C/L levels measured by online TOC analysers. Size exclusion chromatography in combination with organic carbon detection allowed for fractionation of the organic carbon present in the water. Due to the small effective pore size of RO membranes the higher molecular weight (MW) proteins and humic substances were completely removed. Comparison with injected standards shows the MW of the organic matter in the permeate to be less than 400 Da. At low µg/L concentrations but still detected were building blocks which reflect humic substance-like material of lower MW[29]. In accordance with literature[12] low MW acids and neutrals could also be observed. The smallest fraction of the organic matter detected were the low MW neutrals which include mono-oligosaccharides, alcohols, aldehydes, phenols and ketones[29]. Many of the small organic compounds are precursors which effectively react with chlorine to produce DBPs. Methyl ketones, for example, can undergo further chlorine substitution and result in haloketone formation while aldehyde oxidation can result in CH formation[28].

To probe for precursor removal of the DBPs measured, RO feed and RO permeate collected from the full-scale AWTP 2 and permeate produced at pilot-scale for water from AWTP 1 was either chlorinated or chloraminated to reach a residual of 1.5±0.2 mg-Cl_2/L after 24 hours of contact time. Of the DBPs measured, 98-99% of the measured DBP precursors preferentially reacting with chlorine could be removed by RO membrane filtration. This is slightly higher than the removal of measured DBP precursors preferentially reacting with chloramines ranging between 95-96% (Table 2).

Table 2 *Removal of measured DBP precursors (%) by comparing DBP formation potentials in RO feed and RO permeate. Error indicates the standard deviation of triplicate experiments*

AWTP 1		AWTP 2	
HOCl	NH_2Cl	HOCl	NH_2Cl
98.0 ± 0.03	94.7 ± 1.1	99.0 ± 0.8	96.4 ± 0.5

2.7 DBP Formation in RO Permeate Upon Post-Chlorination – Impact of UV/H_2O_2

During full-scale sampling campaigns, CH and TCNM were rejected below the limit of detection by the RO membranes but could be measured after chlorination of AOP treated permeate in both AWTPs. To investigate the impact of UV and UV/H_2O_2 on DBP formation upon chlorination post-membrane filtration under controlled conditions as described in section 2.4.1, RO permeate was irradiated with LP UV at the laboratory scale using a collimated beam unit. DBPs formed from chlorination with either no pre-treatment, pre-treated with UV only, or UV in combination with H_2O_2 are shown in Figure 3.

During UV treatment THM4 yield increased by up to 50% confirming previous work[16, 30] and our observations at full scale. It has been shown that UV irradiation can restructure dissolved organic matter (DOM) into more hydrophilic species[31] which are believed to be more reactive with chlorine[32] resulting in an increase in THM formation. The presence of 5 mg/L H_2O_2 during UV irradiation lead to an additional increase of 16%. A similar increase of 13% after UV/H_2O_2 treatment (at 1000 mJ/cm^2 + 10 mg/L H_2O_2) of media-filtered surface water was observed by Dotson and co-authors[23]. The hydroxylation of organic matter by the •OH radicals during the UV/H_2O_2 process increases the reactivity of DOM with chlorine and was shown to result in higher DBP formation[17, 30].

DHAN formation was not affected by UV irradiation alone but H_2O_2 addition resulted in a 33% lower DHAN formation. This indicates the destruction of DHAN precursors by the •OH radicals produced during the UV/H_2O_2. However, TN was not significantly reduced after UV/H_2O_2 demonstrating that reduced DHAN formation may also be attributed to precursor transformation. In fact, a recent study showed DON transformation after UV/H_2O_2 treatment reducing formation of N-DBPs from model precursors[17].

Similar to DHANs, 1,1,1-TCP was not affected by the UV treatment. However, the addition of 5 mg/L H_2O_2 resulted in almost doubling the formation. •OH radicals promote formation of small saturated alkyl groups in the form of methyl ketones[33] which are known to be precursors of haloketones. The generation of precursors may have contributed to an 84% increase in CH formation after UV/H_2O_2. •OH radicals produced during UV/H_2O_2 have been shown to generate aldehydes[34] which can react with free chlorine and form CH. Aldehydes are also known to be produced during UV irradiation and were proposed to explain the increase in CH in water treated with UV and followed by chlorination[15].

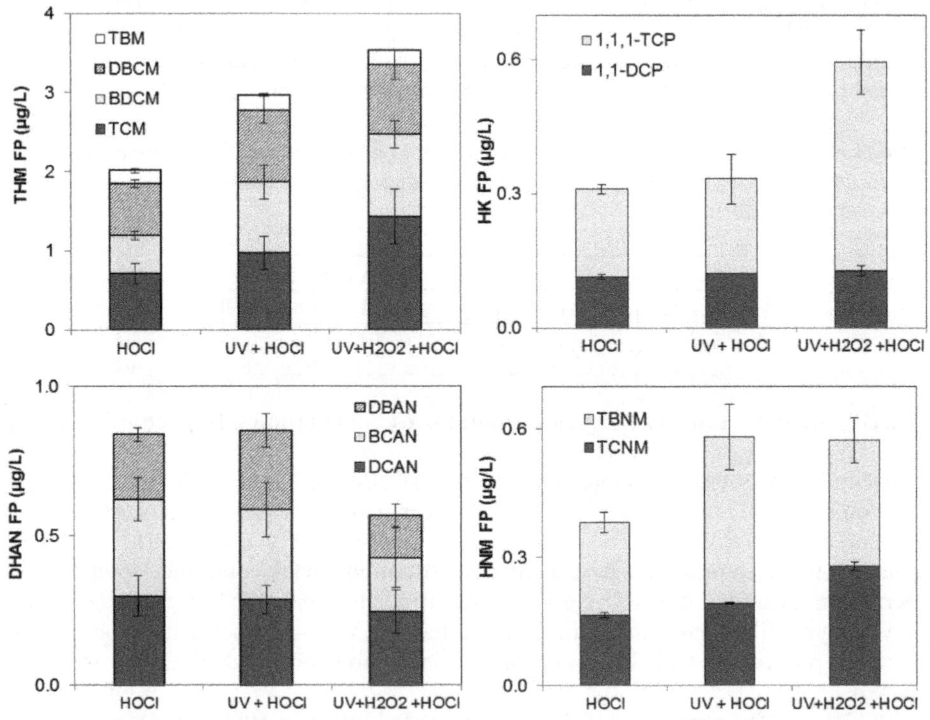

Figure 3 *DBPs formed from chlorination of RO permeate from AWTP2 after either no pre-treatment, pre-treated with UV (625 mJ/cm²) only or UV (625 mJ/cm²) in combination with H_2O_2 (5 mg/L). Error bars indicate the standard deviation between triplicate experimental samples*

TBNM formation increased by around 30% as a result of using UV treatment prior to chlorination while the addition of H_2O_2 in this process caused a slight shift in the TCNM to TBNM ratio without changing their total concentration. A previous study has shown TCNM formation during medium pressure UV treatment to be primarily formed by the nitrating agent •NO_2^- generated during nitrate photolysis[35]. In the RO permeate up to 0.26 mg-N/L of nitrate (Table 1) was present in the RO permeate that could be converted to •NO_2^- en route to HNM production. Nitrate mainly absorbs between 200-240 nm but the LP UV lamp used emits at 254 nm. It has been suggested that the small amount of irradiation absorbed at 254 nm may be sufficient to form nM amounts of TBNM during UV treatment followed by chlorination or chloramination[15] which may explain the 0.2 μg/L increase of TBNM (Figure 3).

3 CONCLUSIONS

This study investigated the fate and formation of DBPs post RO-membrane treatment at full-scale AWTPs. Precursors of the DBPs measured were well removed by the RO membranes with a better removal of precursors that react with chlorine (98-99%) compared to those that react with chloramines (95-96%). The UV/H_2O_2 treatment with subsequent chlorination resulted in increased formation THMs, 1,1,1-TCP, CH and HNMs. UV/H_2O_2 likely promoted the formation of various DBP precursors, such as methylketones, aldehydes and the generation of nitrite radicals through photolysis and subsequent nitration of precursors, which upon chlorination resulted in increased DBP formation. HANs were not affected by UV treatment but the addition of H_2O_2 may have contributed to precursor removal in the form of organic matter transformation. Even μg/L levels of DOC and DON as precursors present in the RO permeate were sufficient to produce TCNM and TBNM at concentrations between 0.3 and 0.4 μg/L after chlorination which is similar to those levels found in occurrence studies for drinking water in the US, Scotland and UK.

ACKNOWLDGEMENTS

The authors gratefully acknowledge the WateReuse Research Foundation (USA), Veolia Water (Australia) and Seqwater's financial, technical, and administrative assistance in funding and managing the project WaterReuse-10-18, within which this manuscript was developed. Dr Maria Jose Farre wants to acknowledge the European Commission for funding project 623711 under the FP7-PEOPLE-2013-IIF - Marie Curie Action: "International Incoming Fellowships" and Dr Wolfgang Gernjak acknowledges funding obtained from the Spanish Government for a Ramon y Cajal Research Fellowship (RYC-2012-12181).

References

1 J. Drewes and S. Kahn, in *Water Resources and Environmental Engineering : Water Quality and Treatment : A Handbook on Drinking Water (6th Edition)*, McGraw-Hill Professional Publishing New York, NY, USA 2011.

2 H. K. Shon, S. Vigneswaran and S. A. Snyder, *Environmental Science and Technology*, 2006, **36**, 327 - 374.

3 A. Matin, Z. Khan, S. M. J. Zaidi and M. C. Boyce, *Desalination*, 2011, **281**, 1-16.

4 H.-C. Flemming, *Experimental Thermal and Fluid Science*, 1997, **14**, 382-391.

5 S. D. Richardson, M. J. Plewa, E. D. Wagner, R. Schoeny and D. M. DeMarini, *Mutation Research*, 2007, **636**, 178-242.

6 D. L. Sedlak and U. von Gunten, *Science*, 2011, **7**, 42-43.

7 K. Doederer, W. Gernjak, H. S. Weinberg and M. J. Farré, *Water Research*, 2014, **48**, 218-228.

8 S. W. Krasner, P. Westerhoff, B. Chen, B. E. Rittmann and G. Amy, *Environmental Science and Technology*, 2009, **43**, 8320–8325.

9 K. L. Linge, J. W. Blythe, F. Busetti, P. Blair, C. Rodriguez and A. Heitz, *Separation and Purification Technology*, 2013, **104**, 221-228.

10 T. Fujioka, L. D. Nghiem, S. J. Khan, J. A. McDonald, Y. Poussade and J. E. Drewes, *Journal of Membrane Science*, 2012, **409-410**, 66-74.

11 K. Doederer, M. J. Farré, M. Pidou, H. S. Weinberg and W. Gernjak, *Journal of Membrane Science*, 2014, **467**, 195-205.

12 R. K. Henderson, R. M. Stuetz and S. J. Khan, *Water Science & Technology*, 2010, **62**, 2747-2753.

13 R. Merlo, J. Wong, V. Occiano, K. Sandera, A. Pai, S. Sen, J. Jimenez, D. Parker and J. Burcham, *Water Environmental Research*, 2012, **84**, 588-595.

14 S. W. Krasner, *Philosophical Transactions of the Royal Society*, 2009, **367**, 4077-4095.

15 B. A. Lyon, A. D. Dotson, K. G. Linden and H. S. Weinberg, *Water Research*, 2012, **46**, 4653-4664.

16 D. A. Reckhow, K. G. Linden, J. K. Hilla Shemer and G. Makdissy, *Journal of American Water Works Association*, 2010, **102**, 100-113.

17 W. Chu, N. Gao, D. Yin, S. W. Krasner and W. A. Mitch, *Environmental Science & Technology*, 2014, **48**, 12190-12198.

18 Public Health Regulation, Office of the Queensland Parliamentary Counsel Revised March 2014, Brisbane, Queensland, 2005.

19 M. Templeton, M. Nieuwenhuijsen, N. Graham, T. Bond, L. Huang and Z. Chen, *Review of the current toxicological and occurence information avialable on nitrogen-containing disinfection by-products*, Drinking Water Inspectorate, 2010.

20 APHA, *Standard methods for the examination of water and wastewater*, American Public Health Association, Washington, DC, 1995.

21 H. S. Weinberg, S. W. Krasner, S. D. Richardson and A. D. J. Thruston, *The Occurrence of Disinfection ByProducts (DBPs) of Health Concern in Drinking Water: Results of a Nationwide DBP Occurence Study*, U.S. Environmental Protection Agency, Athens, GA, 2002.

22 J. R. Bolton and K. G. Linden, *Journal of Environmental Engineering*, 2003, **129**, 209-215.

23 A. D. Dotson, V. S. Keen, D. Metz and K. G. Linden, *Water Research*, 2010, **44**, 3703-3713.

24 B. Chen, W. Lee, P. K. Westerhoff, S. W. Krasner and P. Herckes, *Water Research*, 2010, **44**, 3401-3409

25 J. E. Drewes, C. Bellona, P. Xu, G. Amy, G. Filteau and G. Oelker, *Comparing Nanofiltration and Reverse Osmosis for Treating Recycled Water*, AwwaRF Report 91212, 2008.

26 W. Lee, P. Westerhoff and M. Esparza-Soto, *Journal American Water Works Association*, 2006, **98**, 102-110.

27 A. D. Dotson, P. Westerhoff, B. Chen and W. Lee, in *Disinfection By-Products in Drinking Water*, American Chemical Society, 2008, vol. 995, pp. 274-288.

28 M. L. Trehy, Yost, R. A. and C. J. Miles, *Environmental Science and Technology*, 1986, **20**, 1117–1122.

29 S. A. Huber, A. Balz, M. Abert and W. Pronk, *Water Research*, 2011, **45**, 879-885.

30 C. Cotton, A. Dotson, S. Jousset, K. Linden and J. Collins, 2010.

31 W. Buchanan, F. Roddick, N. Porter and M. Drikas, *Environmental Science & Technology*, 2005, **39**, 4647-4654.
32 P. C. Singer, *Water Sci. Techn.*, 1999, **40**, 25-30.
33 G. Kleiser and F. H. Frimmel, *Science of the Total Environment*, 2000, **256**, 1-9.
34 S. Sarathy and M. Mohseni, *Water Research*, 2010, **44**, 4087-4096.
35 A. D. Shah, A. D. Dotson, K. G. Linden and W. A. Mitch, *Environmental Science & Technology*, 2011, **45**, 3657-3664.

TREATMENT OF BALLAST WATER ON BOARD SHIPS

Barbara Werschkun
Wissenschaftsbüro, Naunynstr. 30, D-10997 Berlin, Germany

1 INTRODUCTION

Shipping is essential for global trade, and ballast water is essential for shipping. Up to 50% of the deadweight tonnage of a ship consists of ballast water. The total quantity transferred around the world every year is estimated as 3-5 billion t.[1] Discharge of ballast water contributes substantially to the spread of aquatic invasive species, with irreversible damage to local ecosystems and economies.[2,3] In 2004, the international community adopted the International Ballast Water Management Convention, which establishes limits for the number of organisms in discharged ballast water (Table 1).[4] To achieve these standards, biocidal treatment of ballast water will be necessary, and ballast water management systems (BWMS) will need to be installed on >50,000 ships once the Convention enters into force. During the last 10 years, >30 BWMS have been developed based on physical and chemical methods for water purification.[5-7]

Treatment with strong chemical oxidants is very effective against most organisms but side reactions between the oxidant and dissolved organic material can generate disinfection by-products (DBPs). Many DBPs have been associated with genotoxicity, carcinogenicity, or toxicity to reproduction in animal experiments as well as epidemiological studies.[8,9] Quantity and speciation of DBPs depend on the composition of the treated water. In ballast water, the DBP pattern can be expected to reflect in particular the high bromide content of natural marine water (up to 65 mg/L). An evaluation of DBP data from the first 10 oxidative BWMS that obtained international approval was published in 2012.[10] This paper continues that earlier work, adding another 16 data sets that have become available in the meantime.

Table 1 *Future ballast water performance standards*

Organism type	*Limit value*
Organisms > 50 µm	<10 individuals/m^3
Organisms 10 - 50 µm	<10 individuals/mL
Vibrio cholerae O1/O139	<1 colony forming unit/100 mL
Escherichia coli	<250 colony forming unit/100 mL
Intestinal Enterococci	<100 colony forming unit/100 mL

2 MATERIAL AND METHODS

This evaluation is based on information given in the guidelines for approval of ballast water management systems and the international procedure for approval of BWMS that make use of active substances.[11,12] All evaluated documents are publicly available from the International Maritime Organization (IMO).[13] Table 2 provides an overview of all BWMS included in the analysis.

Data on DBP formation are usually obtained during the efficacy testing of BWMS in land-based test facilities with water tanks of 200-500 m^3. In order to demonstrate effective disinfection even under challenging conditions, the test water must contain high concentrations of organisms and abiotic matter. For instance, the concentration of dissolved organic carbon is required to be >1 mg/L and >5 mg/L for seawater (>32 psu [practical salinity units]) and brackish water (≤32 psu), respectively. Such values may not necessarily reflect the typical conditions during routine operation of BWMS on board ships. Samples are typically taken after the treated water has been stored for a period of five days. DBP analysis is generally performed in accredited laboratories by gas, liquid, or ion chromatography following internationally accepted protocols.

Table 2 *Overview of evaluated BWMS*

Name	*References*[a]	*Oxidant; dose*	*Pre-Treatment*	*Post-Treatment*
Sedna[b]	57/2/5; 57/2/10	PAA[c]; 21-25 mg/L	filtration	none
Electro-Clean	58/2; 58/2/7	Cl_2; 10 mg/L	none	$Na_2S_2O_3$
OceanSaver	58/2/1; 58/2/8	Cl_2; 2.5 mg/L	filtration	$Na_2S_2O_3$
CleanBallast	59/2; 59/2/16	Cl_2; 2.5 mg/L	filtration	$Na_2S_2O_3$
NK O3	59/2/3; 59/2/16	O_3; 2.5 mg/L	none	$Na_2S_2O_3$
Greenship	59/2/6; 59/2/19	Cl_2; 1 mg/L	hydrocyclone	none
TG	60/2/2; 60/2/12	Cl_2; 20 mg/L	filtration	Na_2SO_3
Special Pipe	61/2/2; 61/2/15	O_3; 3 mg/L	filtration	active carbon
BalClor	61/2/4; 61/2/15	Cl_2; 9.5 mg/L	filtration	$Na_2S_2O_3$
OceanGuard	61/2/7; 61/2/21	Cl_2; 2 mg/L	filtration	$Na_2S_2O_3$
HiBallast	62/2/5; 62/2/18	Cl_2; 9 mg/L	filtration	$Na_2S_2O_3$
Purimar	62/2/6: 62/2/18	Cl_2; 3 mg/L	filtration	$Na_2S_2O_3$
SiCure	62/2/10; 63/2/10	Cl_2; 6 mg/L	filtration	Na_2SO_3
ERMA First	63/2/1; 63/2/11	Cl_2; 10 mg/L	filtration	$NaHSO_3$
MICROFADE	63/2/2; 63/2/11	Cl_2; 2 mg/L	filtration	Na_2SO_3
AquaStar	63/2/3; 63/2/11	Cl_2; 10 mg/L	cavitation	$Na_2S_2O_3$
Neo-Purimar	63/2/6; 63/2/21	Cl_2; 10 mg/L	filtration	$Na_2S_2O_3$
JFE BallastAce	64/2/1; 64/2/7	Cl_2; 10 mg/L	filtration	Na_2SO_3
SmartBallast	64/2/2; 64/2/7	Cl_2; 11 mg/L	none	$Na_2S_2O_3$
Aquarius	65/2/1; 65/2/9	Cl_2; 10 mg/L	filtration	$NaHSO_3$
EcoGuardian	65/2/4; 65/2/19	Cl_2; 9 mg/L	filtration	$Na_2S_2O_3$
SKY	66/2; 66/2/7	PAA[c]; 21-25 mg/L	none	Na_2SO_3
Evonik BWMS	66/2/5; 66/2/10	PAA[c]; 21-25 mg/L	filtration	catalase; Na_2SO_3
Marinomate	67/2; 67/2/4	Cl_2; 10 mg/L	turbulence	$Na_2S_2O_3$
BlueZone	67/2/1; 67/2/4	O_3; 2.5 mg/L	none	$Na_2S_2O_3$
Kurita	67/2/2; 67/2/4	Cl_2; 20 mg/L	none	Na_2SO_3

[a]Cited are the document numbers of the IMO Marine Environment Protection Committee, which are publicly available online (www.imo.org);[12] [b]withdrawn from the market; [c]PAA = peracetic acid.

For the most frequently measured DBPs (trihalomethanes, haloacetic acids, haloacetonitriles, and bromate) the maximum values reported for each BWMS were grouped into for concentration ranges between <1 and >100 µg/L (Tables 3 and 4).

3 RESULTS AND DISCUSSION

As shown in Table 2, most, but not all, BWMS involve some mechanical pre-treatment such as filtration before the chemical oxidant is applied. Almost all BWMS destroy excess oxidant by means of a reducing agent such as thiosulfate, sulfite or bisulfite prior to discharge.

Chlorine is by far the most widely used oxidant in ballast water treatment. Although a few BWMS make use of chemical precursors such as NaOCl, $Ca(OCl)_2$ or sodium dichloroisocyanurate, the majority generate the primary oxidant by seawater electrolysis. The dose ranges employed vary considerably, ranging from 1 to 20 mg/L. For the more recent BWMS, a trend towards dosages around 10 mg/L can be recogized. Due to the high concentration of bromide in seawater, the primary oxidant hypochlorite is immediately consumed under formation of the secondary oxidant hypobromite/hypobromous acid,[14] which becomes decisive for DBP formation. This is reflected by the speciation of DBPs found in chlorine-based BWMS (Table 3). Tribromomethane and tribromoacetic acid are frequently found in the concentration range >100 µg/L, followed by dibromochloromethane, dibromoacctic acid, and dibromoacetonitrile, which are typically between 10 and 100 µg/L. Chlorinated species are rarely found in concentrations >1 µg/L, with the exception of chloroacetic acids, which in some cases were present in unexpectedly high concentrations. A closer look at these cases reveals that the control samples contained equally high concentrations of the same substances, probably due to contamination of the test water from near-by industries. Also unexpected are the rather high levels of bromate detected in some cases.

Table 3 *DBP data for BWMS based on chlorine [µg/L]*

		Number of data sets with				
	c_{max}	$c < 1$	$1 \geq c < 10$	$10 \geq c < 100$	$c \geq 100$	*No data*
Bromate	920.0	2/20	2/20	9/20	4/20	3/20
Tribromomethane	890.0	0/20	0/20	4/20	16/20	0/20
Dibromochloromethane	120.0	0/20	8/20	11/20	1/20	0/20
Bromodichloromethane	43.0	8/20	8/20	3/20	0/20	1/20
Trichloromethane	32.0	12/20	3/20	3/20	0/20	2/20
Tribromoacetic acid	970.0	2/20	2/20	8/20	6/20	2/20
Dibromoacetic acid	120.0	0/20	2/20	16/20	1/20	1/20
Bromoacetic acid	191.0	2/20	11/20	4/20	1/20	2/20
Dibromochloroacetic acid	32.7	4/20	5/20	5/20	0/20	6/20
Bromochloroacetic acid	40.0	4/20	13/20	2/20	0/20	1/20
Bromodichloroacetic acid	28.0	8/20	6/20	3/20	0/20	3/20
Chloroacetic acid	495.0	9/20	4/20	2/20	2/20	3/20
Dichloroacetic acid	5.8	11/20	8/20	0/20	0/20	3/20
Trichloroacetic acid	45.1	11/20	5/20	1/20	0/20	3/20
Dibromoacetonitrile	133.0	5/20	2/20	8/20	1/20	4/20
Bromoacetonitrile	16.0	1/20	2/20	1/20	0/20	16/20
Bromochloroacetonitrile	3.9	5/20	3/20	0/20	0/20	12/20
Chloroacetonitrile	<1	3/20	0/20	0/20	0/20	17/20
Dichloroacetonitrile	6.9	10/20	1/20	0/20	0/20	9/20
Trichloroacetonitrile	1.5	6/20	1/20	0/20	0/20	13/20

The disproportionation of hypobromite to bromide and bromate is usually very slow. However, it has been shown that metal oxides can catalyse this reaction.[15] Elevated levels of bromate may therefore indicate the presence of metal ions due to the corrosive action of hypochlorite and related species in the test system.

Table 4 *DBP data for BWMS based on a) ozone and b) peracetic acid [μg/L]*

	a) Ozone:			*b) Peracetic acid:*		
	NK O3	*Special Pipe*	*BlueZone*	*Sedna*	*SKY*	*Evonik*
Bromate	<1	100	39.8	no data	no data	no data
Tribromomethane	78.5	170	92.8	1.5	1.4	360
Dibromochloromethane	<1	4.2	4.0	<1	<1	36
Bromodichloromethane	<1	<1	4.1	<1	<1	2.3
Trichloromethane	<1	<1	<1	<1	<1	<1
Tribromoacetic acid	no data	no data	5.2	no data	<5	<50
Dibromoacetic acid	7.9	no data	30.6	no data	8	230
Bromoacetic acid	no data	no data	<1	no data	<5	36
Dibromochloroacetic acid	no data	no data	26.9	no data	<5	<50
Bromochloroacetic acid	<1	no data	2.6	no data	<1	<10
Bromodichloroacetic acid	no data	no data	1.0	no data	no data	<10
Chloroacetic acid	11.4	no data	<1	no data	4	<10
Dichloroacetic acid	1.1	no data	<1	no data	<1	<10
Trichloroacetic acid	51.4	no data	<1	no data	no data	<10
Dibromoacetonitrile	no data	no data	673.9	no data	no data	no data
Bromoacetonitrile	no data	no data	23.5	no data	no data	no data
Bromochloroacetonitrile	no data	no data	no data	no data	no data	no data
Chloroacetonitrile	no data	no data	no data	no data	no data	no data
Dichloroacetonitrile	no data	no data	no data	no data	no data	no data
Trichloroacetonitrile	no data	no data	no data	no data	no data	no data

Ozonation of bromine-containing water also produces hypobromite as the major secondary oxidant, which can be further oxidised to bromate in the presence of excess ozone.[16] When looking at the available data from ozone-based BWMS (Table 4), formation of bromate is not consistently observed, even though all systems use similar ozone dosages. A possible reason might be the different composition of the test water, which could in some cases contain additional ozone-depleting components. Much less data are available for DBP formation with ozone than with chlorine. This refers not only to the number of BWMS, but also to the comprehensiveness of the DBP analysis performed for each BWMS. Nevertheless, it becomes apparent that the same chemical species predominate in the DBP mixture (tribromomethane, dibromoacetic acid, dibromoacetonitrile).

Peracetic acid (PAA) is employed in BWMS in form of the commercial preparation PERACLEAN OCEAN (14-17 wt % PAA, 13-15 % H_2O_2, 24-29 % acetic acid). A recent experimental study investigated in detail the chemical reactions associated with the application of this oxidant in saline waters.[17] Initially, PAA reacts with bromide under formation of the secondary oxidant HOBr. Subsequently, excess H_2O_2 rapidly reduces HOBr to bromide and the formation of DBPs is thus prevented. DBP analyses for two PAA-based BWMS seem to confirm these findings (Table 4). A third BWMS, however, presents a completely different picture, with the formation of brominated DBPs being rather similar to chlorine or ozone. The explanation can be found in the overall treatment sequence employed in this specific case. Simultaneously with the injection of PERACLEAN OCEAN into the ballast line, the enzyme catalase is added in order to destroy H_2O_2 (to prevent its discharge into the marine environment). Thus, the lifetime of HOBr is substantially increased and it can play the same role in the formation of DBPs as in the other oxidative BWMS.

4 CONCLUSION

The growing database on DBP formation during ballast water treatment confirms previously suggested patterns. DBP formation under these conditions is largely controlled by bromine. The predominating chemical species are tribromomethane, tribromo- and dibromoacetic acid, dibromoacetonitrile and bromate. All of these are suspected to pose long-term risks to human health. Measures to limit DBP formation that are commonly employed in water treatment plants on the land are hardly feasible for ship-board installation due to spatial and operational constraints. Further efforts should therefore be directed towards monitoring the potential for human exposure and establishing suitable protocols for exposure reduction, in particular in shipping-related workplaces.

References

1 O. Endresen, H. L. Behrens, S. Brynestad, A. B. Andersen and R. Skjong, *Mar. Pollut. Bull.*, 2004, **48**, 615.
2 J. T. Carlton, *Oceanogr. Mar. Biol.*, 1985, **23**, 313.
3 J. T. Carlton and J. B. Geller, *Science*, 1993, **261**, 78.
4 *International Convention for the Control and Management of Ships' Ballast Water and Sediments,* BWM/CONF/36, International Maritime Organization, London, 2004.
5 M. Gregg, G. Rigby and G. M. Hallegraeff, *Aquatic Invasions*, 2009, **4**, 521.
6 *Ballast water treatment technologies and current system availability*, Lloyds Register, London, September 2012.
7 B. Werschkun, S. Banerji, O. C. Basurko, M. David, F. Fuhr, S. Gollasch, T. Grummt, M. Haarich, A. N. Jha, S. Kacan, A. Kehrer, J. Linders, E. Mesbahi, D. Pughiuc, S. D. Richardson, B. Schwarz-Schulz, A. Shah, N. Theobald, U. von Gunten, S. Wieck and T. Höfer, *Chemosphere*, 2014, **112**, 256.
8 S. D. Richardson, M. J. Plewa, E. D. Wagner, R. Schoeny and D. M. DeMarini, *Mutat. Res.*, 2007, **636**, 178.
9 C. M. Villanueva, K. P. Cantor, S. Cordier, J. J. K. Jaakkola, W. D. King, C. F. Lynch, S. Porru and M. Kogevinas, *Epidemiology*, 2004, **15**, 357.
10 B. Werschkun, Y. Sommer and S. Banerji, *Water Res.*, 2012, **46**, 4884.
11 *Guidelines for approval of ballast water management systems (G8)*, Resolution MEPC.174(58), International Maritime Organization, London, 2008.
12 *Procedure for approval of ballast water management systems that make use of active substance (G9)*, Resolution MEPC.169(57), International Maritime Organization, London, 2008.
13 *How and where to find IMO information*, International Maritime Organization, London, 2010.
14 K. Kumar and D. W. Margerum, *Inorg. Chem.*, 1987, **26**, 2706.
15 C. Liu, U. von Gunten and J.-P. Croué, *Environ. Sci. Technol.*, 2012, **46**, 11054.
16 U. von Gunten, *Water Res.*, 2003, **37**, 1469.
17 A. D. Shah, Z.-Q. Liu, E. Salhi and U. von Gunten, *Environ. Sci. Technol.*, 2015, DOI: 10.1021/es503920n.

Disinfection By-products Monitoring and Analysis

TRACING GENOTOXIC N-DBPs AFTER MEDIUM PRESSURE UV WATER TREATMENT BY NITROGEN LABELLING

A.J. Martijn[1], A. Kolkman[2], M. Aljammaz[3], D. Vughs[2], K.A. Baken[2]

[1] PWN Water Supply Company North Holland, P.O. Box 2046, 1990 AA, Velserbroek, The Netherlands
[2] KWR Watercycle Research Institute, P.O. Box 1072, 3430 BB Nieuwegein, The Netherlands
[3] Wageningen University, Division of Toxicology, P.O. Box 8000, 6700 EA Wageningen, The Netherlands

1 INTRODUCTION

Organic contaminant control is a major issue in advanced drinking water treatment. The presence of pesticides, endocrine disruptors and pharmaceuticals has caused PWN Water Supply Company North Holland in The Netherlands to implement multiple barriers in their surface water treatment plants. For improved organics control, a combination of advanced oxidation by medium pressure ultraviolet light in combination with hydrogen peroxide (MP UV/H_2O_2) treatment followed by granular activated carbon (GAC) filtration was implemented in 2005. The design of these processes was based on bench-scale and pilot-scale evaluations. In bench-scale experiments, dose-response relationships were established for these organic pollutants, and degradation at specific process conditions (i.e., 0.54 kWh/m^3 or 600 mJ/cm^2; 6 mg/L H_2O_2) was confirmed at the pilot-scale. Process conditions were based on 80 percent removal of atrazine, a pesticide. For a range of commonly found pesticides, such as 2,4-D, diuron and nicosulfuron, 80 percent degradation was also observed[1].

For medium pressure ultraviolet (MP UV) technology, formation of the inorganic reaction product, nitrite, in nitrate rich waters has been widely reported[2,3]. The formation of harmful organic reaction products by UV based photolysis and advanced oxidation process (AOP) of organic micropollutants in synthetic water was investigated by many researchers[4,5]. However, in natural waters, formation of harmful organic reaction products has not been well documented yet, although Reckhow[6] proposed a mechanism for the formation of nitro organics due to photo nitration. Thorn and Cox[7] show the incorporation of nitrogen into the organic matrix under extreme conditions.

Given the numerous reaction products, which include metabolites and byproducts that are formed by oxidation or hydroxyl radical oxidation of water matrix constituents and targeted organic micropollutants, water samples were tested using a bioassay. In this research, the high throughput Ames fluctuation test was used to screen water samples for MP UV induced genotoxicity without knowing the properties, concentrations, and number of the compounds responsible for a response[8]. No Ames test response was found before and after low pressure UV treatment[9]; however, MP UV/H_2O_2 treatment utilizing synthetic quartz sleeves that did not cut-off UV light at wavelengths below 240 nm caused a significant response[9,10]. Bench scale UV irradiation experiments with reconstituted water containing NOM in the presence and absence of nitrate were conducted. MP UV irradiation of NOM did not show an increase in the Ames test response, while MP UV irradiated water samples with NOM in the presence of nitrate did show a response[11].

Suzuki[12] showed that three aromatic amino acids, tryptophan , phenylalanine and tyrosine, were found to give direct acting and frameshift mutagens by irradiation (at wavelength >300 nm) in aqueous nitrite solution, using the Ames test with and without S9 mix using Salmonella typhimurium TA98 and TA100. Other tested amino acids, threonine, cysteine, glutamic acid, arginine, and proline did not give a response in the Ames test while irradiated in a nitrite aqueous solution. Eutrophic surface water or waste water treatment effluent may well contain amino acids as part of their organic matter. MP UV applications in water treatment may apply the lower wavelengths (i.e., 200 to 240 nm) as well, where the UV absorbance of nitrite and nitrate is even stronger, resulting in higher formation of nitrogenous radicals[13]. Therefore the formation of nitrated aromatic compounds by MP UV photolysis of nitrate under practical water treatment conditions may be the cause for the observed Ames test response because nitroaromatics are well known genotoxic compounds[14].

To quantify the total mutagenic effect of MP UV irradiation of water containing nitrate and NOM, the preceding studies converted the Ames test response into 4-nitroquinoline oxide (4-NQO) equivalents, a known genotoxic compound commonly used as a positive control in the Ames test. This showed that MP UV treatment resulted in the formation of 120 ng/L 4-NQO equivalents in the water extracts[15,16]. To evaluate the hazard, a preliminary risk evaluation which is based on the Threshold for Toxicological Concern (TTC)[17,18] was applied. The TTC approach establishes a human exposure threshold value below which no appreciable health risk occurs[19]. The TTC concept may be applied when the presence of a new contaminant in food is observed, for which no toxicological information is available. It could also be useful in setting priorities for testing large groups of functionally similar chemicals to which exposure is generally very low, such as flavourings[17] or deriving target values for drinking water contaminants[18]. The proposed TTC for genotoxic compounds in food is 75 ng/L[19], whereas for drinking water, a target value of 10 ng/L was derived[18]. However, the TTC prescribes a compound specific risk assessment when nitro-aromatic compounds may be present.

Only after identification and quantification of formed nitrated organics from the organic water matrix by MP UV treatment under practical conditions, can risk assessment for these compounds be carried out to quantify the possible health impact. Given the complex nature of the natural organic matter, the randomness of involved (nitro) radical oxidation processes and the low concentration of the formed compounds, an innovative approach was developed to identify and quantify the formed nitro-organic compounds. Current analytical techniques are not able to detect the change in inorganic nitrogen balance due to the nitrate photolysis, or to detect the increase of dissolved organic nitrogen content due to nitro- and / or nitroso radical reactions with the organic water matrix. The hypothesis is that inorganic nitrogen, originating from nitrate, is incorporated in organic molecules from the natural organic matter. The proposed mechanism of nitration or nitrosation is via nitro- and nitrosoradical formation from nitrate and potentially nitrite photolysis, as described by Mack[2] and Goldstein[13].

To overcome these analytical drawbacks, reconstituted water using nitrate with nitrogen isotope ^{15}N and Pony Lake NOM from the International Humic Substances

Society was irradiated with MP UV light. The same experiment was conducted with nitrate with nitrogen ^{14}N. Nitrated and / or nitrosated organic compounds that were formed due to nitrate photolysis were detected via a mass balance difference, using LC-MS-Orbitrap in combination with statistical software. To identify the detected masses and associated chemical composition of the formed nitrogenous reaction products, commercially available standards are used[20,21]. This paper illustrates the iterative process between screening results, identifying an effect and chemical analytics, all of which is necessary to evaluate the potential risk.

2 METHODS AND RESULTS

2.1 Ames test response full scale MP UV/H_2O_2 treatment at wtp Heemskerk

The Ames test is an in vitro test using a biological assay to assess the genotoxic potential of chemical compounds. Since the 1970s, this test has been used as a standard method for detecting potential genotoxic and carcinogenic compounds. A bacterium is used in the assay as an indicator for DNA damage[22]. The bacterium used in this test is Salmonella typhimurium which is genetically modified, preventing the bacterium from synthesizing histidine (his), which is required for growth. The mutagenic potential is assessed by exposing his-organisms to different concentrations of chemicals by which the genotoxic effect of the chemical compounds gives the bacteria the ability to grow without histidine, and form colonies (reverse mutation). To perform the Ames test on water samples, sample concentration is required. For this research, samples were concentrated up to a factor of 20,000, using OASIS HLB solid phase extraction. The Ames fluctuation test was obtained from Xenometrix GmbH (Allschwil, Switzerland), uses a liquid culture instead of Agar plates and uses small sample quantities to obtain faster results. A mixture of rat liver enzymes is used in Ames test to mimic metabolic activation, referred to as +S9[8]. All Ames tests were performed including a procedure control and a positive control, using the protocol described by Heringa et al.[9]

The Ames test response was measured in concentrated water samples from a full scale MP UV/H2O2 treatment installation of PWN (Figure 1)[10], which had five UV reactors in series. Each UV reactor was equipped with sixteen 12 kW MP UV lamps in quartz sleeves allowing utilization of the full spectrum. The influent of the MP UV/H_2O_2 treatment installation was conventionally pretreated eutrophic surface water from the IJssel Lake, (DOC: 3.5 mg C/L; NO_3: 12 mg NO_3/L), with 6 mg/L H_2O_2. The applied UV dose was 600 mJ/cm^2, equalling 0.54 kWh/m^3 in this water type. Each UV reactor added one-fifth of the total delivered UV dose. Figure 1 illustrates that with increasing UV dose, the formation of genotoxic compounds increases. The Ames test response in the 20,000 fold concentrated water samples after the full treatment train exceeded the positive control, indicating a hazard, which required further investigation.

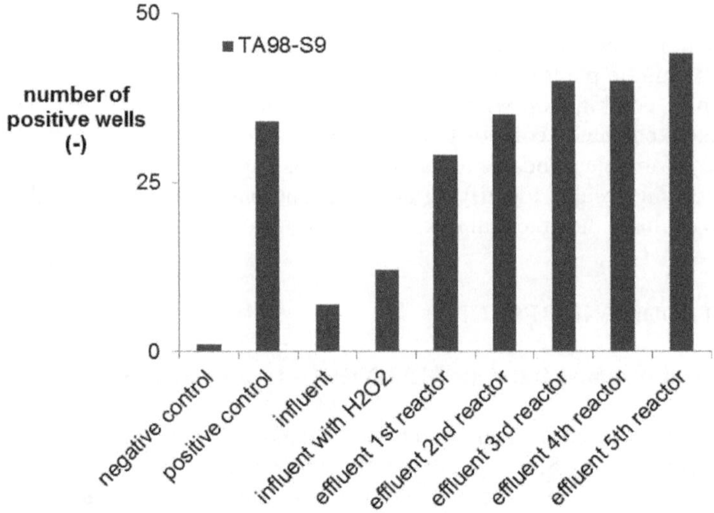

Figure 1 *Ames test response (TA98-S9) at several stages in the PWN wtp Heemskerk MP UV/H_2O_2 treatment train (0.54 kWh/m^3) (November 2010) in concentrated water samples (cf 20,000). Reprinted with permission of International Ozone Association. 'UV and UV/H_2O_2 Treatment: The Silver Bullet for By-product and Genotoxicity Formation in Water Production', Martijn, A.J., Kruithof, J.C., Ozone: Science & Engineering, 2012, Taylor & Francis Ltd, reprinted by permission of the publisher (Taylor & Francis Ltd, http://www.tandfonline.com).*

2.2 Water quality parameters

2.2.1 Natural Organic Matter. The size exclusion chromatography-liquid chromatography-organic carbon detection-organic nitrogen detection (LC-OCD-OND) method is a powerful analytical tool to characterize organic matter and to observe relative differences in dissolved organic carbon (DOC). The SEC broadly groups the organics into five fractions: biopolymers, humics, building blocks, LMW acids and LMW neutrals (in the order of retention time). Two detectors, organic carbon detector (OCD) and ultraviolet detector (UVD) detect the organics. The OCD spectrum is used to determine the total mass of organic carbon, whereas the UVD spectrum detects only the UV adsorbing species (i.e., double bond carbon)[23]. The DOC concentration in the raw water source shows little seasonal fluctuation and is between 5.5 and 6.0 mg C/L. The main cause for fluctuation in the raw water DOC concentration is the algal related component, represented in Figure 2 by the LC-OCD-OND fraction biopolymers. Also after pretreatment, the fluctuation in biopolymers is still visible, although less attenuated.

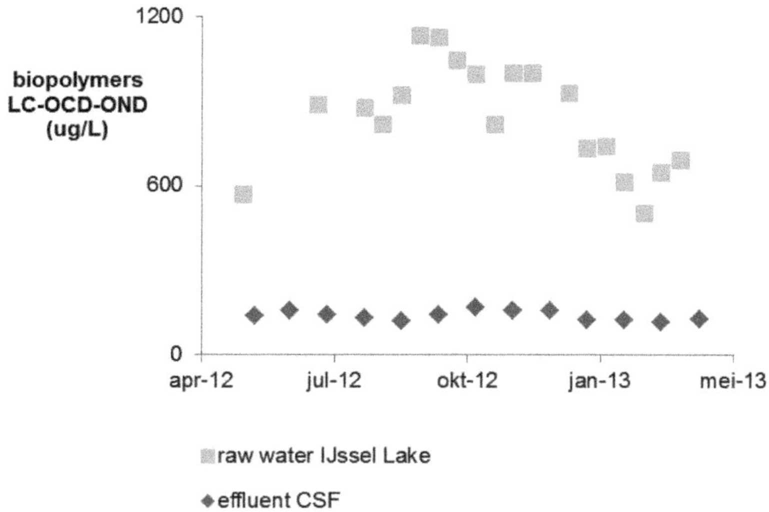

Figure 2 *Typical seasonal fluctuation of the fraction biopolymers (LC-OCD-OND method) in natural organic matter in raw water from the IJssel Lake and effluent from the conventional pretreatment.*

Relating the character of the natural organic matter to the findings of Suzuki et al.[12], amino acids in eutrophic surface water may come from biological activity, such as algae blooms. The biopolymer fraction in the LC-OCD-OND analytical method represents the high molecular weight organic carbon, typically not of terrestrial or aquatic humic substances. The seasonal fluctuation in the raw water follows the trend of algae blooms (data not shown). When the samples for the full scale Ames test data (Figure 1) were taken, the biopolymer concentration was at a maximum in the raw water.

2.2.2 Nitrite Formation by Nitrate Photolysis. Nitrate photolysis with its stable end product nitrite is an important MP UV byproduct, as well as an indication for the involved nitrogenous radical formation. Nitrate has two absorption bands in the UV region, one in the near UV region from 260 nm to 350 nm with its maximum at 300 nm and a much more intensive band below 260 nm with its maximum at 200 nm[24]. The emission spectrum of MP UV lamps shows a substantial emission in the wavelengths (<240 nm) where nitrate has a high molar absorption coefficient (Figure 3)[16]. Especially at these emitted lower wavelengths, UV contributes strongly to the nitrite formation[2].

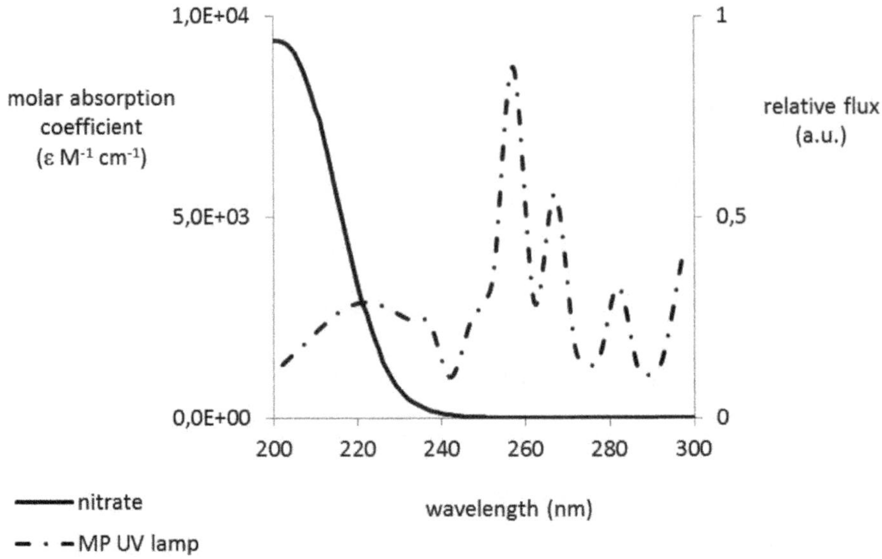

Figure 3 *Emission spectrum of a medium pressure UV lamp and the molar absorption coefficient between 200 and 300 nm of nitrate in demineralized water*

The major reaction product of nitrate photolysis is peroxynitrite, and it produces nitroso-, nitro- and hydroxyl radicals as intermediates[13]. These radicals may cause reaction product formation by interaction with organic compounds. The stable inorganic end product of nitrate photolysis by UV light is nitrite. However, the impact of UV photolysis of nitrate on harmful organic reaction product formation and the possible health impact is not (well) documented yet, although the hazard of nitrate photolysis was predicted[2,6]. When the Ames test results presented in Figure 1 were obtained, the nitrate level was 12 mg NO_3/L and 380 ug NO_2/L was analysed in the effluent of the sampled full scale MP UV/H_2O_2 treatment train.

2.3 MP UV bench scale experiments using a collimated beam apparatus

To further investigate the mechanisms involved in the relationship between nitrate photolysis, the presence of natural organic matter and the formation of genotoxic compounds, all as measured by the Ames test, bench scale experiments were performed. The irradiations for bench scale experiments were performed in a collimated beam apparatus equipped with a 3 kW medium pressure Hg lamp (Trojan Rl-01) with a well-defined spectral output. The light passed through a synthetic quartz window before reaching the solution, allowing utilization of the low wavelengths radiation. The irradiance readings were taken with an IL research Radiometer, with a SED240/W detector calibrated every 2 nm within the 200-320 nm range. Absorption spectra were recorded in a 1 cm path length quartz cell, with a Hach-Lange DR5000 spectrophotometer. The irradiation time was accurately monitored for each sample and the UV dose delivered to the solution was calculated using the UV dose calculation method developed by Bolton and Linden[25]. The collimated beam experiments have been performed under conditions to guarantee representative results for the full scale UV/H_2O_2 installation. For the irradiations, 60 x 35 mm crystallizing dishes were employed with a sample volume of 55 mL. The irradiation path length in the sample was approx. 1.95 cm. The crystallizing dish was placed on a stirring plate, and the solution was continuously stirred with a small magnetic stir bar.

2.4 Ames test response in reconstituted water containing NOM with and without nitrate

Collimated beam experiments with synthetic water containing IHSS Pony Lake NOM, in the presence and absence of nitrate, were carried out under MP UV photolysis conditions (UV dose 600 mJ/cm^2). IHSS Pony Lake NOM is an example of a microbial end member environment. Microbiologically derived NOM, such as from Pony Lake, has a higher aliphatic carbon content and is rich in organic nitrogen, attributable to carbohydrate, protein and lipid precursor materials[26]. The IHSS Pony Lake NOM content was 2.5 mg/L C and the nitrate concentration was 2.71 mg NO_3/L. At the MP UV photolysis condition, 565 μg NO_2/L was formed.

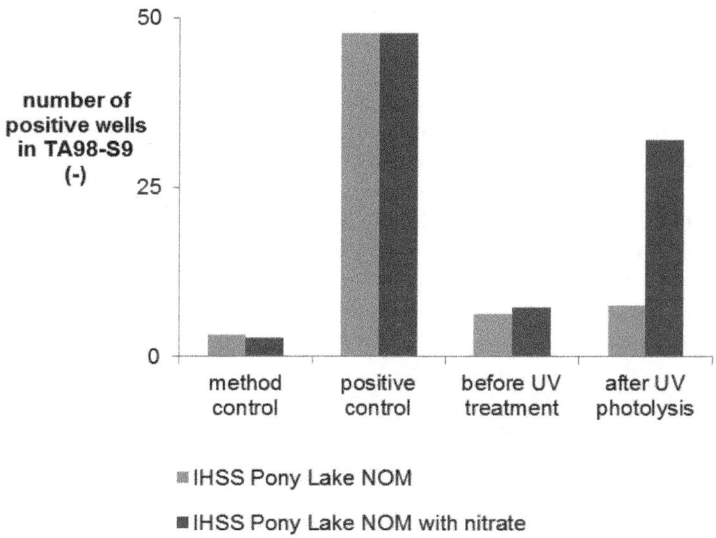

Figure 4 *Ames test response (TA98-S9) in concentrated water samples in reconstituted water containing 2.5 mg C/L IHSS Pony Lake NOM with and without nitrate (12 mg NO3/L), before and after MP UV treatment (600 mJ/cm2)*

The Ames test results (Figure 4) show no effect of MP UV irradiation (600 mJ/cm^2) when only NOM is present. In the presence of both NOM and nitrate, MP UV irradiation with a UV dose of 600 mJ/cm^2, a substantial response in the Ames test is observed, indicating formation of genotoxic compounds. The Ames test response observed after MP UV irradiation may be caused by multiple nitrated and nitrosated compounds at different concentrations. In order to quantify the total mutagenic effect of MP UV irradiation of surface water containing nitrate and NOM, the Ames test response was converted into 4-nitroquinoline oxide (4-NQO) equivalents, a known genotoxic compound commonly used as positive control in the Ames test. This showed that MP UV treatment of surface water resulted in the formation of 120 ng/L 4-NQO equivalents in the water extracts[15]. Conversion of the Ames test response in 4-NQO equivalents for the Ames test results in Figure 4 resulted in formation of 575 ng/L 4-NQO equivalents[16].

The formation of a large number of similar, unknown genotoxic compounds at very low concentrations, from which toxicity data are lacking, requires a preliminary risk assessment. The TTC concept provides a preliminary risk assessment and prescribes a compound specific risk assessment when the concentrations exceed the level that is considered to pose a negligible human health risk. The observed 4-NQO levels after MP UV treatment (600 mJ/cm^2) of reconstituted water from bench scale experiments and of pretreated surface water from a full scale MP UV installation, exceed the level that is generally considered to pose a negligible risk and justify further study in the chemical identity of the formed genotoxic compounds.

2.5 Reaction product research

2.5.1 Model Compound Study. Formation of nitrated aromatic compounds by MP UV photolysis is most probably the cause of the observed Ames test response. Machado and Boule[27] showed the formation of nitrophenol via photonitration of phenol in an aqueous solution by excitation of nitrite and nitrate ions. Suzuki et al.[12] identified tryptophan as amino acid most susceptible for formation of genotoxic compounds via MP UV photolysis of nitrate in aqueous solution. To illustrate the incorporation of inorganic nitrogen in the organic water matrix, MP UV collimated beam experiments were conducted with the organic model compounds, phenol and tryptophan. For both tryptophan and phenol, Figure 5 illustrates the formation of nitrite and the decrease of the nitrate concentration. The original inorganic nitrogen decreases with the increase of UV dose, indicating formation of nitrated organic compounds. Although formation of a range of hydroxylated and nitrated reaction products was observed in the model compound experiments with phenol and tryptophan, only 2-nitrophenol, 4-nitrophenol and 4-nitrocatechol could be identified, using Waters ACQUITY UPLC and column ACQUITY UPLC@ BEH C18 1.7μm (2.1* 50mm).

The results of the model compound experiments illustrate the complexity of identifying the formed nitrated reaction products, requiring a new experimental and analytical approach.

2.5.2 Labelled Nitrogen to Detect Nitrogen Containing Reaction Products. Based on the hypothesis that the formation of mutagenic compounds after MP UV water treatment results from the reaction of photolysis products from nitrate with (photolysis products of) NOM, we developed an innovative approach to detect these N-DBPs[21]. The approach is based on labeling the newly formed N-DBPs with a stable isotope originating from ^{15}N labeled nitrate (Figure 6). To detect the expected low concentrations of formed N-DBPs, concentration and sample pretreatment using solid phase extraction (SPE) and analysis by liquid chromatography coupled to high resolution mass spectrometry (LC HR-MS) was applied. HR-MS has proven to be excellently suited for screening samples for the presence of organic contaminants at low concentrations, e.g. in environmental analysis[28,29,30], sewer epidemiology[31], food safety, and doping research. Moreover HR-MS enables chemical formula assignment and identification of unknown components[32]. This approach enables the accurate and sensitive detection of the stable isotope tag in the formed N-DBPs in artificial water samples treated with MP UV in laboratory studies.

The raw MS data files from the artificial water samples were analyzed using software tools (Sieve, Thermo Scientific) to detect differences between sets of triplicate samples from the ^{14}N and ^{15}N MP UV treated samples, and the control samples. The MS data was screened for the presence of the nitrogen containing reaction products using Xcalibur software.

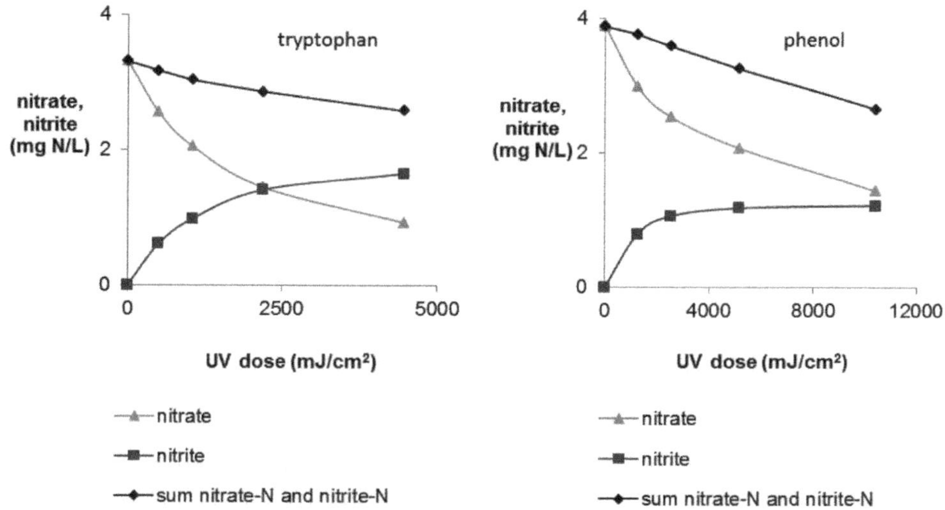

Figure 5 *Nitrate and nitrite concentration and total inorganic nitrogen content after MP UV CB experiments with phenol (0.26 mmole/L) and nitrate (0.40 mmole/L) and tryptophan (0.20 mmole/L) and nitrate (0.33 mmole/L) for a range of UV doses*

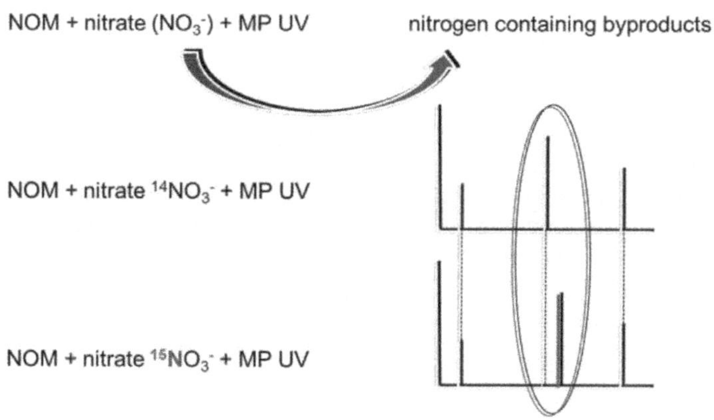

Figure 6 *Stable isotope labeling strategy for tracing N-DBPs. N-DBPs that are formed after MP UV when nitrate is present in its stable isotope labelled version, resulting in a peak in the mass spectrum with a $\Delta m/z$ of 0.99704 compared to the normal N-DBP. Background ions will be present in both spectra with the same mass*

Collimated beam experiments, using reconstituted water containing IHSS Pony Lake NOM (2.2 mg/L C) and nitrate (10.9 mg NO_3/L) were performed. Nitrate was added to ultrapure water as $K^{14}NO_3$, as the heavy stable isotope ($K^{15}NO_3$), and as a 1: 1 mixture ($^{14}N/^{15}N$). Concentrations of NOM and nitrate and experimental MP UV conditions were chosen at a level that positive responses in the Ames fluctuation test were expected, based on previous experiments [11], and thus potentially mutagenic compounds are presumably formed. Control samples were also prepared, i.e. artificial water containing NOM and ^{14}N nitrate, which were not treated with MP UV [21].

Application of this method showed that nitrogen originating from nitrate is indeed incorporated in N-DBPs and successfully resulted in detection of 84 newly formed nitrogen containing reaction products. Concentrations of 1 up to 135 ng bentazon-d6 equivalents (eq.)/L for individual nitrogen containing reaction products were found in reconstituted water containing nitrate and IHSS Pony Lake NOM. The summed concentration after MP UV treatment of the formed 84 nitrogen containing reaction products was about 1.2 μg bentazon-d6 eq./L and 69 ng atrazine-d5 eq./L for negative and positive ionization, respectively. Using commercially available standards, only 4-nitrophenol, 4-nitrocatechol and 2-methoxy-4,6-dinitrophenol could be identified[21]. Identification of the remainder of the formed nitrogen containing reaction products is future work.

2.6 Ames test response and reaction product identification in full scale samples

Water samples of the influent and effluent of the full scale MP UV/H_2O_2 treatment, and of the effluent of the post treatment, were taken at wtp Heemskerk. Sample preparation for both Ames test (TA98 with- and without S9) and LC-HR-MS analysis was performed (Figure 7). Of the 84 nitrogen containing reaction products discovered in the artificial water samples, in total 22 of these compounds were also detected in the full-scale MP UV treatment samples treatment. The three nitrogen containing reaction products whose identities were confirmed, namely 4-nitrophenol, 4-nitrocatechol and 2-methoxy-4,6-dinitrophenol, were also present in treated full-scale water samples. These findings link the results obtained in a MP UV laboratory setting to a full-scale drinking water treatment plant, and show that N-DBPs formed in a laboratory setting are relevant for full-scale MP UV/H_2O_2 processes[21].

The similar trends of the chemical screening results and the mutagenic response indicate that (part of) the 22 nitrogen containing reaction products present in the full scale water may contribute to the observed mutagenicity. The mutagenic potency is as yet unknown for the majority of these compounds. The detected Ames fluctuation test response may thus represent the total effect of a mixture of mutagenic (such as 2-methoxy-4,6-dinitrophenol) and non-mutagenic (such as 4-nitrophenol and 4-nitrocatechol) nitrogen containing reaction products.

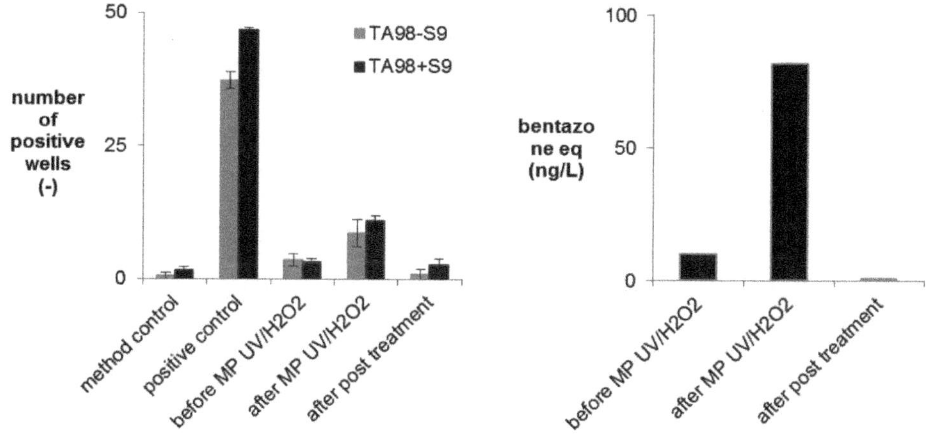

Figure 7 *Ames test response (TA98 with and without S9) before and after MP UV/H2O2 treatment and after post treatment and LC-HR-MS identified nitrogen containing reaction products, expressed as bentazone equivalents before and after MP UV/H2O2 treatment and after post treatment in concentrated water samples (cf 10,000) (March 2013). Reprinted (adapted) with permission from (Kolkman, A.; Martijn, A.J.; Vughs, D.; Baken, K.A.; Van Wezel, A.P., Tracing Nitrogenous Disinfection Byproducts after Medium Pressure UV Water Treatment by Stable Isotope Labeling and High Resolution Mass Spectrometry, Enviromental Science & Technology, 2015, 49 (7), pp 4458–4465). Copyright (2015) American Chemical Society.*

3 CONCLUSION AND DISCUSSION

In surface water, an important source for drinking water, numerous organic micropollutants are present. Regulatory requirements, maintaining customer confidence and restricting a (possible) health impact require barriers for organic micropollutants in surface water treatment processes. Advanced oxidation, based on the in-situ formation of hydroxyl radicals, is a robust technology to oxidize organic micropollutants. However, advanced oxidation processes may form harmful byproducts from NOM and inorganic water matrix constituents.

Water matrix related byproduct formation is a known effect of oxidation and advanced oxidation technologies in water treatment. Disinfection byproducts from chlorination and their health impact is extensively studied since the 1970's[33]. Formation of the carcinogen bromate by ozonation of bromide containing water types for drinking water production is currently limiting the applicability of ozone based processes, both for disinfection and advanced oxidation.

For medium pressure UV based processes, formation of nitrite by nitrate photolysis in nitrate rich water types was reported. Furthermore, recently formation of genotoxic compounds in MP UV water treatment was reported, however, only indirectly in bioassays on concentrated water samples. It was shown that MP UV nitrate photolysis in the presence of natural organic matter resulted in the formation of genotoxic compounds.

For purposes of preliminary risk assessment, the measured effect in the Ames test bioassay was converted in 4-NQO equivalents, resulting in levels of 575 ng/L 4-NQO equivalents in reconstituted water. Model compound studies using phenol and tryptophan into the nature of possible compounds confirmed the incorporation of nitrogen in natural organic matter via nitrate photolysis.

From the preliminary risk assessment based on the TTC concept it was concluded that identification of the genotoxic compounds causing the Ames test response was necessary. In labelled nitrogen experiments with complex natural organic matter, formation of 84 nitrated organic molecules was found, but only three could be identified. Based on an internal standard method, a total bentazone and atrazine equivalent of 1.2 ug/L bentazone and 69 ng/L atrazine equivalents were found.

Compared to the effect derived 4-NQO equivalents the analytically found summed N-DBPs equivalents are in the same order of magnitude. This method is very powerful and can be used to further identify the genotoxic nitrogen containing reaction products that are responsible for the observed Ames test response.

References

1. J.C. Kruithof, A.J. Martijn, *WST: Water Supply*, 2013, **13**, p 130.
2. J. Mack, J.R. Bolton, *Journal of Photochemistry and Photobiology A: Chemistry*, 1999, **128**, p 1.
3. C.M. Sharpless, K.G. Linden, *Environmental Science and Technology*, 2001, **35**, p 2949.
4. H. Shemer, K.G. Linden, *Journal of Photochemistry and Photobiology A: Chemistry*, 2007, **187**, p 186.
5. S. Esplugas, D.M. Bila, L.G.T. Krause, M. Dezotti, *Journal of Hazardous Materials*, 2007, **149**, p 631.
6. D.A. Reckhow, K.G. Linden, J. Kim, H. Shemer, G. Makdissy, *Journal of American Water Works Association*, 2010, **102**, p 100.
7. K.A. Thorn, L.G. Cox, *Journal of Environmental Quality*, 2012, **41**, p 865.
8. S. Flückiger-Isler, M. Baumeister, K. Braun, G. Gervais, N. Hasler-Nguyen, R. Reimann, J. van Gompel, H.-J. Wunderlich, G. Engelhardt, *Mutation Research*, 2004, **558**, p 181.
9. M.B. Heringa, D.J.H. Harmsen, E.F. Beerendonk, A.A. Reus, C.A.M. Krul, D.H. Metz, G.F. IJpelaar, *Water Research*, 2011, **45**, p 366.
10. A.J. Martijn, J.C. Kruithof, *Ozone: Science & Engineering*, 2012, **34**, p 92.
11. A.J. Martijn, M.G. Boersma, J.M. Vervoort, I.M.C.M. Rietjens, J.C. Kruithof, *Desalination and Water Treatment*, 2014, **52**, p 6275.
12. J. Suzuki, T. Ueki, S. Shimizu, K. Uesugi, S. Suzuki, *Chemosphere*, 1985, **14**, p 493.
13. S. Goldstein, J. Rabani, *J. Am. Chem. Soc.*, 2007, **129**, p 10597.
14. H.S. Rosenkranz, R. Mermelstein, *Journal of Environmental Science and Health*, 1985, **3**, p 221.
15. A.R. Van Rompay, A.J. Martijn, E.J.M. Penders, P.M. Baggelaar, *The 18th International Congress on In Vitro Toxicology ESTIV2014*, 2014, Noordwijk, The Netherlands.
16. A.J. Martijn, J.C. Kruithof, R.A.M. Hughes, R.A. Mastan, A.R. Van Rompay, J.P. Malley jr., *Journal - American Water Works Association*, 2015, **107**, p 120.

17. S. Barlow, in *Threshold of Toxicological Concern (TTC)*, ILSI Press, Washington, 2005.
18. M.N. Mons, M.B. Heringa, J. van Genderen, L.M. Puijker, W. Brand, C.J. van Leeuwen, P. Stoks, J.P. van der Hoek, D. van der Kooij, *Water Research*, 2013, **47**, p 1666.
19. R. Kroes, C. Galli, I. Munro, B. Schilter, L.-A. Tran, R. Walker, G. Wurtzen, *Food and Chemical Toxicology*, 2000, **38**, p 255.
20. D. Vughs, *Proceedings 2014 International Mass Spectrometry Conference*, 2014, Geneva, Switzerland.
21. A. Kolkman, A.J. Martijn, D. Vughs, K.A. Baken, A.P. van Wezel, *Environmental Science and Technology*, 2015, **in press**.
22. J. McCann, B.N. Ames, *Ann. N.Y. Acad. Sci.*, 1976, **271**, p 5.
23. S.A. Huber, A. Balz, M. Abert, W. Pronk, *Water Research*, 2011, **45**, p 879.
24. K.S. Krishnan, A.C. Guha, *Proceedings of the Indian Academy of Sciences*, 1934, **1**, p 242.
25. J.R. Bolton, K.G. Linden, *Journal of Environmental Engineering*, 2003, **129**, p 209.
26. D.M. McKnight, E.D. Andrews, S.A. Spaulding, G.R. Aiken, *Limnol. Oceanogr.*, 1994, **39**, p 1979.
27. F. Machado, P. Boule, *Journal of Photochemistry and Photobiology A: Chemistry*, 1995, **89**, p 73.
28. E.L. Schymanski, H.P. Singer, P. Longrée, M. Loos, M. Ruff, M.A. Stravs, C. Ripollés Vidal, J. Hollender, *Environmental Science and Technology*, 2014, **48**, p 1811.
29. C. Moschet, I. Wittmer, J. Simovic, M. Junghans, A. Piazzoli, H. Singer, C. Stamm, C. Leu, J. Hollender, *Environmental Science and Technology*, 2014, **48**, p 5423.
30. T.L. ter Laak, L.M. Puijker, J.A. van Leerdam, K.J. Raat, A. Kolkman, P. de Voogt, A.P. van Wezel, *Science of the Total Environment*, 2012, **427**, p 308.
31. L. Bijlsma, E. Emke, F. Hernández, P. De Voogt, *Analytica chimica acta*, 2013, **768**, p 102.
32. E.L. Schymanski, J. Jeon, R. Gulde, K. Fenner, M. Ruff, H.P. Singer, J. Hollender, *Environmental Science and Technology*, 2014, **48**, p 2097.
33. S.D. Richardson, M.J. Plewa, E.D. Wagner, R. Schoeny, D.M. DeMarini, *Mutation Research - Reviews in Mutation Research*, 2007, **636**, p 178.

THE IMPACT OF ON-LINE THM MONITORING ON DRINKING WATER PLANT DESIGN, TOTAL LIFE COSTS, AND REGULATORY STANDARDS AND REQUIREMENTS

Rick Bacon and Harmesh Saini
Aqua Metrology Systems, Sunnyvale, California USA

1 INTRODUCTION

Increasingly stringent local, state and national regulations have spurred the need for better and quicker information to monitor and track the composition of waters. Physical, chemical and biological parameters of water have historically been monitored using routine grab samples. This approach collected only a relatively small data set, potentially missing important events occurring between sampling schedules. Moreover, the infrequent data sets from manual sampling offered little insight for utilities to proactively manage, control and mitigate contaminants such as disinfection by-products (DBPs) across their applications. There are four primary species of Trihalomethanes (THMs) - chloroform ($CHCl_3$), bromodichloromethane ($CHCl_2Br$), dibromochloromethane ($CHClBr_2$) and bromoform ($CHBr_3$).

The automation of sampling, analysis and reporting available through the growing suite of commercially viable online water quality instruments offers great potential for onsite or remote process control automation, as well as for making management decisions on water quality at the source, intake, discharge, treatment plant and distribution system. Since water quality parameters are frequently changing, a firm and immediate understanding of water quality is necessary in order to anticipate problems and maintain optimal treatment conditions.

A step-by-step water quality analysis enables operators to maximize efficiency and this type of productivity-driven information has to be almost real-time in order to be useful. The high frequency and continuous data obtained through online instruments is a powerful tool to anticipate problems and warn of contamination in source waters, throughout treatment processes and in distribution systems.

Online instrumentation, once a niche segment of the broader water quality analysis market, is poised for significant growth. The global market for water analysis instrumentation is forecast to reach $1.86 billion by the year 2017, as noted in the research report "Water Analysis Instrumentation: A Global Strategic Business Report" by the Global Industry Analysts Inc.[1] While laboratory-based water analysis instruments represent the largest product segment in the water analysis instrumentation market, online systems are emerging as the fastest growing market segment with a compounded annual growth rate of 4.6% over the analysis period.[1]

The substantial growth in the online instrumentation market stems from a variety of drivers.

- Regulatory: more stringent regulations are being put into effect and results from external analysis are typically returned up to 10 business days later; preventing the immediate implementation of preventative actions since the underlying water quality parameters have changed by the time analysis is received. Online instruments allow operators to ensure critical water quality parameters are in compliance with required target values and also to check for deviations from stable values. Online instruments also afford operators and water quality managers the ability to effectively monitor treatment systems in real-time and proactively mitigate the impact of a potential regulatory breach through the timely adjustment of contaminant remediation processes.
- Public: heightened public awareness and concern over water quality in natural ecosystems and drinking water treatment facilities increases the accountability for operators and water quality managers to ensure their ability to better anticipate issues or problems instead of simply reacting to them after the fact; thus increasing the demand to implement real-time water quality analysis across their operations.
- Economics: the high frequency data provided through online instruments enable operators and water quality managers to optimize process controls and treatment schemes. By maximizing efficiency, potential capital and operational cost savings can be realized by reducing energy and material consumption throughout the treatment process and limiting reliance on personnel to conduct manual sampling and external or internal laboratory staff to undertake the analysis.
- Consolidation: with increased consolidation of water utilities, stand-alone entities now working together across broad geographical areas will rely on online instruments to provide remote monitoring and controls. Additionally, the creation of consecutive water systems resulting from the consolidation of utilities will use online instruments to ensure optimal water quality parameters are in place at handover points across networks and distribution systems.

In the absence of regulatory drivers calling for the implementation of online water quality instruments, operators and water quality managers have been installing online analysers and monitors because they are convinced of their ability to improve operational efficiency and maximize optimal water quality.

This paper is based on the insights gained from high frequency online THM data (nearly 0.5 million records) from drinking water plants in the UK, Spain and across the USA, and many conversations with managers and water quality personnel in drinking water utilities and state regulators in the USA. This data is being accumulated in real-time 24/7 from 40 online monitors that are taking readings with a frequency of every 2-4 hours.

2 ONLINE MONITORING TECHNOLOGY

Technological advances in water quality instrumentation have led to the development of online THM analysers to proactively manage THM mitigation strategies. THM analysers employ two essential steps to determine THM values. First, extraction and concentration of THMs from a water sample using a head-space, purge and trap (P/T) or membrane permeation approach. Second, separation and detection of THMs using a gas chromatography (GC)/electron capture detector, e-nose sensors, colorimetric or fluorescence meter.

The online analyser detailed in this paper combines the P/T and colorimetric techniques into one THM instrument while eliminating the pitfalls commonly associated with each approach. P/T-GC techniques often require the use of gas (helium or argon), external manual calibration, and operation by skilled technicians. In colorimetric methods, concentrated THMs react with alkaline pyridine or its derivative to form a red colour or a fluorescent product for measurement via a colorimetric or fluorescence detector. Pyridine produces a hazardous and smelly waste; and usually THM speciation is not possible through a traditional colorimetric approach using the Fujiwara chemical reaction.[4]

The online analyser detailed in this paper is based on the 100-year old Fujiwara reaction specific for THM detection. However, the methodology has been modified to allow for robust and reproducible analysis, inclusive of THM speciation via an instrument that does not require operation by a skilled technician. Following a "purge-and-trap" sampling method, desorption into a chemical mixture is used to generate a coloured product and time-resolved spectrophotometric analysis for detection and determination of speciated THM levels. The online monitor uses the following analytical methodology (Figure 1):

(1) A 250-mL sample is abstracted from the on-line water supply and used to fill the purge vessel (container)
(2) The water in the container is purged with air and the volatile compounds are retained on the trap containing an adsorbent material, thus extracting the THMs from the water sample via purge-and-trap
(3) Once the purging time is completed, the trapped analytes (THMs) are thermally desorbed and dissolved into a proprietary chemical reagent mixture.
(4) The solution is heated and generates a red color, whose intensity is measured spectrophotometrically. The reagent mixture produces a reaction with the THMs that is based on the Fujiwara reaction, which has been modified to minimize possible interferences from other compounds present in the sample. A derivative of pyridine and an aqueous base is employed, thus converting the four THMs into the same colored product. Note- each of the four THMs has a unique kinetic profile in the formation and decomposition of the colored reaction-intermediate.
(5) The absorbance of the reaction mixture is measured at two or three different times and a proprietary algorithm calculates the speciation concentration of the individual THMs.

The analyser reports values for the THM dominating species of chloroform or bromoform, depending upon the makeup of sampling water, as well as Total THM and has an accuracy specification of ±10%. It has also demonstrated a relative standard deviation of 3% or less on $CHCl_3$ and approximately 1% or less on the total THM. Results are delivered through communication options including 4-20 mA current loop, USB download from the monitor, and cloud-based data service.

The online THM analyser is programmed with a default standard sampling schedule of every 4 hours; average sampling time of 90 – 110 minutes. A more or less frequent sampling schedule can be programmed to meet application specific needs. The online analyser uses three reagents and two on-board calibration standards. The system runs a daily validation and can be configured to run a full calibration more frequently than once a week if required. Onboard reagents and THM calibration standards compensate for variations in the colorimetric reaction, ambient temperature fluctuations and aging of the

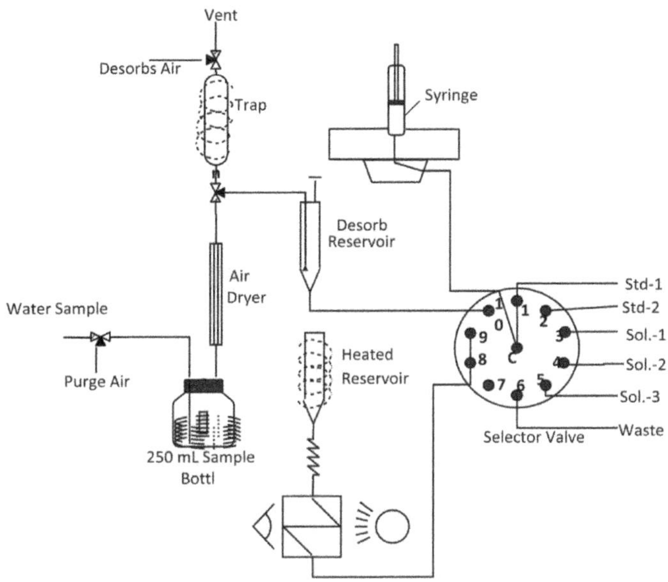

Figure 1 *Protocol of analytical method for online THM monitor (AMS THM-100™)*

trap. Automated calibrations coupled with quarterly maintenance and servicing, result in minimal instrument drift over extended periods. Reagent packs typically last 3 to 4 months based on standard sampling schedules of 6 samples per day.

Software continually monitors the performance of internal run-time parameters and alerts the user to the need for more frequent calibration if needed. A service package is offered through the manufacturer to further minimize onsite operational support. The manufacturer also monitors the health of the analyser remotely, 24/7, and works with the client to ensure the instrument remains online and working under optimal conditions. There are analysers working in the field for in excess of 3 years with a 99% uptime.

3 COMPARISON BETWEEN ONLINE AND LABORATORY METHODS

How does the precision and accuracy of this online monitor compare with that of laboratory based methods used to assess compliance with regulatory standards? The accuracy and reliability of the monitor has been validated by many utilities that include Aguas de Barcelona, Scottish Water, Severn Trent, American Water and United Water.

Independent testing highlights the precision and accuracy of the results provided by a specific online THM analyser. The specified bias for USEPA methods 524.2 and 551.1 for THMs and individual compounds is 80 – 120% with a relative percent difference of ±20%. In the case of the online THM analyser presented in this paper with an accuracy specification of ±10%; the instrument demonstrated a relative standard deviation of 3% or less on $CHCl_3$ and Br-THMs and approximately 1% or less on the Total THM value during independent testing.

Figure 2 and Table 1 show the results provided by 5 US –EPA accredited laboratories that each received 3 spiked samples from the same mother sample for testing. These results are compared with the results of analysis of the same sample on 3 online analysers. While the results from each the laboratories are within the specified limits of the EPA Standard Methodology they took between 7 and 13 days to be returned.

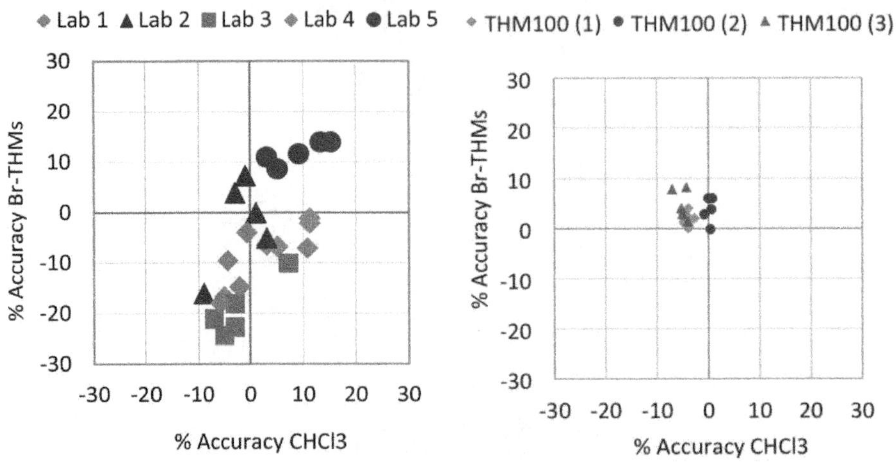

Figure 2 *Data Reliability: Laboratories versus Online (Spiked Samples)*
Reprinted from Journal AWWA 105 (11) by permission. Copyright © 2013 The American Water Works Association [2]

Table 2 *THM Online Monitor vs. US-EPA THM Laboratory Methods*

	US-EPA 524.2 Method			US-EPA 551.2 Method			THM-100 Online Monitor		
	CHCl3	Br-THM	THM	CHCl3	Br-THM	THM	CHCl3	Br-THM	THM
Average* (µg/L)	50.1	28.5	78.6	52.0	28.0	80.0	47.9	30.9	78.8
Std. Dev. (µg/L)	2.7	1.7	4.1	3.3	1.9	5.2	0.4	0.7	0.5
% Accuracy	1.3	-5.1	-1.1	5.2	-6.7	0.7	-3.1	2.9	-0.9
% RSD	5.4	6.2	5.2	6.4	6.6	6.4	0.8	2.3	0.6
* Data based upon Spike values of Chloroform 49.4 µg/L, Br-THM 30.0 µg/L and THM 79.4 µg/L									

Reprinted from Journal AWWA 105 (11) by permission. Copyright © 2013 The American Water Works Association [2]

3.1 Selection criteria for online THM water quality monitors

A summary of online THM monitors as compared to onsite and offsite laboratories can be found in Table 2. The precision of online THM monitors can be equivalent to or better than that of onsite or offsite laboratories. While online analysers provide timely detail regarding THM formation in treated water, the efficacy of commercially available online systems is varied. As a result, it is important to evaluate any online THM analyser against a specific set of criteria to determine if it is suitable to meet application specific needs.

Before evaluating the specific operating parameters of a given online THM analyser, it is important to confirm its overall performance and accuracy. Validation of performance is best obtained through independent testing that comparatively measures the online analyser against test results from accredited external laboratories such as the Environmental Laboratory Approval Program or state-specific programs. Without independent testing, it is difficult to validate the accuracy of an online THM analyser.

Table 2 *Online THM Monitor vs. Onsite/Offsite Laboratories*

Parameter	Online Monitor	Onsite Lab	Offsite Lab
Real-time continuous THM monitoring	X	X	
Fully automated, no human intervention required	X		
Accuracy of a certified lab	X	X	X
Repeatability, more then laboratory techniques	X		
Same-day turnaround time	X	X	
Speciation of THMs	X	X	X
Operating cost savings; no lab technician required	X		
Sampling cost reduction	X		
Treatment cost savings; adjust for varying THM level	X	X	
Suitable for regulatory reporting		X	X
Useful for pilot studies because of quick turnaround	X	X	
Useful for Stage 2 DBPR compliance	X	X	X
Useful for contract compliance at consecutive water system handover points	X		

* *THM online monitor, THM-100 by Aqua Metrology Systems*

Although online THM analysers use a colorimetric process to report real-time Total THM values, not all commercially available systems can provide measurement values for THM species[3]. Most online THM monitors will detect chloroform, bromoform, bromodichloromethane, and dibromochloromethane, and report the result as total THM. The online monitor presented in this paper measures for the chloroform or bromoform species as well as Total THM.

The benefit of using an online THM analyser is the ability to have an automated instrument capable of providing real-time, continuous THM results. However, operational conditions might require that the analyser also accept multi-point manual sample analysis. For example, if more than one location throughout a distribution system required continuous THM monitoring, yet capital and operational costs were limited, it would be beneficial to have the ability to manually grab multi-point samples from remote locations and test them at a central location housing the online THM analyser.

4 IMPLICATIONS OF THE ROUTINE USE OF ONLINE THM MONITORS IN DRINKING WATER

In the USA the regulatory limit for total THMs (sum of chloroform, bromoform, bromodichloromethane and dibromochloromethane) is 80 μg/l based on the running quarterly average of specific measurement points. In Figure 3, the high frequency data from a New Jersey utility shows THM levels at the point of transfer over a 14 week period for water it imports from a local wholesaler[5]. The wholesaler claimed that the water it was shipping was within regulatory limits. The circles indicate when the wholesaler took samples of the water it was selling to the retailer and used this to argue it was in compliance. Until the retailer had access to this high frequency data neither it nor the local EPA was any the wiser and it was under huge pressure from the local EPA to invest heavily to upgrade its storage reservoirs in order to address the problem of THMs in its distribution system. Based on high frequency data from an online monitor installed at the transfer point the retailer now immediately ceases purchases of water until THM levels drop to 60ppb and has avoided significant capital and operating expenditures.

In this case it will be the wholesaler that faces remediation costs. However there are numerous cases of expensive THM remediation projects that failed to deliver the promised results because they were based on designs that solved for levels of THM derived from infrequent and unrepresentative analytical results that did not capture much higher underlying THM levels. Equally, there are several examples of utilities that were overspending significantly on THM remediation by over-blending to dilute THM precursors with water from a Reverse Osmosis plant, over-treating with chemicals, or excessively air-stripping because they were ignorant of the impact of these processes on THM levels. The annual cost of overspend in some cases ran to in excess of €1m.

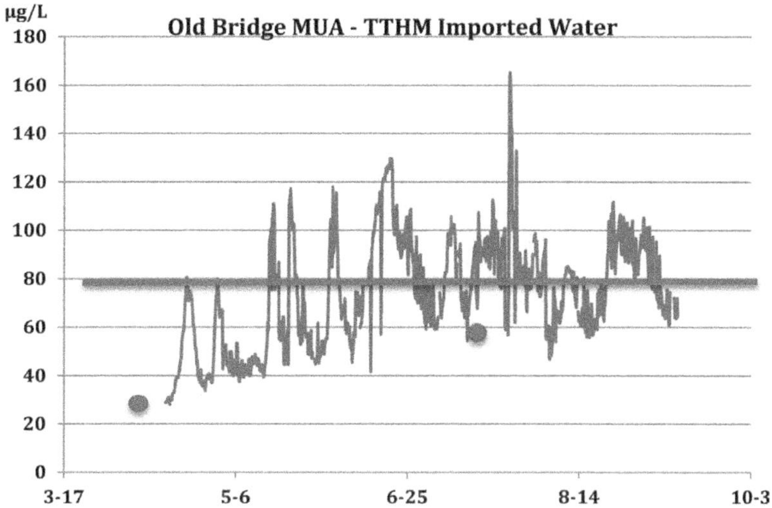

Figure 3 *High Frequency Real-time THM data shows the true picture*

5 CONCLUSION

Fluctuations in THM levels from source water or process control changes are traditionally difficult for water utilities to characterize based on the frequency of quarterly grab samples and the turnaround time of external laboratory analysis. As a result, utilities often conservatively operate their facilities while waiting to receive analytical results so as to ensure they remain in compliance with regulatory requirements. This approach often leads to overtreatment, which in turn increases operational expenses through additional spending on chemicals, sludge removal, processing, power and other resources.

Automated THM monitors help utilities to ensure regulatory compliance by providing immediate results on TTHM values throughout the water treatment process and prompting alterations to THM mitigation strategies. As a result, online instruments are poised to be an integral tool aiding in management decisions on water quality from source to distribution.

References

1. Research report “Water Analysis Instrumentation: A Global Strategic Business Report” by the Global Industry Analysts Inc.
2. Saini, H.; West, M.; Wang, Q.; Garvey, J.; Mui, R. *Journal AWWA.,2013,* ***105****:11:28.*
3. Takahashi, Y.; Onodera, S.; Morita, M.; Terao, Y. *Journal of Health Science,2003,* ***49*** *(1):1-7.*
4. Emmert, G.L.; Brown, M.A.; Simone Jr, P.S.; Geme, G.; Cao, G., 2007. Method of Real Time Measurement of THMs and HAAs in Distribution Systems Part 2. AwwaRF and AWWA. Denver, Col.
5. Roy, M.; Mui, R.; *Water & Wastes Digest, March 2014, 12.*

CHANGES IN INORGANIC NITROGEN RATIO ([NH_4^+-N]/[NO_3^--N]) DURING OZONATION OF DRINKING WATER AND ITS APPLICATION FOR MICROPOLLUTANT REMOVAL PREDICTION

G.A.D. De Vera[1], M.J. Farré[1,2], W. Gernjak[1,2] and J. Keller[1]

[1]The University of Queensland, Advanced Water Management Centre (AWMC), Queensland 4072, Australia
[2]ICRA, Catalan Institute for Water Research, Scientific and Technological Park of the University of Girona, H_2O Building, Emili Grahit 101, 17003 Girona, Spain

1 INTRODUCTION

Ozonation is considered as a treatment process of choice for the disinfection and oxidation of micropollutants and organic matter. However, the selection of appropriate ozonation conditions is critical to ensure its effective application. For example, working at conditions with high ozone CT may improve disinfection, but it can also increase the formation of disinfection byproducts (DBP) such as bromate.[1] It is therefore important for water utilities to characterise the ozonation process to ensure proper disinfection, while avoiding DBP formation.

One way to characterise ozonation is to calculate the R_{CT}, or the ratio of $^{\bullet}OH$ and O_3 exposures, as proposed by Elovitz and von Gunten.[2] In this approach the contribution of both ozone and hydroxyl radicals ($^{\bullet}OH$) in the overall oxidation of a water sample is determined from the decay of ozone and *para*-chlorobenzoic acid (pCBA), a $^{\bullet}OH$ probe. Characterising the ozonation process is especially important for disinfection as CT values are dependent on ozone exposures. If the ozonation conditions promote ozone decay ($^{\bullet}OH$ pathway), then the required CT value to achieve a certain log removal of microorganism may not be obtained.[1] Variations in the exposures of ozone and $^{\bullet}OH$ during pre-ozonation can also impact the transformation of organic matter, which may subsequently affect DBP formation during final chlorination.[3] The degree of oxidation by ozone and $^{\bullet}OH$ also has huge implications on the micropollutant removal, as functional groups in these compounds may be more reactive towards ozone or $^{\bullet}OH$.[4]

The R_{CT} values of ozonation systems, however, are not easily accessible since it will require water utilities to monitor ozone and pCBA decay by spectrophotometry and liquid chromatography, respectively. This led to studies that aimed to have a real-time measurement of ozonation characteristics, for improved control of the process. Most measured the CT_{10} parameter,[5] which is the ozone concentration at the reactor outlet, multiplied by the time required for the first 10% of a tracer to reach that outlet. Others operated using automated dosing systems, that are based on ozone decay constants obtained from ozone concentrations measured at certain points in the ozone contactors.[6, 7] However, these require process control systems that may not be available for use in other treatment plants. It also still requires bench-scale experiments to measure pCBA decay to determine $^{\bullet}OH$ exposures. As there is not a currently available method that can measure R_{CT} online, it will be useful to evaluate alternative methods which can easily measure such parameters.

In our previous work,[8] we have demonstrated that R_{CT} can be indirectly measured via simple colorimetric inorganic nitrogen measurements. Such an approach offers opportunities for on-line measurement of the extent of oxidation by $^{\bullet}OH$ and O_3 since higher nitrate concentrations were observed for conditions at higher ozone exposure and higher ammonium at $^{\bullet}OH$ dominant conditions. We have demonstrated that R_{CT} and $[NH_4^+\text{-N}]/[NO_3^-\text{-N}]$ are directly related to each other, i.e. samples with higher R_{CT} have higher $[NH_4^+\text{-N}]/[NO_3^-\text{-N}]$. This relationship is illustrated in Figure 1,[8] and observed due to the different reaction mechanisms of O_3 and $^{\bullet}OH$ towards dissolved organic nitrogen species, such as amino acids. As a major pool of organic nitrogen, amino acids like glycine can be oxidised to produce NH_3 and NO_3^- during ozonation. This was presented in the study of Berger, et al.[9] where they found NH_3 for $^{\bullet}OH$ reactions and NO_3^- for ozone reactions to be the main products (Figure 2). It is then plausible that similar reactions take place for fractions of dissolved organic nitrogen, causing the changes in inorganic nitrogen levels.

This present study therefore aims to confirm the observed relationship between R_{CT} and inorganic nitrogen by (a) monitoring the changes in NO_3^--N and NH_4^+-N in actual water samples before and after ozonation; and (b) using predicted R_{CT} values in assessing micropollutant removal.

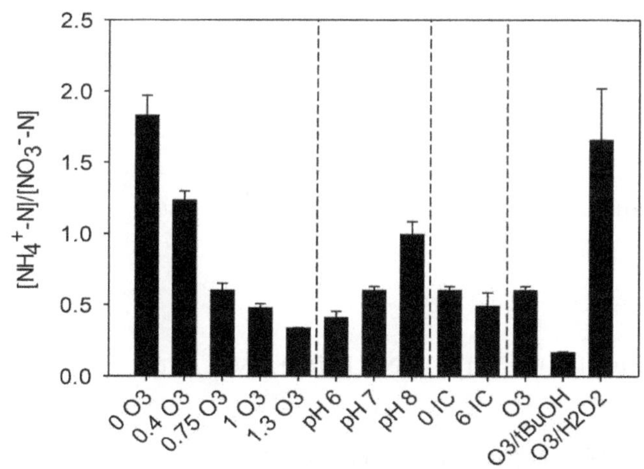

Figure 1 *$[NH_4^+\text{-N}]/[NO_3^-\text{-N}]$ after ozonation of a coagulated water sample at different transferred ozone dose (0.4-1 mg/mg DOC), pH (6-8, 1 mM phosphate), inorganic carbon (IC) (0-3 mg/mg DOC), and in the presence of 10 mM tert-butanol and 1 mg H_2O_2/mg O_3. Baseline conditions: DOC = 20 mg/L, C/N = 28.6 mg/L, ozone dose = 0.75 mg/mg DOC, pH = 7 (1 mM phosphate), inorganic carbon = 0 mg/mg DOC, temperature = 22 ^{0}C.* [8]

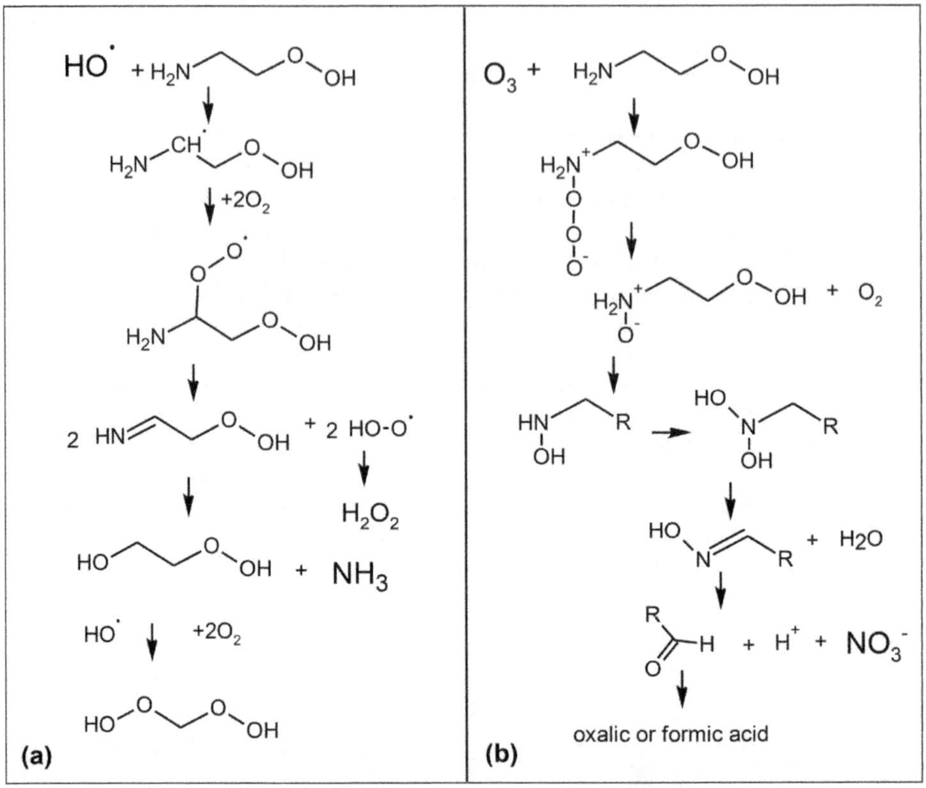

Figure 2 *Reaction pathways proposed by Berger, et al.*[9] *for the oxidation of glycine by (a)* $^{\bullet}OH$ *and (b)* O_3 *that produced* NH_3 *and* NO_3^-*, respectively.*

2 METHOD AND RESULTS

2.1. Changes in $[NH_4^+\text{-N}]/[NO_3^-\text{-N}]$ before and after pre- and intermediate ozonation

In November 2014, samples were taken from the Landers Shute drinking water treatment plant, Australia. This plant has a capacity of 140 ML/d and utilises both pre-coagulation ozonation and inter-stage ozonation in its treatment train.

To investigate changes in $[NH_4^+\text{-N}]/[NO_3^-\text{-N}]$ after ozonation, inorganic nitrogen concentrations were measured directly before and after the pre-ozonation and inter-stage ozonation processes. For this purpose, a Lachat QuikChem8500 Flow Injection Analyser was used for quantification and the results are shown in Figure 3.

It was observed clearly that after treatment at both pre-ozone and inter-ozone stages, nitrate and ammonium levels increased. These changes in the inorganic nitrogen concentration were likely caused by the reaction of ozone and $^{\bullet}OH$ with dissolved organic nitrogen. To confirm this, batch ozonation experiments were conducted in the laboratory, using water samples before the pre-ozone and inter-ozone stages. The experiments were performed under the same conditions as the treatment plant. Although there were relatively small deviations from the $[NH_4^+\text{-N}]/[NO_3^-\text{-N}]$ levels in the actual treatment plant with the ones obtained in the laboratory, it is still apparent that concentrations of nitrate and ammonium after ozonation at pre-ozone and inter-ozone stage of actual water samples, agreed well with the ones observed in the laboratory batch experiments. This was observed even in the presence of background nitrate concentrations produced during ozone generation. It should be noted that having a nitrate-free ozone solution only led to a higher $[NH_4^+\text{-N}]/[NO_3^-\text{-N}]$ ratio, but did not necessarily affect the increasing or decreasing trend of the ratio with changes in $^{\bullet}OH$ or O_3 exposures.

In addition, the nitrate levels of water samples taken before and after ozonation, at various treatment plants in the South East Queensland region, in 2013 were also found to increase after the ozonation process (Figure 4). This indicates that changes in inorganic nitrogen after ozonation happen for most advanced water treatment plants. This approach however applies only for source waters with minimal background nitrate concentration.

All of these results strongly suggest that the inorganic nitrogen species produced from oxidation of dissolved organic nitrogen may indeed be used to assess the performance of the ozonation process in a treatment plant.

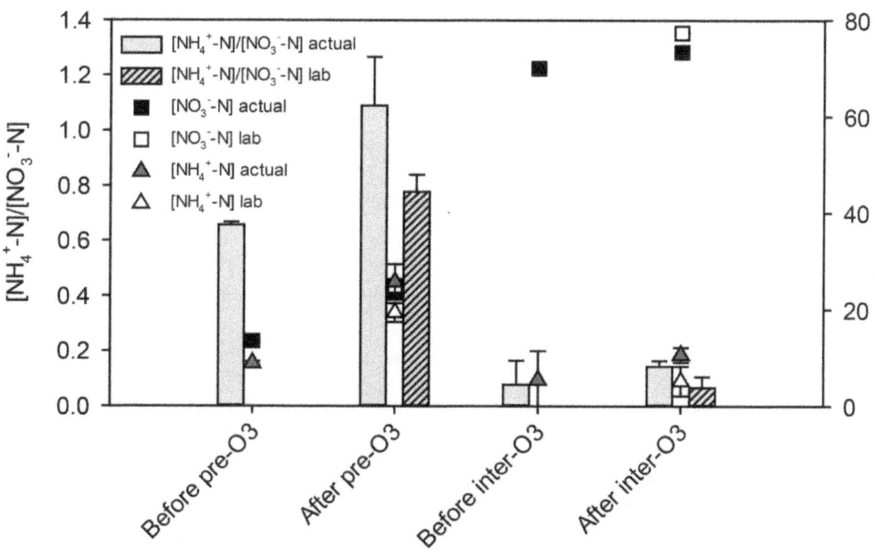

Figure 3 *$[NH_4^+\text{-N}]/[NO_3^-\text{-N}]$ ratio measured before and after pre-ozonation and intermediate-ozonation in Landers Shute drinking water treatment plant in comparison with batch samples ozonated in the laboratory. Conditions: pre-ozone dose = 2.1 mg/L, pH=7.4, intermediate ozone dose = 0.5 mg/L, pH = 6.8. Error associated with the results is the standard deviation (n=3). The detection limit for inorganic nitrogen is 1 μg/L.*

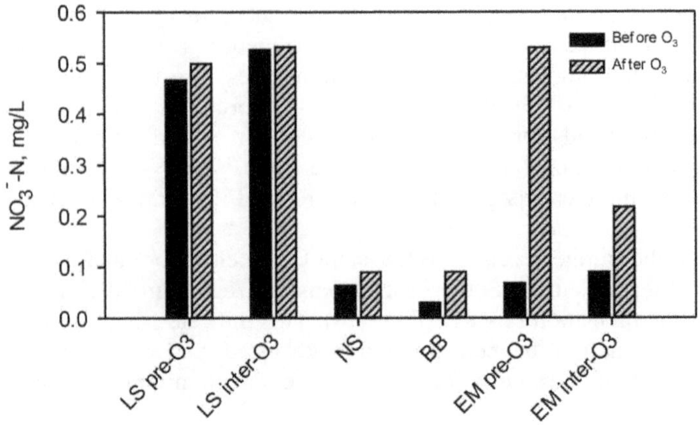

Figure 4 *NO_3^--N concentrations before and after ozonation at different treatment plants in South East Queensland region in 2013. LS = Landers Shute; NS = Noosa; BB = Banksia Beach; EM = Ewan Maddock*

2.2. Prediction of micropollutant removal from $[NH_4^+$-N]/[NO_3^--N] ratio

One of the applications of R_{CT} is for improved prediction of micropollutant removal in ozonated waters.[4] If the rate constants of the reaction with O_3 and $^{\bullet}OH$ with the micropollutant (MP) are known, then their removal can be predicted as expressed in equation (1).[2]

$$-ln\left(\frac{[\mathrm{MP}]}{[\mathrm{MP}]_0}\right) = k_{O_3}\int[O_3]dt + k_{\cdot OH}R_{CT} \qquad (1)$$

In previous work,[5] we demonstrated that R_{CT} and $[NH_4^+$-N]/[NO_3^--N] of ozonated samples are directly proportional. If this relationship holds true for a particular water sample with a relatively constant characteristic, R_{CT} values can be regularly obtained from the inorganic nitrogen ratio. This could be useful for routinely checking the ozonation process and also in predicting micropollutant removal.

In this study, three micropollutants with different reactivity with ozone and $^{\bullet}OH$ were selected to assess whether the R_{CT} calculated from the inorganic nitrogen ratio can be used to predict micropollutant removal. The following compounds were spiked into a coagulated water sample prior to adding ozone: ibuprofen (IBP), carbamazepine (CBZ), and atrazine (ATZ). The second-order rate constants for these compounds are shown in Table 1 and were taken from the study of Lee, et al.[4] These constants, together with the predicted R_{CT} and O_3 exposure in Table 2, were used to calculate the expected concentration of the micropollutants ([MP]) after all ozone had reacted.

The concentrations of the individual micropollutants were determined using a Shimadzu ultra-high performance liquid chromatography - mass spectrometry (LC/MS) system coupled with a 4000 QTRAP MS equipped with a turbo ion spray source (Applied Biosystems-Sciex, Foster City, CA, U.S.A.). Carbamazepine and atrazine were detected and quantified using the ESI (+) mode whereas ibuprofen was quantified using the ESI (-) mode. Liquid chromatography was performed using an Alltima C18 Column (250 × 4.6 mm, particle size 5 μm) and gradient elution with eluent A being acetonitrile (95%) + HPLC grade water (4.9%) + formic acid (0.1%) and eluent B being HPLC grade water (99%) + acetonitrile (0.9%) + formic acid (0.1%). The compound dependent MS parameters are shown in Table 3. Settings for source-dependent parameters were as follows: curtain gas (CUR) = 30V; nitrogen collision gas (CAD) = high; ion source gases GS1 and GS2 = 62 V, source temperature (TEM) = 700 ^{0}C, and ion spray voltage = 5500 V.

Figure 5 shows the % removal of micropollutants in a coagulated water sample after treatment with 1.2 and 0.6 mg O_3/mg DOC. At the higher ozone dose, it can be seen that the experimental concentrations for all of the micropollutants are nearly identical with the predicted concentrations. At 1.2 mg O_3/mg DOC, both the experimental and predicted % micropollutant removal were >99% elimination. At a lower dose of 0.6 mg O_3/mg DOC, the experimental % removal for ATZ and IBP were somehow lower than the predicted % removal. This may be due to the possible slight overestimation of ozone exposures, since the exposure at rapid ozone decay was only estimated from the initial ozone dose and the first ozone concentration measurement at 30 s. Also, the presence of 0.5 μM of micropollutant may somehow alter the kinetics of ozone decay especially when a lower dose was used. Nevertheless, it was observed that the removal of micropollutants was still predicted well, indicating the R_{CT} values obtained from $[NH_4^+$-$N]/[NO_3^-$-$N]$ can be applied for micropollutant prediction.

Table 1 *Second-order rate constants of the reaction of ibuprofen (IBP), carbamazepine (CBZ), and atrazine (ATZ) with ozone and* $^\bullet OH$[4]

Rate constant, $M^{-1}s^{-1}$	Ibuprofen	Carbamazepine	Atrazine
k_{O3}	9.6	$3.0x10^5$	6
$k_{\bullet OH}$	$7.4x10^9$	$8.8x10^9$	$3.0x10^9$

Table 2 *R_{CT} and O_3 exposure values used in the predicting micropollutant removal*

Sample[a] (O_3 dose, mg/mg DOC)	Measured $[NH_4^+$-N]/ $[NO_3^-$-N]	Predicted R_{CT}[b]	Predicted O_3 exposure[c], Ms
CBZ (0.6)	0.70	$3.31x10^{-9}$	0.04
CBZ (1.2)	0.29	$6.82x10^{-11}$	0.09
IBP (0.6)	0.57	$2.32x10^{-9}$	0.05
IBP (1.2)	0.29	$7.74x10^{-11}$	0.09
ATZ (0.6)	0.66	$3.05x10^{-9}$	0.04
ATZ (1.2)	0.24	$-2.91x10^{-10}$	0.10

[a]*[micropollutant] = 0.5 μM;* [b]*Calculated from the linear relation of R_{CT} and $[NH_4^+$-$N]/[NO_3^-$-$N]$: $R_{CT}=8x10^{-9}*[NH_4^+$-$N]/[NO_3^-$-$N]-2x10^{-9}$; $R^2=0.99$;* [c]*Calculated from the logarithmic relation of ozone exposure and $[NH_4^+$-$N]/[NO_3^-$-$N]$: O_3 exposure=-0.059Ln($[NH_4^+$-$N]/[NO_3^-$-$N]$)+0.0148, $R^2=0.93$.*

Table 3 *MS parameters used to quantify IBP, CBZ and ATZ*

MP	Q1 Mass (Da)	Q3 Mass (Da)	Dwell(msec)	DP[a]	CE[b]	CXP[c]
IBP	205	161	100	-52	-11.5	-10
CBZ	237.2	193.3	62	61	47	12
ATZ	216.13	174.2	62	81	27	10

[a]*DP=declustering potential;* [b]*CE=collision energy;* [c]*CXP=cell exit potential*

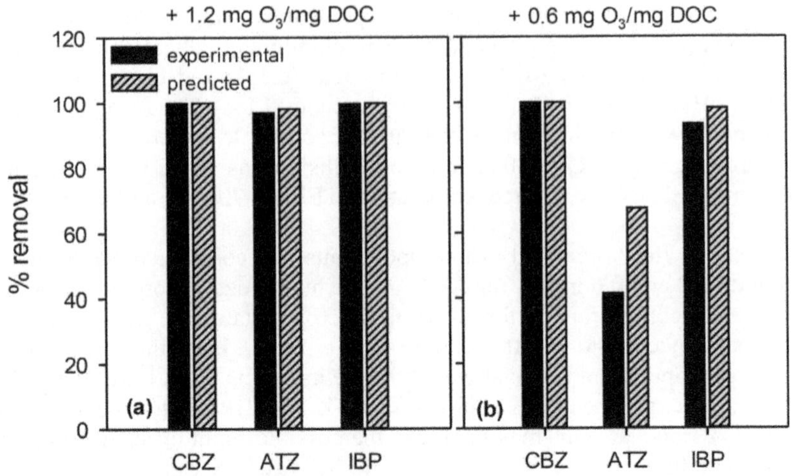

Figure 5 *Comparison of experimental and predicted concentrations of CBZ (carbamazepine), ATZ (atrazine), and IBP (ibuprofen) after ozonation at (a) 1.2 mg O_3/mg DOC and (b) 0.6 mg O_3/mg DOC. Concentrations of micropollutants were predicted based on known rate constants of the micropollutant with O_3 and $^{\bullet}OH$, R_{CT}, and ozone exposure (equation 1). Conditions: Sample = coagulated water, DOC of water sample = 20 mg/L, spiked micropollutant concentration = 0.5 μM, pH =7, inorganic carbon = 0 mg/mg DOC, temperature = 22 $\pm$ 1 ^{0}C.*

3 CONCLUSION

This study demonstrated that monitoring changes in inorganic nitrogen during ozonation may be useful in characterising the ozonation process. From the R_{CT} values calculated using the $[NH_4^+\text{-N}]/[NO_3^-\text{-N}]$ ratio, micropollutant removal after ozonation was predicted reasonably well. The use of $[NH_4^+\text{-N}]/[NO_3^-\text{-N}]$ may also have several applications in drinking water treatment, including the assessment of disinfection and organic matter removal efficiency. In our future work, dissolved organic nitrogen oxidation mechanisms will be studied using model nitrogen-containing compounds. It is also aimed to test this hypothesis using online systems to look at changes in NO_3^- and NH_4^+ during ozonation, at different conditions and precursor concentration.

Acknowledgements

This study was funded by a tailored collaboration with Seqwater (Australia) and Water Research Foundation (project WRF #4484). G.A.D De Vera is grateful to AusAID for the PhD scholarship. Dr Maria Jose Farre wishes to acknowledge the European Commission for funding project 623711 under the FP7-PEOPLE-2013-IIF - Marie Curie Action: "International Incoming Fellowships" and Dr Wolfgang Gernjak acknowledges funding obtained from the Spanish Government for a Ramon y Cajal Research Fellowship (RYC-2012-12181). The authors would also like to thank Deb Gale and other Seqwater staff who were involved in the sampling at treatment plants. We also would like to thank Dr. Beatrice Keller-Lehmann for her help with flow injection and HPLC analyses and Mr. Ali Farhat for the LC/MS analysis.

References

1. U. von Gunten, *Water Res.*, 2003, **37**, 1469-1487.
2. M. S. Elovitz and U. von Gunten, *Ozone-Sci. Eng.*, 1999, **21**, 239-260.
3. G. Kleiser and F. H. Frimmel, *Sci. Total Environ.*, 2000, **256**, 1-9.
4. Y. Lee, D. Gerrity, M. Lee, A. E. Bogeat, E. Salhi, S. Gamage, R. A. Trenholm, E. C. Wert, S. A. Snyder and U. von Gunten, *Environ. Sci. Technol.*, 2013, **47**, 5872-5881.
5. K. L. Rakness, I. Najm, M. Elovitz, D. Rexing and S. Via, *Ozone-Sci. Eng.*, 2005, **27**, 335-350.
6. J.-c. Jeon, C.-H. Jo, S.-B. Kwon, K.-H. Lee, T.-M. Hwang and Y. Chung, *Desalin. Water Treat.*, 2011, **33**, 337-350.
7. H.-P. Kaiser, O. Köster, M. Gresch, P. M. J. Périsset, P. Jäggi, E. Salhi and U. von Gunten, *Ozone: Science & Engineering*, 2013, **35**, 168-185.
8. G. A. D. De Vera, M. J. Farre, W. Gernjak and J. Keller, *Submitted*, 2014.
9. P. Berger, N. Karpel Vel Leitner, M. Dore and B. Legube, *Water Res.*, 1999, **33**, 433-441.

THE BEHAVIOUR OF HALOACETIC ACIDS IN DISTRIBUTION ZONES IN SCOTLAND

Georgia Agori[1], Kevin Snaddon[2], Emma H. Goslan[1]

[1] Cranfield University, Cranfield, Bedfordshire, MK43 0AL
[2] Scottish Water Juniper House, Heriot Watt Research Park, Avenue North, Edinburgh, EH14 4AP

1 INTRODUCTION

Disinfection is one of the most important processes for producing safe drinking water and controlling waterborne diseases. Chlorination is the most common disinfection strategy applied by water treatment facilities throughout the world because of its economical availability, effectiveness against a wide range of bacteria, viruses and other pathogenic organisms and due to its property of maintaining an adequate residual in the distribution system to prevent re-growth of pathogens [1]. However, since the mid-1970s public health concerns have been raised about the disinfection process [2] when researchers found potentially harmful by products known as disinfection by-products (DBPs). DBPs are formed by the reaction of chlorine with naturally occurring organic matter (NOM) like humic substances and inorganic ions such as bromide/iodide present in water [3, 4, 5]. Since then, more than 600 DBPs have been found in treated drinking water [6], with trihalomethanes (THMs) and haloacetic acids (HAAs) being the major DBPs detected in chlorinated water on a weight basis [1, 2].

DBPs are considered potentially carcinogenic and more recently, toxicological studies have shown a link between long-term exposure to DBPs and increased risk of cancer and adverse reproductive and developmental outcomes [2, 7, 8]. Therefore the control of DBP formation has become an issue for the drinking water sector. The objective of utilities is to provide treated and disinfected drinking water which protects consumers from pathogens while reducing DBP formation in order to adhere to present and forthcoming governmental standards.

A significant amount of research has been conducted on THMs in distribution but less attention has been given to the fate of HAAs. The concentrations of HAAs are often found to be equal to or greater than the concentration of THMs. Hence HAAs are a significant DBP in chlorinated drinking water and their monitoring in the distribution system is essential [9, 10]. In surface waters HAA are present at ng/L to μg/L concentrations, in drinking water distribution systems in μg/L and in treated wastewater in μg/L to mg/L [11]. The main HAAs of concern in drinking water are the nine chlorinated and/or brominated species that include monochloroacetic acid (MCAA), dichloroacetic acid (DCAA), trichloroacetic acid (TCAA), monobromoacetic acid (MBAA), dibromoacetic acid (DBAA), tribromoacetic acid (TBAA), bromochloroacetic acid (BCAA), bromodichloroacetic acid (BDCAA) and dibromochloroacetic acid

(DBCAA) [12] with MCAA, TCAA, DCAA being the dominant compounds for most systems [13,14]. Especially for the UK, studies show that the dominant compounds of various water distribution systems of UK are TCAA and DCAA [7, 15]. The sum of MCAA, MBAA, DCAA, TCAA and DBAA concentrations is known as HAA_5. DCAA has been classified as a group B_2 chemical (i.e., a probable human carcinogen) by the Integrated Risk Information System of the United States Environmental Protection Agency (US EPA) and TCAA as a group C compound (possible human carcinogen) [16]. Furthermore, both of these compounds have been proven to be animal carcinogens [17].

The formation, control, and health effects of HAAs, THMs and other DBPs in drinking water are issues of international concern and a significant amount of research on them has been conducted in the last decades. Thus many developed countries, as well as the World Health Organization, have established maximum acceptable concentrations for DBPs in drinking water. The only regulated HAAs at this time are HAA_5, because occurrence data on tribromoacetic acid, dibromochloroacetic acid and bromodichloroacetic acid, collectively referred to as HAA_3, were limited and until recently none of the published methods for analysing HAAs could quantify HAA_3 accurately and in a reproducible manner [18, 19]. The US EPA has set the regulation in 1998 for HAA_5 in drinking water to be 60 μg/L with a limit of 80 μg/L for THMs [18, 20] and from 2005 was based on a locational running annual average [21, 22]. Health Canada has fixed recommendations at 80 μg/L in 2008 for HAA_5 and WHO has set up guidelines for chlorinated HAAs (MCAA: 20 μg/L; DCAA: 50 μg/L; TCAA: 200 μg/L), but not for brominated HAAs [23]. However it is important to consider HAA_9 when analysing DBPs, because researchers have shown that HAA_9 levels can be 20-50% higher than HAA_5 levels in waters with high bromide concentrations and because these brominated species might have greater effects on human health due to being potentially more carcinogenic than their chlorinated counterparts [2, 18, 23, 24]. In Taiwan the limit for THMs is 80 μg/L and there is no current regulation for HAA, however a limit of 60μg/L has been proposed. In Japan total THMs are regulated at 100μg/L. In addition MCAA, DCAA and TCAA are regulated at 20, 40, and 200μg/L respectively. In China the limits for THMs, DCCA and TCAA are 100, 50 and 100μg/L respectively [2]. In Australia and New Zealand the regulated limits of MCAA, DCAA and TCAA in drinking water are 150, 100 and 100 μg/L respectively [22, 25]. On the other hand, currently there is no limit in the European Union (EU), including the UK, for HAAs. Only THMs are currently covered by legislation in the EU and the regulation limits for total THM concentration in drinking water are 100 μg/L. An MCL of 80μg/L for total HAAs (HAA_9) however is suggested in a recent revision of the Drinking Water Directive [13, 15, 22]. Furthermore a recent report to the UK Drinking Water Inspectorate has recommended that HAAs should be included as a high priority (the highest category) in the list of regulatory chemical parameters to be routinely monitored [7].

However the monitoring of HAAs in the distribution system is complicated and several parameters influence their formation and speciation: the type of disinfection scenario and chemicals (chlorination only, chlorination/chloramination, or ozonation/chlorination), water treatment processes, nature of water source, bromide content, pH of chlorination and distribution, temperature, concentration of total and free chlorine, residence time, distribution system characteristics and levels of total organic carbon (TOC)[1, 3, 7, 9, 22]. The kinetics of HAAs, their formation and stability remain largely unidentified. Current knowledge supports that their formation occurs faster than THMs, particular species have shown tendencies to decompose at elevated pH values, and some species are known to biodegrade in the absence of chlorine residual [26]. The stability of HAAs plays a key role and it is an important factor that may impact their measurement [27].

In order to comply with the THMs standard an increasing number of water systems in UK are converting to chloramines (combined chlorine) as a secondary disinfectant in their treatment plants. Generally in the UK, the water is exposed to free chlorine for 30 minutes before ammonia is added at a weight ratio of 3:1 to 4:1 of Cl_2:N [4]. Different studies show that chloramination produces lower concentrations of THMs and HAAs compared to chlorination and chloramination suppress the formation of trihalogenated HAAs [28, 29, 30].

HAA occurrence in chlorinated water may vary significantly based on season and location in the distribution system. These variations are due to changes in raw and treated water qualities as well as operational parameters related to chlorination [15, 31]. Many full-scale and bench-scale studies are available regarding the seasonal and spatial variations of THMs and HAAs in the distribution systems of several regions around the world such as the US, Canada, Australia, Taiwan, China, France, England and Turkey [6, 10, 18, 20, 22, 32, 33, 34]. During warm months there is often a higher concentration of NOM, a higher than average temperature and so an increased rate of microbial growth. For these reasons chlorination doses can be higher during treatment, resulting in substantially higher levels of DBPs during this period [1, 3].

Spatial changes in HAAs in the distribution system are also important because treated water prior to consumption can spend an extensive amount of time within the distribution system. Thus it is important to investigate the possible growth of DBPs in distribution [35]. Some studies have shown that the behaviour of THMs and HAAs in the distribution system is different. THMs were observed to increase in the distribution system and become stable in the system extremities, whereas HAAs decrease as water residence time increases within the distribution system after an initial increase [3, 36]. In warm water the decrease in HAA over time is higher than observed in cold waters where microbial activity is minimal [3, 9]. Biodegradation is considered to be responsible for the decrease in HAAs along the distribution system [9, 13]. Several studies have shown the involvement of microorganisms in the degradation of HAAs in the absence of disinfectant residual (e.g high residence time locations) [9, 37]. However biodegradation is likely the major loss process occurring in drinking water distribution systems [38] although other degradation reactions have been reported: hydrolysis (decarboxylation) and abiotic degradation by iron pipes and the resulting corrosion products [12, 38, 39], sorption or reaction within the pipe surface. Thus pipe surface is a crucial factor affecting all of the processes above [39].

This study aims to investigate the relative occurrence and speciation of HAAs in distributed water in Scotland and evaluate the impact of seasons on their concentration and speciation. Most of the zones used surface water and only 13 % of the 298 zones used ground water during the study period. The principal type of disinfectant used by most works was chlorination and only 14 % of zones are supplied by chloraminated water.

The spatial variability of HAAs concentrations in two distribution zones using different disinfection strategies was also studied during July 2014 and the impact of the two most common disinfection strategies on HAAs formation and fate along the distribution system was investigated. Information gathered on the distribution systems such as residence time and pipe material is combined with the analytical data to gain an understanding of each system. A novel aspect of this work is the use of flow cytometry (FCM) to enumerate the viable bacteria

2 METHODS

2.1 Sampling Strategy

A four season sampling programme was undertaken between September 2013 and August 2014. HAA data (n=1690 samples) was collected from 298 distribution zones in Scotland. Sample locations in the distribution system were random houses. The samples were collected by Scottish Water operators and were analysed at Scottish Water laboratories. The samples are analysed for HAAs, THMs, free and total chlorine, TOC and pH by Scottish Water laboratories. The results of free and total chlorine, TOC and pH are not presented in this chapter.

Two distribution zones were selected to determine the behaviour of HAAs in distribution. Drinking water samples were collected from the two distribution systems during the summer season in July 2014. These two zones were chosen depending on the concentration of HAAs (HAA_5 >20ug/l), their accessibility and the disinfection method used in the treatment works. However as it was difficult to calculate the water age, the distances from the WTWs are given for 5 or 6 different sampling locations. That involved some network tracing to find out the locations which are the furthest in distance from the WTW. By doing this we assumed that the further away the postcode is, the greater the water's age. Furthermore as the pipe material may affect the fate of HAA in the distribution system, information about the distribution mains was gathered.

The pipe material of the chloraminated distribution system was a mixture of ductile iron, spun and cast iron and mixed density polyethylene (PE) (High Performance PE, Medium Density PE) with Unplasticised Poly Vinyl Chloride (UPVC). The pipe material of the chlorinated distribution system was a mixture of ductile, spun and cast iron, Asbestos Cement and mixed density PE (High Performance PE, Medium Density PE) with PVC and UPVC. The percentage of iron material of the pipes for each system is presented in Tables 1 and 2. The samples were analysed for HAAs, THMs, free and total chlorine, TOC, pH and bacteria counts by flow cytometry by Scottish Water laboratories. The results of TOC and pH are not presented here.

Table 1 *Chlorinated distribution system sampling locations*

Sample point	Iron material %	Distance from WTW (m)
1	92.06	33.54
2	61.32	40.91
3	47.45	41.21
4	63.08	42.09
5	51.08	54.91

Table 2 *Chloraminated distribution system sampling locations*

Sample point	Iron material %	Distance from WTW (m)
1	100.00	6.42
2	65.57	10.73
3	84.15	16.54
4	91.35	19.23
5	93.44	25.41
6	94.92	32.87

2.2 Sample Collection

Before sampling water was run for 5 minutes to clear any standing water in the supply line. Water pH, total and free residual chlorine were measured in situ. Samples for measuring TOC, THMs, flow cytometery and HAAs were collected and transported to the laboratory following the appropriate sampling and storage practices. For HAA analysis water samples were collected in 150 mL amber glass bottles with PTFE-lined screw caps. A dechlorination solution (sodium thiosulfate) was added to the bottles to avoid further DBP formation during the time between sample collection and analysis (10%W/V). Samples were stored in a calibrated refrigerator at 5±3°C prior to extraction. Samples were extracted within 5 days of the sampling date.

2.3 Analytical Methods

HAAs were analyzed using a modified form of USEPA Method 552.3. A known volume of sample was measured out and internal standard is added. The sample was then acidified using sulphuric acid and sodium sulphate added, then extracted using MTBE. The Haloacetic acids were partitioned into the organic layer and converted into their methyl esters to enhance the volatility of the analytes, by the addition of acidic methanol followed by heating for 2 hours. The organic layer containing the methylated haloacetic acids was separated from the acidic methanol and neutralised by adding sodium bicarbonate solution. The organic layer was concentrated by evaporation under gentle nitrogen flow then analysed on a Perkin Elmer Gas Chromatograph Clarus 600C Mass Spectrometer (GC-MS). The compounds of interest were detected using the MS in Selected Ion Recording (SIR) mode. Quality assurance procedures included analysis of field blanks and the use of internal standard in each sample and standard.

The THM concentrations reported were provided by the Scottish Water Organic Laboratories for samples taken on the same sampling day. A known volume, (5ml), of sample was transferred to a headspace vial. This vial was then heated for 25 minutes to drive the THMs into the headspace. A portion of the headspace was then injected into a gas chromatograph where the solvents of interest were separated prior to mass spectrometric detection using a capillary column. Measurement of free and total chlorine in potable water was carried out on site using Hach Colorimeter. Flow cytometry (FCM) is used for the quantification of microbial cells in drinking water. Water samples were collected alongside routine microbiology samples with an aliquot of the sample being transferred in to a 50mL centrifuge tube. The sample was transferred into 2mL micro centrifuge tubes and mixed with CYBR Green stain or Propridium Iodide stain. The stained sample was then passed through the Accuri C6 flow cytometer for analysis.

2.4 Data Analysis

Sample values below the detection limit of the methods were assigned the detection limit for statistical analysis. Basic analyses were conducted using spreadsheet calculations (MS Excel) and for more elaborate statistical procedures we used the (free) statistical software R (R Foundation for Statistical Computing, Vienna, Austria).

3 RESULTS AND DISCUSSION

3.1 Four-season Occurrence Survey of HAAs in Scotland

Most of the 298 zones used surface water with only 13% using ground water. The principal type of disinfectant used was chlorination with only 14% of zones employing chloramination. The individual and total HAA_5 levels and the total THM levels found in our study are reported (Table 3). Average HAA_5 concentrations in distribution systems were about 50% lower than total THMs.

Compared with a previous study [7] undertaken in the UK where a mean concentration of total HAA_6 ranged from 35.1-94.6 (μg/L) and a maximum concentration of 244 (μg/L) was observed, the total HAA_5 concentrations reported in this study were relatively low (median value of 11.5 μg/L , with a maximum value of 150.4 μg/L and a minimum value <5.3 μg/L) Our results had lower median concentration compared with a more recent study [15] in one utility system in England and its distribution network, where the levels of HAA_9 were ranging from 11.1-20.6 μg/L with a maximum value of 41 μg/L. However, our maximum HAA_5 concentration was much higher compared to that study. Moreover our results come from 298 distribution zones not only from one zone. Further to this, a study carried out in Scotland [4] showed median HAA_9 concentration of 20 μg/L and more recently the same group carried out a study [40] of HAA_9 measured in final samples of 9 WTWs of England on a seasonal basis and found levels ranging from 0.3-28.9 μg/L but overall the median value was 12.5 μg/L which is similar to that observed in our study. During the period of study, of HAA_5, DCAA and TCAA were the main species where the sum of these two species represented over 83% of the total HAA_5 (41 % and 42 % respectively). Generally non-brominated HAAs dominated in the waters of the distribution zones under study.

Furthermore (Figure 1) most systems complied with the maximum acceptable concentration limit set for the US (HAA_5: 60μg/L) In <1 % of the zones under study, the average HAA_5 concentration exceeds 60μg/L.

Table 3 *Individual and total HAA and total THM levels (μg/L)*

	TCAA (μg/L)	DCAA (μg/L)	BCAA (μg/L)	DBAA (μg/L)	MCAA (μg/L)	MBAA (μg/L)	HAA_5 (μg/L)	THM_4 (μg/L)
Q1	1.40	2.20	0.90	1.10	1.00	0.80	6.00	12.10
Minimum	1.10	1.30	0.90	1.10	1.00	0.80	5.3	3.20
Median	5.30	4.50	1.20	1.10	1.00	0.80	11.45	24.50
Maximum	82.00	65.70	22.90	34.30	3.70	5.30	150.40	153.20
Q3	10.00	8.70	2.20	1.30	1.00	0.80	19.40	43.80
Mean	7.53	6.50	1.94	1.68	1.06	0.85	15.47	30.77
Std	8.29	6.17	1.76	2.14	0.23	0.24	13.29	24.14

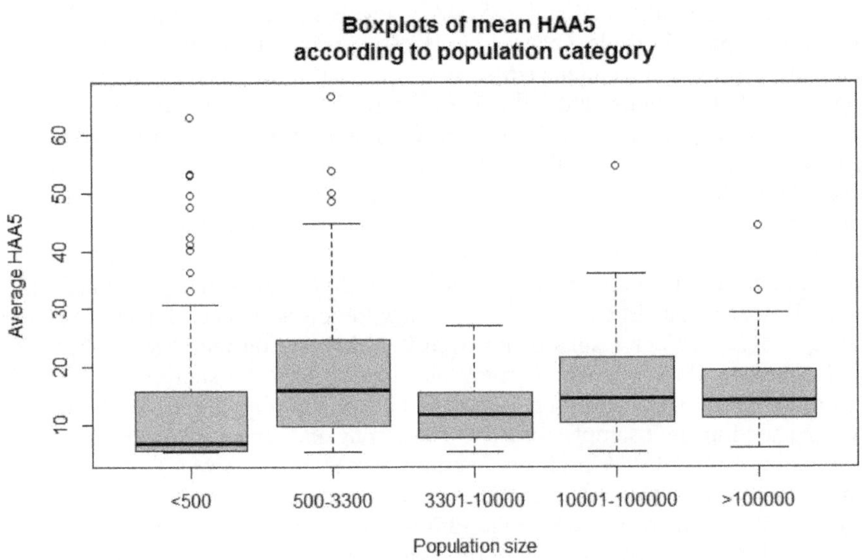

Figure 1 *Annual average HAA_5 (μg/L) according to population category*

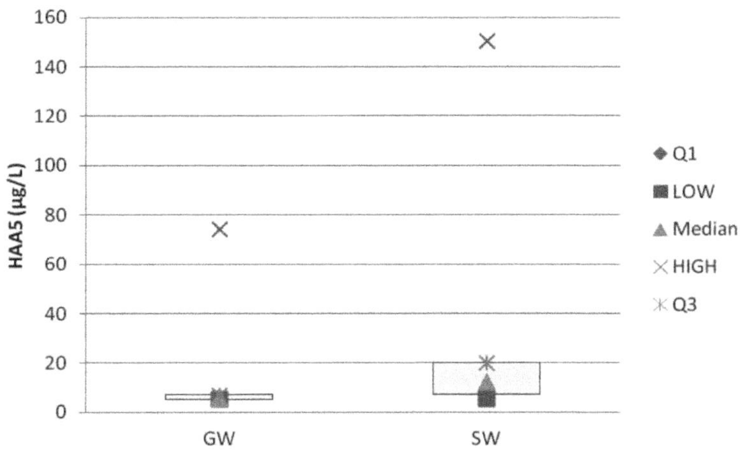

Figure 2 *HAA_5 distribution according to water source (SW:surface, GW: ground water)*

The plants with a mean HAA_5 concentration higher than 60µg/L were small systems with a service population less than 3300 people. The very small WTW (<500 people) appear to have the lowest concentration of HAA_5, on a median basis.

The highest median concentration of HAA_5 appears in networks of 500-3300 residents where there is also the widest range of concentrations. However there is a very weak correlation between the population served and the annual average HAA_5.The Spearman correlation coefficient (r_s) was 0.25. These low concentrations of HAAs are due to the fact that most systems in large cities tend to implement more effective DBP control strategies and the NOM has been efficiently removed [17]. The high levels found are thought to be due to the non-optimized removal of organics at these works.

We also detected differences in HAAs levels by type of source water and disinfectant used. When water type is considered, it is shown that ground water produces lower levels of HAAs than surface water (Figure 2). The results agree with several other studies that found that ground water supplied systems experienced very low levels of total HAAs [5, 7, 22]. Surface reservoirs or river waters have relatively high TOC levels and are influenced by algal blooms. The groundwater tends to contain less NOM and hence less potential to form DBPs [7]. For all systems supplied by both ground water and surface water, DCAA and TCAA were the dominant species but their concentrations were higher in surface water supplied zones.

The HAA_5 distribution according to the disinfection practice is shown (Figure 3). Although the chloraminated waters have lower HAA_5 levels compared to chlorinated waters, the difference between the medians is not that pronounced. Moreover all the works using chloramination meet the USEPA HAA_5 maximum contaminant limit whereas some chlorinated samples exceeded it. Chloramination is shown to control the formation of DBPs in the distribution system, however the 30 minutes contact time of free chlorine at the works before ammonia is added is likely sufficient to form higher levels of THMs and HAAs before water enters the system [4]. The significant impact of chloramination on HAAs speciation has been reported. Dihalogenated and trihalogenated HAAs are generally present in equal amounts in chlorinated water but dihalogenated HAAs are always the dominant group when using chloramination [4]. The work here agrees with the literature (Figure 4).

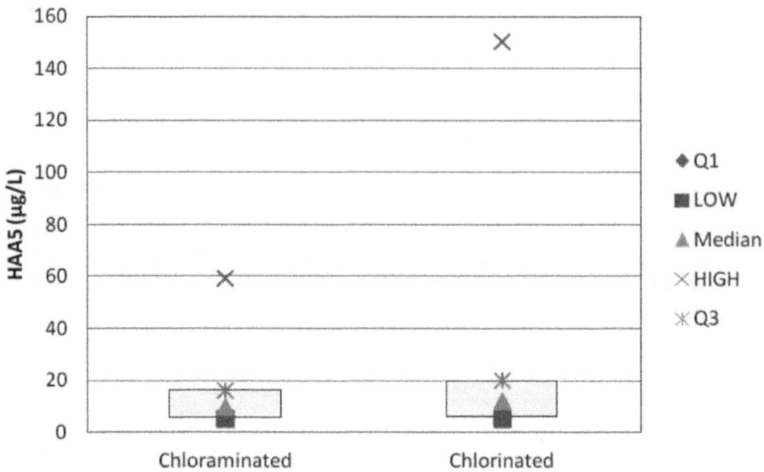

Figure 3 *HAA_5 distribution according to disinfection strategy used*

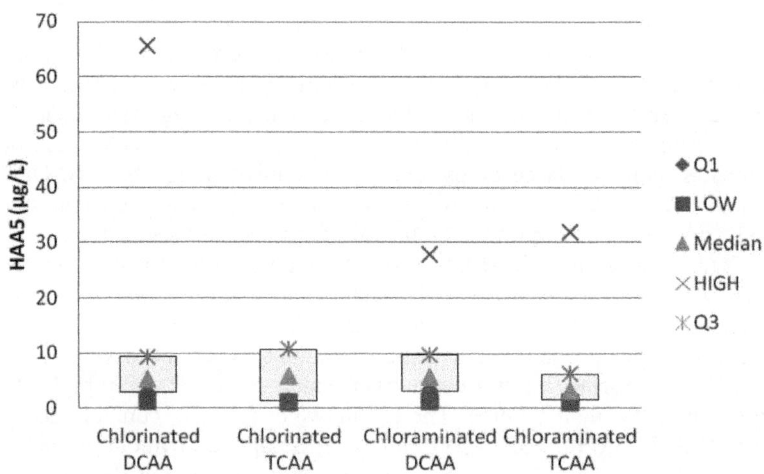

Figure 4 *Distribution of DCAA and TCAA in chlorinated and chloraminated zones*

Many full-scale and bench-scale studies are available regarding the seasonal and spatial variations of THMs and HAAs in the distribution systems of several regions around the world such as the USA, Canada, Australia, Taiwan, China, France, England and Turkey [6, 10, 18, 20, 22, 32, 33, 34]. In most of the studies THMs concentrations were higher during summer, but HAAs seasonality is less well defined [20]. In Canada the seasonal variations were very pronounced and the highest HAAs concentrations were measured during summer and spring and for THMs summer and autumn [32]. Though in Turkey a study showed higher HAA_9 average concentrations in spring and autumn and lowest during summer. A Chinese study found similar results to the Canadian study. A study in Taiwan showed higher HAA_5 during summer [22]. In Scotland a study [4] found higher HAA_9

concentrations during summer as well and lower during spring and winter. The seasonal variation of HAA_5 in our study is presented (Figure 5). The highest values were observed in summer but the variation in the median over the seasons was less than 5 μg/L. The highest median and the widest range of HAA_5 concentrations were observed in autumn and summer (13.2 and 14.5 μg/L respectively) while the lowest were obtained in winter and spring season (10.4 and 9.7 respectively). During autumn and summer, concentrations of HAAs ranged from <5.3-85.6 μg/L and <5.3-150.4 μg/L respectively. This is possibly because the water temperature is higher in summer and autumn, promoting the formation reactions of HAAs. Additionally more DBPs precursors may be present, because higher TOC levels were also observed during summer and autumn. One possible explanation for some high winter HAA concentrations is that during winter there is less water demand so low plant flow. Therefore there may have been longer detention time in service reservoirs and longer free chlorine contact time [4]. We could also assume that during the warm months there is a greater degradation of HAAs along the distribution systems, explaining the less pronounced seasonal variation of HAAs [3]. The seasonal variations of DBPs and therefore of HAAs are related to changes in NOM quantity and characteristics of the water sources and consequently variations in operational parameters concerning the disinfection. The climate of Scotland is temperate and tends to be very changeable, but not normally extreme. The changes between summer and winter are generally relatively moderate, rather than extreme hot or cold. Hence in our study the seasonal changes in surface water quality and water temperature are not as extreme in countries such as Canada over the year and do not necessitate large operational variations in water treatment.

In addition the behaviour of DCAA and TCAA are not the same during seasons. The median data (Table 4) show almost no seasonal variation for the DCAA however the seasonal variation of TCAA is more pronounced and higher concentrations were detected during summer and autumn. A possible explanation is the degradation of DCAA in warm waters due to the presence of biofilm into the distribution system [3].The analysis of HAAs fate according to location will allow better observation of HAA biodegradation. Similarly with DCAA there is no seasonal variation for BCAA and DBAA (data not presented).

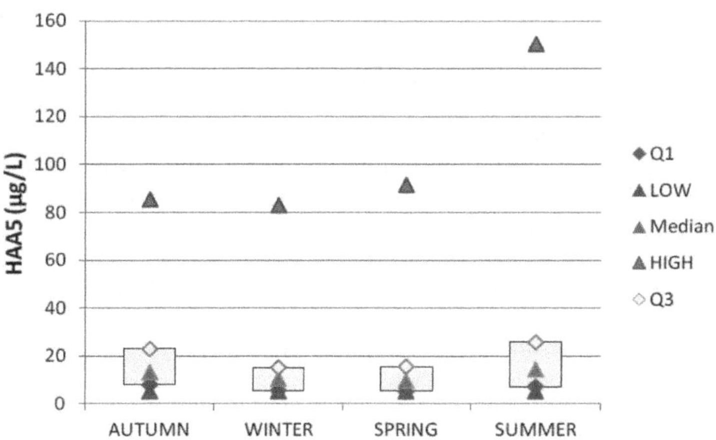

Figure 5 *HAA5 seasonal variations for all zones*

Table 4 *Seasonal variations of DCAA and TCAA*

	DCAA				TCAA			
	AUTUMN	WINTER	SPRING	SUMMER	AUTUMN	WINTER	SPRING	SUMMER
Q1	2.60	2.00	1.90	2.50	2.60	1.30	1.10	1.90
LOW	1.30	1.30	1.30	1.30	1.10	1.10	1.10	1.10
Median	4.45	4.80	4.20	4.80	7.40	4.35	3.90	7.60
HIGH	35.90	46.20	36.10	65.70	62.70	43.50	54.10	82.00
Q3	10.05	7.80	7.90	10.10	13.03	7.63	7.48	12.90
Mean	7.03	5.83	6.04	7.47	9.86	5.97	6.08	9.10
Std	6.34	5.12	5.61	7.67	9.75	6.23	7.19	9.54

3.2 Spatial Variability of HAAs Concentrations in two Distribution Zones

In both zones HAAs increase and decrease along the distribution (Figures 6 and 10).

3.2.1 Fate of HAAs in Chloraminated Distribution Zone.

There were favourable conditions such as low residual chlorine (Figure 7) and warm water for the biodegradation of HAAs. Thus biodegradation can be one explanation for the decrease of HAAs observed at sampling points 4 and 6. Zhang et al. found that trihalogenated HAA species are more stable than mono and dihalogenated species [13]. Here DCAA decreased substantially but TCAA decreased at a slower rate. The ratio TCAA:DCAA was stable from the first to the third sampling location but on the fourth location their ratio increased to 2 and on the last sampling point was 1.2. Furthermore the other HAAs concentrations were below their detection limits. Concerning the TOC concentrations, the value didn't vary significantly in the distribution system and its concentration ranged from 1.3 to 1.9 reaching the lowest value at the last sampling point. The pH didn't vary significantly along the distribution system with values ranging from 7.7 to 8.4 (data not presented).

The flow cytometry (Figure 8) showed a decline in the levels of both total and intact cells. At sampling points 4 and 6 there was a significant rise in intact levels reinforcing our hypothesis of HAAs biodegradation. However the substantial increase of intact cell levels on the second sampling point, probably associated with the decrease in total chlorine didn't cause any HAA biodegradation. However the factors that affect HAAs biodegradation in the distribution system are not clearly understood. Although bacteria could be viable, they might not have been metabolically active regarding HAA biodegradation [41]. Lack of degradation could be observed if there is lack of HAA degrading organisms, or due to environmental factors like pH, temperature, micronutrient limitation, dissolved oxygen limitation, or big pipe diameter [42]. Furthermore in that part of the distribution system the iron proportion is 61%, much less than in the rest of the distribution system. That may play a role in the formation of the biofilms on the pipe surface. Growth of bacteria on surfaces to form biofilms is possible on any type of surface but it might be more favourable on corroded iron. Although large portions of the distribution pipes are composed of iron there was no evidence of dehalogenation. Moreover most iron water pipes have been in place for decades and are corroded so absorption to iron minerals is a potential loss pathway for HAAs. Of note was the large number of dead cells found from the flow cytometry results which could have contributed to HAA formation.

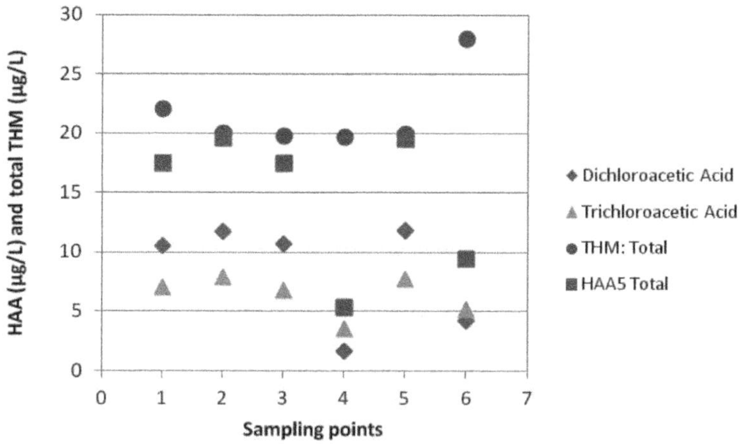

Figure 6 *Spatial variability of THM total and HAAs species along the distribution system of the chloraminated zone*

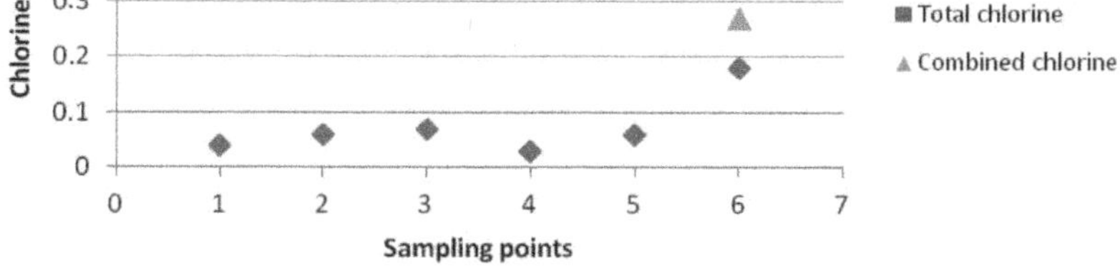

Figure 7 *Spatial variability Chlorine residual along the distribution system of the chloraminated zone*

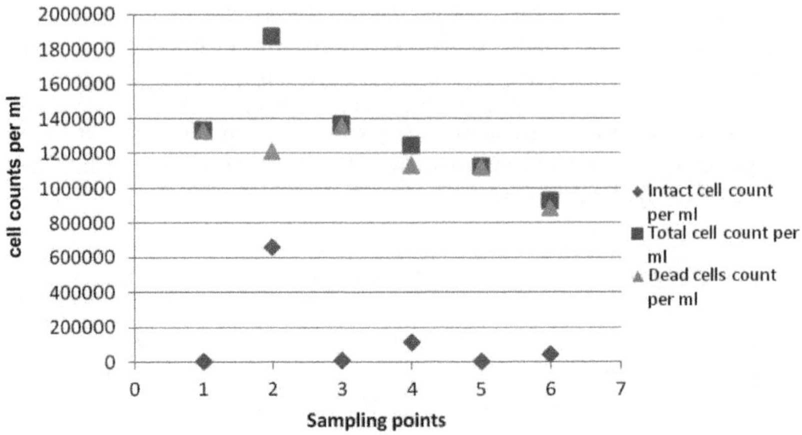

Figure 8 *Spatial variability of cells counts along the distribution system of chloraminated zone.*

Free chlorine (HOCl) is a powerful oxidant that can react with organic compounds like the dead cells and create more DBPs including HAAs. In addition to free chlorine, combined chlorine (chloramines) reacts with organic matter to form DBPs, but at a much slower rate [2].

3.2.2 Fate of HAA in the Chlorinated Distribution System.

The first fall in HAA_5 was due to the decrease in DCAA and TCAA. Both decreased to almost the same degree at 43 % and 45 % respectively. However during the second fall in HAA_5 TCAA was stable and slightly increased on the last sampling point of the distribution system, but DCAA decreased 70 % from the third to the forth sampling point. From the nearest to the furthest point TCAA decreased 5 % and DCAA 57 %. Furthermore BCAA was the only other HAAs species whose concentration varied along the distribution system and its behaviour was similar to that of DCAA. It is known that dihalogenated HAAs are likely to have formed through similar chemical pathways and could be why they have similar behaviour [9, 22]. All the other compounds were found at concentrations lower than their detection limits. A possible explanation for the drops in both HAA_5 and TCAA and DCAA is the relatively low levels of NOM detected and chlorine residual (Figure 9). Furthermore the level of dead cells is quite low (Figure 11) compared to the number of dead cells detected in the chloraminated zone.

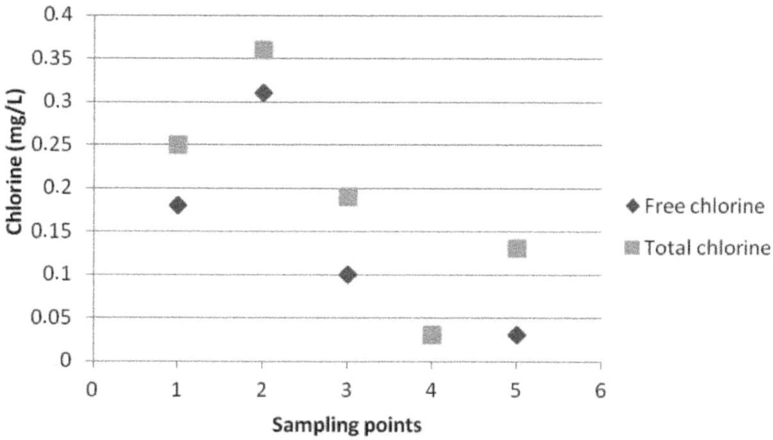

Figure 9 *Spatial variability of total and free chlorine along the chlorinated distribution system*

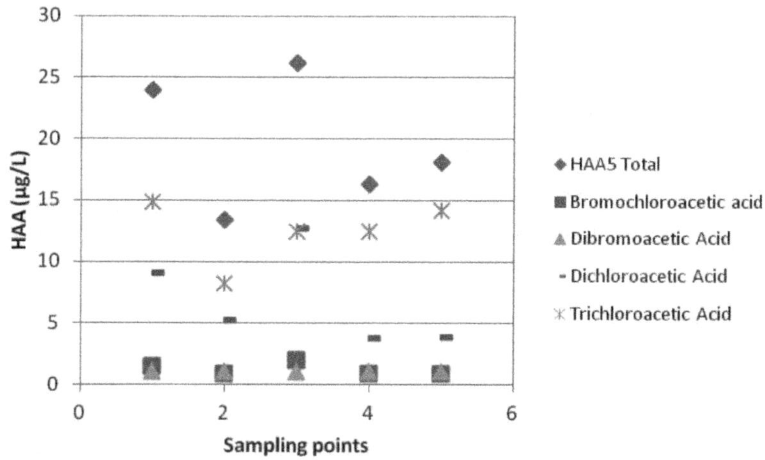

Figure 10 *Spatial variability of HAAs species along the chlorinated distribution system*

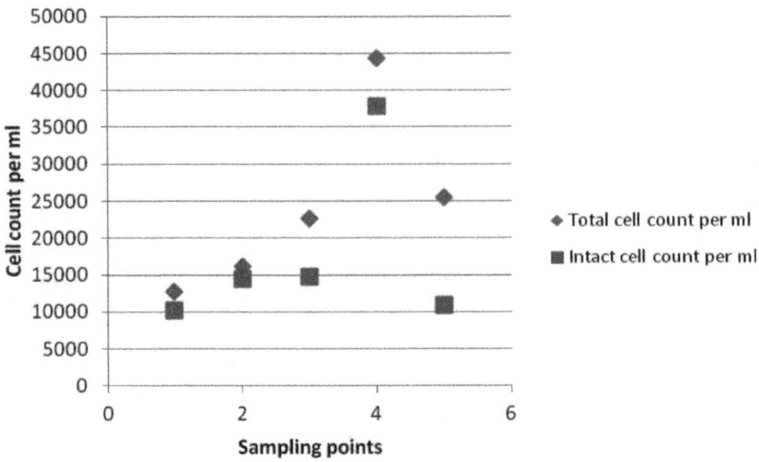

Figure 11 *Spatial variability of total cell and inact cells along the chlorinated distribution system*

In addition abiotic reduction of HAAs is possible in the iron pipes. The water travels through a pipeline consisting of 92% iron to reach a pipeline consisting of 63% iron on the second sampling point. Studies have shown that dehalogenation occurs in the presence of iron distribution pipes with TCAA degrading to DCAA [12, 43]. This may have occurred here as the TCAA:DCAA ratio and TXAA:DXAA (as BCAA was observed as well) was observed to decrease through the first to the third points of the distribution. On the second point of HAAs decreasing the most probable reason is biodegradation. In this case TCAA stays stable and DCAA and BCAA decreased substantially.

4 CONCLUSIONS

The work carried out showed that the formation of HAAs, their speciation and their fate in a distribution system is complex and depends on several factors including season, characteristics of water, treatment processes, water distribution pipe materials. Distribution systems are dynamic in nature and there may be significant temporal and spatial variations in DBPs concentrations. Thus the exposure of the public to HAAs may vary spatio-seasonally along a given distribution system. Given the complex nature of the distribution systems and the number of the factors that affect the formation and fate of HAAs and their interactions along a distribution system, it is important that the DBP precursors are reduced as much as possible at the treatment works

ACKNOWLEDGEMENTS

The authors would like to thank everyone in Scottish Water for laboratory support and for providing useful data for the project. Financial support for this project was provided by Scottish Water.

REFERENCES

1. P. Brett and L. Barron, *Journal of Chromatography A*, 2004, **1046**, 1–9.
2. T. Karanfil, S.W. Krasner, P. Westerhoff and Y. Xie, *ACS Symposium Series of American Chemical Society*, 2008, **1**, 2-19.
3. M.J., Rodriguez, J.B. Serodes, and P. Levallois, *Water Research*, 2004, **38**, 4367–4382.
4. E.H. Goslan, S.W. Krasner, M. Bower, S.A. Rocks, P. Holmes, L.S. Levy and S.A Parsons, *Water Research,*2009, **43**, 4698-4706.
5. C.M.Villanueva, M. Kogevinas, M. and J.O. Grimalt, *Water Research*, 2003, **37**, 953–958.
6. C.M Shanks, J.B. Serodes and M.J. Rodriguez, *Water Research*, 2013, **47**, 3231-3243.
7. E. Malliarou, C. Collins, N. Graham and M.J Nieuwenhuijsen, *Water Research*, 2005, **39**, 2722–2730.
8. M.J. Nieuwenhuijsen, R. Smith, S. Golfinopoulos, N. Best, J. Bennett, G. Aggazzotti, E. Righi,G. Fantuzzi, L. Bucchini, S. Cordier, C.M. Villanueva, V. Moreno, C. Vecchia, C. Bosetti, T. Vartiainen, R. Rautiu, M. Toledano, N. Iszatt, R. Grazuleviciene and M. Kogevinas, *Journal of Water and Health,* 2009, **07**, 185–207.
9. P.C. Singer, *Water Science and Technology: Water Supply*, 2002, **2,** 487–492.
10. V. Uyak, K. Ozdemir and I. Toroz, *Science of the total environment*, 2008, **390,** 417-424.
11. B.M. McRae, T.M. LaPara and R.M. Hozalski, *Chemosphere*, 2004, **55**, 915–925.
12. L. Zhang, W.A. Arnold and R.M. Hozalski, *Environmental Science and Technology*, 2004, **38**, 6881-6889.
13. P. Zhang, T.M. Lapara, E.H. Goslan, Y.F. Xie, S.A. Parsons and R.M. Hozalski, *Environmental Science and Technology,* 2009, **4**, 3169-3175.
14. C.Pluchon, J.B. Sérodes, C. Berthiaume, S.J. Charette, Y. Gilbert, G. Filion, J. Fournier-Larente, M. Rodriguez and C. Duchaine, *Water Science & Technology: Water Supply, 2013,* **13**, 447–461.
15. Y.Zhang, D. Martinez, C. Collins, N. Graham, M.R. Templeton, J. Huang, and M. Nieuwenhuijsen, *Journal of Water Supply: Research and Technology—Aqua* , 2011, **60**, 275-285.
16. C. Chen, S. Chang and C. Wang, *Journal of Chromatographic Science,* 2009, **47**, 67-74.
17. T.D Lekkas and A.D Nikolaou, *Environmental Engineering Science*, 2004, **21**, 493-506.
18. P.C. Singer, *Water Science and Technology: Water Supply*, 2004, **4**, 151–157.
19. Y. Ma and C. Chiang, *Journal of Chromatography A,* 2005, **1076**, 216-219.
20. S. Parvez, Z. Rivera-Núñez, A. Meyer and J.M. Wright, *Environmental Research*, 2011, **111**, 499–509.
21. U.S. Environmental Protection Agency.Stage 2 Disinfectants and Disinfection Byproduct Rule (Stage 2 DBP rule). http://water.epa.gov
22. Y. Zhang, C. Collins, N. Graham, M.R. Templeton, J. Huang and M. Nieuwenhuijsen, *Water and Environment Journal,* 2010, **24**, 237–245.
23. S.Chowdhury, *Water International,* 2013, **38**, 61–77.
24. T. Bond, J. Huang, N. Graham, M.R. Templeton, *Science of the Total Environment, 2014,* **470–471**, 469–479.
25. S.K. Golfinopoulos, A.D. Nikolaou, *Desalination*, 2005, **176**, 13-24.
26. L. Liang and P.C. Singer, *Sci. Technol.*, 2003, **37**, 2920-2928.

27. X. Zhang and R.A. Minear, *Water Research*, 2002, **36**, 3665–3673.
28. Y. Qi, C. Shang and I.M.C. Lo, *Water Research*, 2004, **38**, 2375–2383.
29. T.K. Nissinen, I.T. Miettinen, P.J. Martikainen and T. Vartiainen, *Chemosphere, 2002,* **48**, 9–20.
30. P.C. Singer, A. Obolensky and A. Greiner, *Water Works Assoc.*, 1995, **87**, 83-92.
31. M.J. Rodriguez, J.B. Sérodes, P. Levallois and Proulx, François, *J. Environ. Eng. Sci.*, 2007, **6**, 355-365.
32. N. Knight, K. Watson, S. Carswell, E. Comino, E. and G. Shaw, *Air, Soil and Water Research,* 2011, 1–17.
33. V. Uyak, K. Ozdemir, I. Toroz, *Science of the Total Environment*, 2007, **378**, 269-280.
34. S. Li, X. Zhang, W. Liu, L. Cao and Z.Wang, *Journal of Environmental Science and Health, Part A: Toxic/Hazardous Substances, 2001,* **A36**, 475–481.
35. L.A. Rossman, R.A. Brown, P.C. Singer and J.R. Nuckols, *Wat. Res.,* 2001, **35**, 3483–3489.
36. A.M. Evans, J.M. Wright, A. Meyer and Z. Rivera-Nunez, *Water research*, 2013, **47**, 6130-6140.
37. W. Bayless and R.C. Andrews, *Journal of Water and Health*, 2008, **06.1,** 5-22.
38. A. S. Grigorescu, T.M Lapara, R.M Hozalski, *Rom. J. Biochem., 2010,* **47**, 165–177.
39. W. Wang and L. Zhua, *Journal of Hazardous Materials* 2010, **174,** 40–46.
40. S.A. Parsons and Goslan, E.H. 2011, *Cranfield University study.*
41. H. Tung and Y.F Xie, *Water research*, 2009, **43**, 971-978.
42. P. Zhang, R.M Hozalski, L.H Leach, A.K Camper, E.H Goslan, S.A Parsons, Y.F Xie, and T.M LaPara, *Federation of European Microbiological Societies, Microbiol Lett*, 2009, **297**, 203–208.
43. R.M Hozalski., W.A Arnold, C.L Chun, T.M LaPara, J.Y Lee, C.R Pearson, P Zhang, *ACS Symposium Series of American Chemical Society,* 2008, **23**, 335-345.

FACTORS AFFECTING THE DETECTION OF DISSOLVED ORGANIC NITROGEN VIA ELECTRODIALYSIS PRETREATMENT

Baiyang Chen, Anbang Zhu and Yueting Sui

Harbin Institute of Technology Shenzhen Graduate School, Shenzhen Key Laboratory of Water Resource Utilization and Environmental Pollution Control, Shenzhen 518055, China

1 INTRODUCTION

Dissolved organic nitrogen (DON) is ubiquitously present in the environment, and recently a growing concern of DON has been related to the formation of nitrogenous disinfection byproducts (N-DBPs) [1-3], which are shown to be more toxic than some regulated, carbonaceous DBPs (C-DBPs) [4, 5]. Therefore, a better understanding of the occurrence and control of DON becomes important.

However, existing methods for DON analysis in water are based on an indirect, subtractive process that takes dissolved inorganic nitrogen (DIN) away from total dissolved nitrogen (TDN), which may accumulate and exaggerate errors of each nitrogen species' measurement. As a result, researchers often obtained highly variable or negative DON values with this subtraction process, especially when the ratio of DIN to DON exceeds 0.6 [6, 7]. So, a tool to enable direct and more accurate measurement of DON is desired; and the tool may serve as a key to enhance our understandings of DON occurrence and environmental implications.

Electrodialysis (ED) is one of the options for sample pretreatment [8], and recently it has been applied by us [9] and other researchers to enable reliable DON quantification [10]. The process employs ion exchange membranes (IEMs) to separate DIN from DON and meanwhile take advantage of the electrical force to facilitate the process. In addition, separation of DIN and DON is also related to the distinct difference in molecular weight (MW) between DIN and DON [11]. In early study, researchers have coupled ED with reverse osmosis to successfully recover natural organic matter (NOM) from freshwater and seawater [12, 13]. However, although researchers has reported the effectiveness of ED on NOM recovery and DON analysis, the factors influencing the electrodialysis process for DON measurement are not well-unknown so far, which may hinder common use of it by more researchers.

So, this study aims to evaluate these factors to optimize the electrodialysis efficiency. The factors are generally classified into three categories: reactor configuration, electrolyte, and target compound property. Specifically, it includes the electricity voltage, electrolyte flow rate, electrolyte type and concentration, initial sample concentration and salinity, presence of competing ions, membrane specific area, and electrode materials.

2 MATERIALS & METHODS

The designed ED unit is made by nitrogen-free plexiglass in a rectangle shape. The length of a typical ED unit is 60 mm, and the width of the stack is 40mm. Each unit has three stacks, including two side modules to hold electrodes and contain the electrolyte solution, and one middle module to hold the sample water (Figure 1). A picture of our experimental equipment is shown in Figure 1. The anion membrane (AMI-7001) and cation membrane (CMI-7000) used in this experiment were purchased from Membranes International Inc. (USA). The power input of the electrodialysis unit was provided by a DC transformer (MCH-305DII, MEILI). The electrode material is aluminum, while another electrode made of copper was also evaluated.

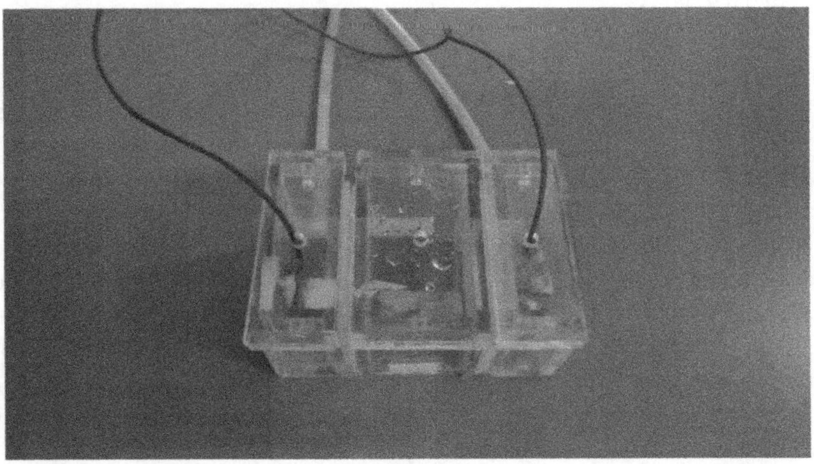

Figure 1 *An electrodialysis unit evaluated by this study.*

The chemicals evaluated include three types of DIN (i.e., ammonia, nitrite, and nitrate) and a series of DON model compounds with varying molecular weights (MW), including urea (60 g/mol); glycine (75 g/mol), glutamic acid (147 g/mol), tryptophan (204 g/mol), tripeptide glutathione (307 g/mol); protein bovine serum albumin (BSA, 66,000 g/mol), and three representative humic substance isolates from Leonardite, Elliot, and Nordic Rivers as purchased from the International Humic Substance Society (IHSS). Before experiments, organic compounds were soaked in ultrapure water for at least 1 day. Usually, the original concentrations of DIN were 100mg-N/L while those of organic nitrogen were set at 10mg-N/L. From an earlier study we already know that an average 88% (ranging from 78% ~100%) recovery of DON compounds can be obtained [9], therefore further improvement is limited with such a high recovery level. So, this study focused on factors affecting DIN removal instead of DON recovery. Nitrate was selected as a representative DIN because it has the highest MW of all three DIN species (i.e., ammonia, nitrite, and nitrate), with slower removal kinetics than ammonia by ED, and is more universally present than nitrite in natural and engineering waters.

Ammonia was detected by a spectrophotometer (DR 3900, Hach Co., Ltd) using EPA method 350.2 with its method detection limit (MDL) at 0.02 mg-N/L. Nitrite was analyzed using the same spectrophotometer but based on EPA method 354.1, with a MDL of 0.003 mg-N/L. Nitrate was measured with an ultraviolet (UV) spectrophotometer (Model UV765, Shanghai Precision & Scientific Instrument Co., Ltd) according to the American Public Health Association (APHA) Method 4500-NO_3^- B (MDL = 0.1mg/L) and an ion chromatograph (IC2010, Tosoh Inc., Japan) according to EPA method 300.0, which has a MDL of 0.005 mg-N/L. TDN analysis was based on APHA method 4500-N C, using UV spectrophotometer (UV765), with a MDL of 0.1 mg-N/L. Conductivity was recorded using an electrode meter (SX-650, San-Xin Instrumentation, Inc., China), and pH was monitored using a pH meter (pH100, Extech Instruments Corporation, USA). To ensure quality assurance and control, the limits of relative standard deviation (RSD) of all measured parameters were less than 10%. In case of violation, more analyses were conducted.

3 RESULTS & DISCUSSION

3.1 Influence of Reactor Configuration

Electric force is a driving force for separation of DIN and DON in ED process. Higher voltage input surely leads to more rapid removal of DIN, as shown in Figure 2. The major concern regarding excessive voltage input is the phenomena of concentration polarization, which may cause unexpected huge pH change for samples and retard the DIN removal kinetics. So, in general, the electricity input cannot increase as much as desired. In this study, the final voltage was set as 30V where no side effects were observed.

Electrode material is another important consideration in industrial desalination processes using ED. In order to avoid frequent replacement of electrode as a result of corrosion or coating of materials, inert material such as Ti and RuO_2 are often applied. However, for analytical chemistry purposes, it is unnecessary to apply these expensive materials. Instead, more consideration was given to costs and efficiency of DIN removal. Figure 2 shows a comparison of aluminum and copper electrodes in the removal of nitrate. Both materials can scarify themselves to reduce the formation of oxygen and chlorine gas at the anode chamber so as to maintain relatively constant pH conditions [9]. Copper appears more effective than aluminum in eliminating nitrate under the same operational conditions. However, copper can easily lose electrons to form precipitative solid, i.e., $Cu(OH)_2$, at cathode side, which may block the membrane. In addition, it exhibited blue color at the anode side, making both sides appear either dirty or muddy. So, the use of copper is not preferred finally.

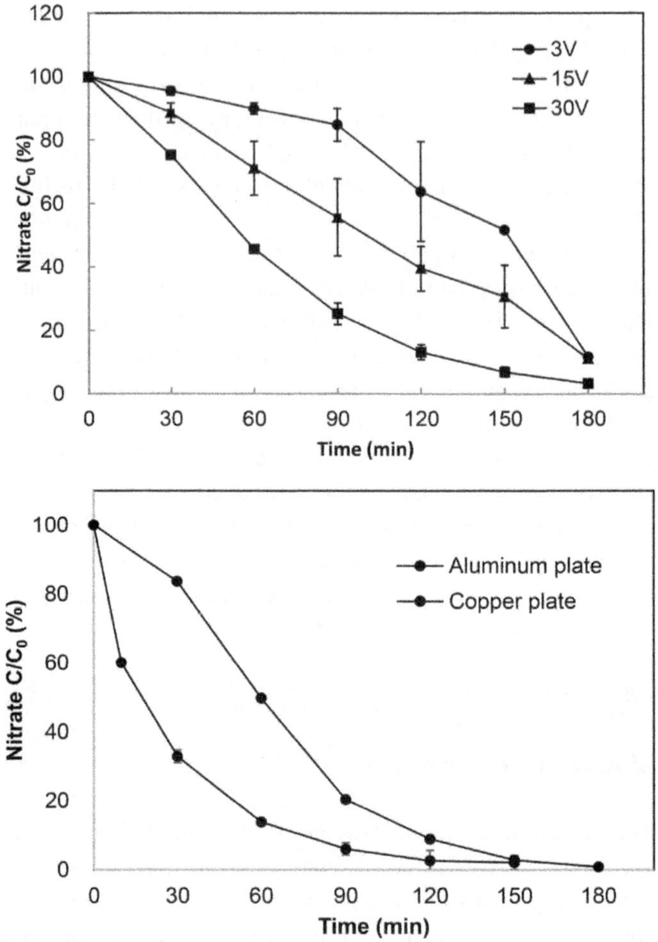

Figure 2 *Effects of power input (top) and electrode material (bottom) on the DIN removal efficiency by ED*

Figure 3 presents the effect of specific membrane area on nitrate removal. Specific membrane area is defined as the area of ion-exchange membrane per unit of sample volume. In this experiment, two sets of ED units were examined. One unit has an opening of 5 cm^2 to treat 100 ml water sample, and the other unit has a 10 cm^2 muzzle area to treat 50 ml water sample; both have the same electrolyte chamber sizes and operating conditions. The results indicate that the unit with higher specific area (0.2 cm^2/cm^3) had much greater nitrate removal rate than that with low specific area (0.05cm^2/cm^3), suggesting that as long as the membrane muzzle is sufficient a small-scale ED unit may enable more rapid pretreatment of water samples.

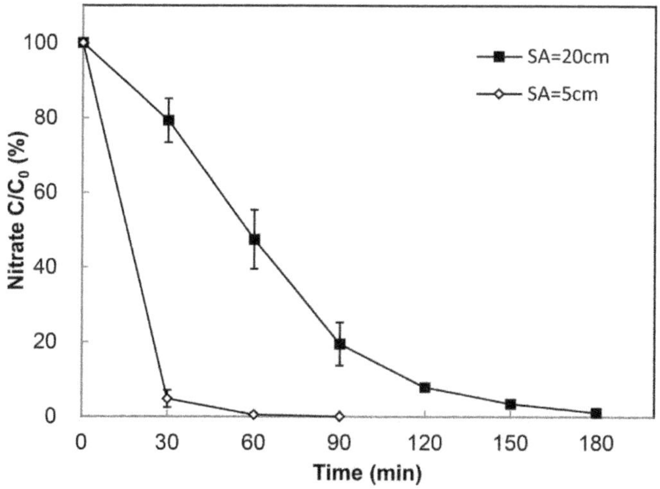

Figure 3 *Effect of IEM specific area (SA: volume of water per area of membrane) on the DIN removal efficiency by ED*

3.2 Influence of Electrolyte

In order to increase electric current and facilitate ion movement, some electrolyte using NaCl were added to both anode and cathode side chambers. In order to maximize the DIN concentration gradient between sample and electrolyte solution, the electrolyte solutions were continuously refreshed to carry away the permeated DIN. However, based upon this study, it appears that the flow rate of carrying media is incapable of affecting the DIN removal significantly (Figure 4). This may imply that the imposed electric current is strong enough to prevent DIN backflow even though its concentration in the side chamber is larger than that in the sample chamber at the final phase of the process. In this context, the use of a metering pump for electrolyte replenishment seems unnecessary.

In contrast, the use of electrolyte appears necessary. An increase of electrolyte concentration from zero to 0.25M in the side compartment has facilitated the nitrate removal considerably (Figure 4). However, further increase of electrolyte concentration added just a few additional benefits in DIN removal when the concentration increased from 0.25M to 0.50M.

3.3 Influence of Compound

In an earlier study, we have identified several important structural and compositional properties that may affect the fate and transport of nitrogen compounds in ED process [9]. For example, because nitrous acid has a low acid-dissociation constant (pKa) of 3.3, nitrite can dominate in nonionic form at $pH < 3.3$ conditions, making its removal slower

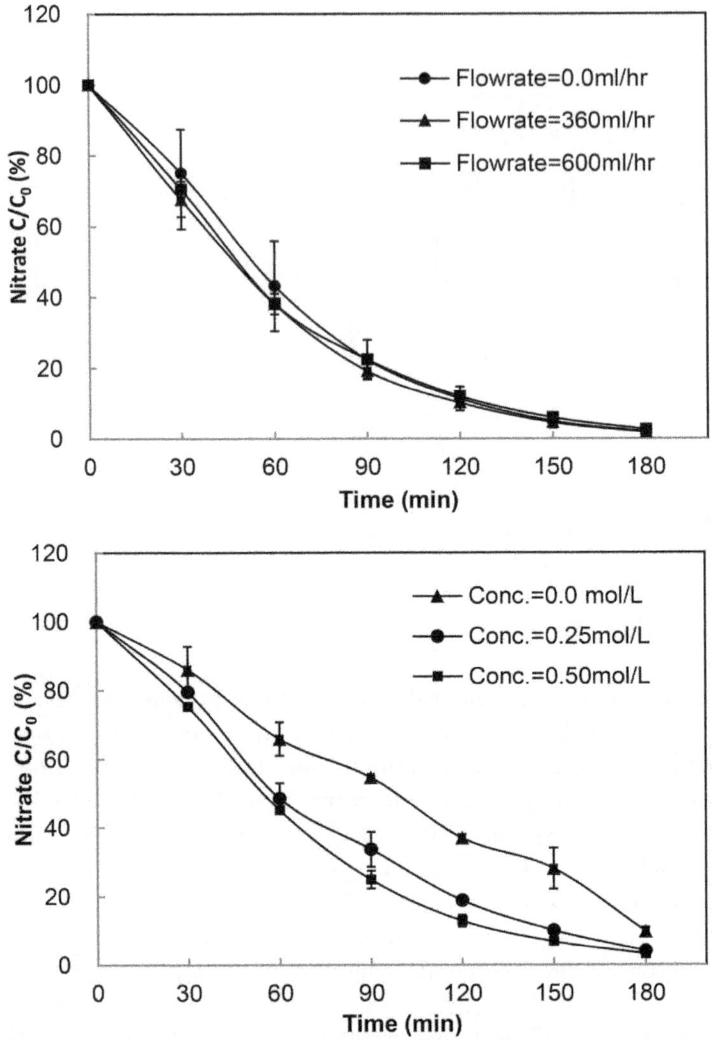

Figure 4 *Effects of electrolyte flowrate (top) and concentration (bottom) on the DIN removal efficiency by ED*

than nitrate and ammonia. In contrast, because certain amino acids, such as glutamic acid and glutathione, exist in ionic form in normal water, their recovery level is lower than other DON compounds. In addition, the aromatic fraction of organic compound is more likely to adsorb to membrane/chamber and therefore result in DON loss. Consequently, the humic substance extracted from Leonardite exhibited lower recovery of DON than those from Nordic lake and Elliott Soil. The information to be provided herein hence may supplement aforementioned data.

Figure 5 shows the effects of initial nitrate concentration and presence of other competing anions on nitrate removal. For a wide range of nitrate concentrations, ED pretreatment maintained a stable yet considerably high efficiency. This assures us that the ED process can function as well in low DIN-containing water as in nitrogen-rich samples. The presence of salt and typical ions like chloride, sulfate, and phosphate, did not alter the pretreatment efficiency dramatically either in this study. This indicates that competition of ions in crossing the membrane may not be a big concern for DIN removal. Instead, the type of DIN species is likely to be a more important factor in the ED process. As shown from Figure 6, ammonia was eliminated faster than nitrate, indicating that low molecular weight (MW) compounds have more potential to be removed in this process.

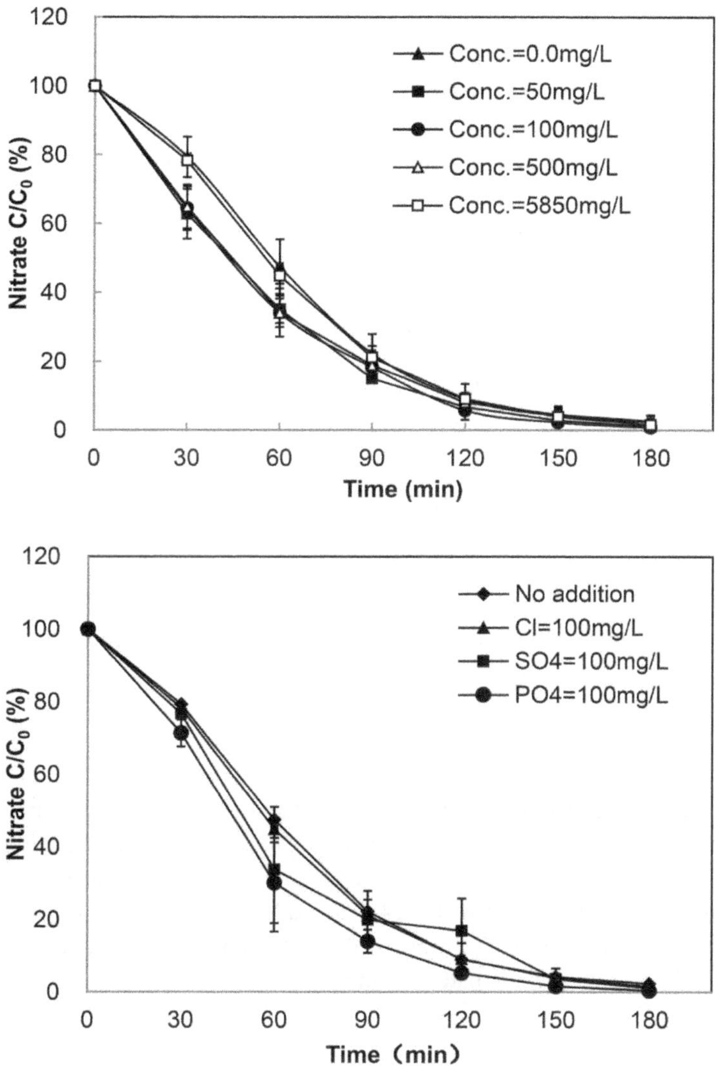

Figure 5 *Effects of initial concentration of nitrate (top) and competing ions (bottom) on the DIN removal efficiency by ED*

4 CONCLUSIONS

Overall, this study evaluated a series of factors that may impact the ED efficiency on DIN removal, which is the key for DON measurement. Several important conclusions are obtained as follows.

1. Power input can facilitate the ED process; and as long as concentration polarization phenomena does not occur, the input voltage should be set as high as possible;
2. Copper electrode is preferred over aluminum in removing nitrate in terms of efficiency; however, copper ion is easily precipitated at high pH condition and makes the electrolyte solution suspended with solids. So, in practice aluminum was applied with a relatively long reacting time to compensate the efficiency;
3. ED configuration is crucial, with an increase of specific area of IEM leading to a significant higher removal rate of DIN and thus lowering the reaction time by half;
4. In the electrolyte chamber, the electrolyte replacement rate exhibited a minor effect on DIN removal, showing that nitrate permeation is not slowed down when the nitrate concentration gradient between two sides of IEM becomes small, which suggests that it is perhaps unnecessary to feed fresh electrolyte continuously. In contrast, higher electrolyte concentration results in higher DIN removal;
5. In the sample chamber, the initial concentration of nitrate and the presence of anions in water seem to be insignificant in DIN removal; however, the type of DIN and DON are critical in either DIN removal efficiency or DON recovery.

Ultimately, we have designed an ED unit capable of separating DON from DIN by >99% within 90 minutes with a consumption of water for only 50 ml. Although the experiments are not comprehensive or sufficient to cover all aspects, the results may provide a glimpse of important points to help optimize the ED process. An automated, easy-to-use tool may be developed by carefully considering these factors and their influences.

ACKNOWLEDGEMENTS

The study was financially supported by the National Natural Science Foundation of China (Grant No: 51278144), Shenzhen Science & Technology R&D Funding (Grant No: JCYJ20120613150442560), and the Shenzhen Funding for Returned Oversea Chinese (Project No: KQCX20130627094615414). In addition, we are also grateful of the contributions from colleagues like Zhong Tang, Xiaoqi Guo, Shengcun Ma, etc. The method of DON measurement is patent pending now.

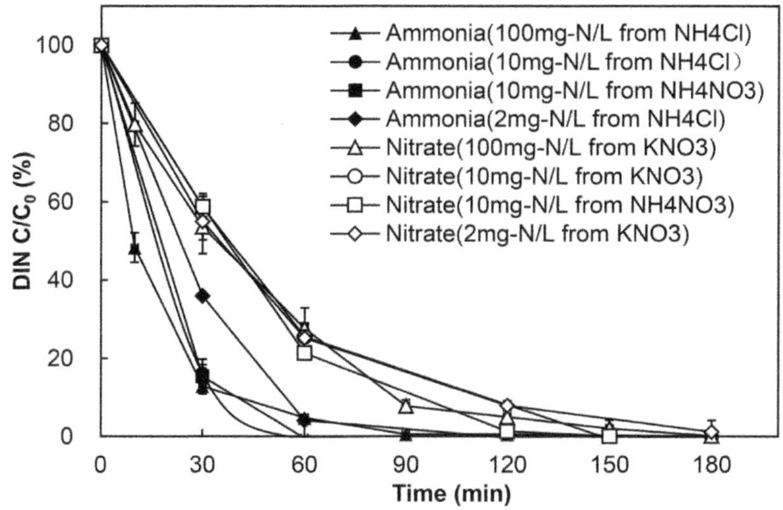

Figure 6 *Effect of type and concentration of DIN on the DIN removal efficiency by ED (Adapted with permission from Zhu etc. 2015. Copyright (2015) American Chemical Society)*

REFERENCES

1. W. Lee, P. Westerhoff and J. P. Croue, *Environ Sci Technol*, 2007, **41**, 5485.
2. X. Yang, C. Shang and P. Westerhoff, *Water Res*, 2007, **41**, 1193.
3. S. W. Krasner, P. Westerhoff, B. Chen, B. E. Rittmann, S.-N. Nam and G. Amy, *Environ Sci Technol*, 2009, **43**, 2911.
4. M. J. Plewa, E. D. Wagner, P. Jazwierska, S. D. Richardson, P. H. Chen and A. B. McKague, *Environ Sci Technol*, 2004, **38**, 62.
5. M. G. Muellner, E. D. Wagner, K. McCalla, S. D. Richardson, Y.-T. Woo and M. J. Plewa, *Environ Sci Technol*, 2007, **41**, 645.
6. W. Lee and P. Westerhoff, *Environ. Sci. Technol.*, 2005, **39**, 879.
7. J. Vandenbruwane, S. De Neve, R. G. Qualls, J. Salomez and G. Hofman, *Sci. Total Environ.*, 2007, **386**, 103.
8. B. Chen, P. Westerhoff, L. Zhang, A. Zhu, X. Yang and C. Wang, *Crit. Rev. Env. Sci. Technol.*, 2014, **45**, 249.
9. A. Zhu, B. Chen, L. Zhang and P. Westerhoff, *Anal. Chem.*, 2015, **87**, 2353.
10. K. Chon, Y. Lee, J. Traber and U. von Gunten, *Water Res*, 2013, **47**, 5381.
11. E. M. Perdue and J. D. Ritchie, in *Treatise on Geochemistry*, eds. D. H. Editors-in-Chief: Heinrich and K. T. Karl, Pergamon, Oxford, 2003, pp. 273.
12. J. F. Koprivnjak, E. M. Perdue and P. H. Pfromm, *Water Res*, 2006, **40**, 3385.
13. B. K. Gurtler, T. A. Vetter, E. M. Perdue, E. Ingall, J. F. Koprivnjak and P. H. Pfromm, *J. Membr. Sci.*, 2008, **323**, 328.

COMPUTATIONAL CHEMISTRY INVESTIGATION OF THE BROMIDE-CATALYZED FORMATION OF NDMA FROM *N,N*-DIMETHYLSULFAMIDE DURING OZONATION

Daniela Trogolo,[1] *Brijesh K. Mishra,*[1] *Michèle B. Heeb,*[2] *Urs von Gunten,*[2,3] *J. Samuel Arey* [1,3]

[1] Environmental Chemistry Modeling Laboratory, EPFL, Station 2, 1015 Lausanne (CH)
[2] Laboratory for Water Quality and Treatment, EPFL, Station 2, 1015 Lausanne (CH)
[3] Eawag, Swiss Federal Institute of Aquatic Science and Technology 8600 Dübendorf (CH)

1 INTRODUCTION

N,N-dimethylsulfamide (DMS), an environmental metabolite of the fungicide tolylfluanid, was found to lead to the formation of *N*-nitrosodimethylamine (NDMA) during ozonation of bromine-containing drinking water treatment.[1] A previous experimental study proposed a detailed NDMA formation mechanism from DMS, suggesting several reactive intermediates that could not be measured experimentally.[2] In this present work, we aim to further assess the NDMA formation mechanism using quantum chemistry modeling.[3]

2 COMPUTATIONAL METHODS

2.1 Model Chemistry Selection

All calculations were performed using the Gaussian09 program.[4] The M05[5] density functional theory model chemistry was applied to geometry optmization and vibrational frequency analysis calculations, whereas energy single point calculations were carried out with the computationally more expensive B2PLYPD[6,7] functional on gas phase M05 geometries, used to confirm the M05 energy values for all the stationary structures. All calculations were conducted using the Dunning aug-cc-pVTZ[8–11] basis set. The M05 and B2PLYPD functionals were selected for their ability to handle potentially difficult electronic structures of ozone, halogen oxides, and nitroso compounds that are characterized by moderate-to-high multireference character.[12,13]

2.2 Aqueous Gibbs Free Energy of Reaction

1 Molar standard state free energies of reaction in aqueous phase were computed as follows. For reaction steps b, c, f, g, j, and n, all species were modeled in aqueous phase with the SMD[14] implicit solvation model, using the following formula:

$$\Delta G_{aq,rxn} = \sum_i \nu_i(E_{g,elec}(A_i) + G_{g,therm}(A_i) + \Delta G_{SMD}(A_i)) + \Delta G_{standardstate} \quad (1)$$

where $E_{g,elec}(A_i)$ is the total gas phase electronic energy of species A_i, and $G_{g,therm}(A_i)$ is the thermal contribution to the Gibbs free energy coming from translations, rotations, and vibrations in gas phase.$\Delta G_{SMD}(A_i)$ is the SMD estimate of free energy of transfer from the gas phase into aqueous solution, ν_i is the stoichiometric coefficient of species A_i for the reaction step, and $\Delta G_{standardstate}$ accounts for any relevant standard state changes of the solute.[15]

For the remaining reactions, we expected that one or more molecules of the solvent (i.e water) play a direct role in the molecular mechanism. For these cases, the solvation effects were considered by applying the SMD implicit solvation model and one or more explicit water molecules. The introduction of water molecules and consequent formation of clusters of water was treated by taking advantage of the cluster-continuum approach developed by Bryantsev.[16] In our study, explicit water molecules were generally associated to bromine atom, nitrogen atoms, and to the oxygen atoms of the SO_2 fragment. For these reactions, the free energies of reaction were computed with this formula:

$$\Delta G_{aq,rxn} = \sum_i \nu_i(E_{g,elec}(A_i(H_2O)_n) + G_{g,therm}(A_i(H_2O)_n) + \Delta G_{SMD}(A_i(H_2O)_n)) + \Delta G_{standardstate} + \Delta G_{clustercorrection} \quad (2)$$

where the terms $E_{g,elec}$, $G_{g,therm}$, and ΔG_{SMD}in eq 2 are determined for each solute + explicit solvent cluster $A_i(H_2O)_n$. The term $\Delta G_{standardstate}$accounts for standard state changes of the solutes, whereas the $\Delta G_{clustercorrection}$ term describes the water cluster thermodynamic contributions, as explained by the cluster-continuum approach.[16]

2.2 Aqueous Gibbs Free Energy of Activation and Aqueous Reaction Rate Constants

The Gibbs free energy of activation was calculated for all reactions for which no experimental rate constant was available. The free energy of activation for a given reaction can be estimated as follows:

$$\Delta G^{\ddagger}_{aq,rxn} = E^{\ddagger}_{g,elec} + \Delta G^{\ddagger}_{SMD} - \sum_i (E_{g,elec}(R_i) - \Delta G_{SMD}(R_i)) + \Delta G_{clustercorrection} + \Delta G_{standardstate} \quad (3)$$

where the superscript ‡ indicates the computed property of the transition structure, and R_i indicates the preceding stationary structure(s).

Reaction rate constants can be estimated using the relationship between free energy of activation and kinetic constant as expressed by the Eyring-Polanyi equation in the transition state theory (TST) approach.[17,18]

$$k_{TST} = \frac{k_B T}{h} e^{\frac{-\Delta G^{\ddagger}_{rxn}}{RT}} \quad (4)$$

where k_B is the Boltzmann constant, h is Planck's constant, T is temperature, R is the molar gas constant, and $\Delta G^{\ddagger}_{aq,rxn}$ is determined based on eq 3.

3 RESULTS

The bromide-catalyzed mechanism of NDMA formation from DMS during ozonation processes was investigated computationally.[3] Figure 1 reports a detailed scheme of the reaction pathways.

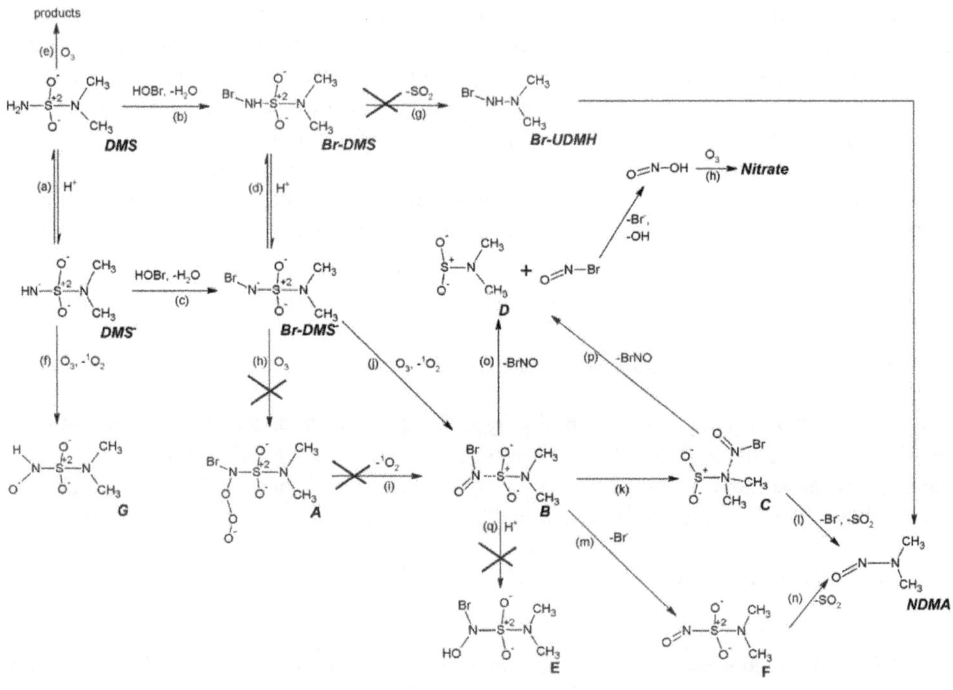

Figure 1. *Molecular mechanism of NDMA formation from DMS and other hypothesized pathways that were found to be less important.*[3]

Gibbs free energies of reaction were used to assess the thermodynamic feasibility of hypothesized reactions, whereas kinetic information was deduced from the free energy of activation for each key step (Table 1).

Ozone reaction with naturally occurring bromide produces HOBr, which can react with DMS (step b) and also with deprotonated DMS⁻ anion (step c). According to quantum chemistry calculations and stopped-flow measurements,[19] the bromination of the *N,N*-dimethylsulfamide species arises predominantly from the HOBr reaction with deprotonated DMS⁻ anion to form the intermediate Br-DMS⁻, even though DMS is the predominant species at neutral pH.

Table 1. *Estimates for Aqueous Gibbs Free Energies of Reaction and of Activation for the Elementary Reaction Steps of the NDMA Formation Pathway.*[3]

Reaction step	Number of explicit waters	$\Delta G_{aq,\, rxn}$ [kcal/mol]	$\Delta G^{\ddagger}_{aq,rxn}$ [kcal/mol]
b	0/3	-8	18
c	0	-18	Not available
f	0	2	Not available
g	0	-14	36
j	0	-31	11
k	5	-4	4
l	5	-26	<1
m	5	6	13
n	0	-38	3
o	5	-3	3
p	5	1	5

The electron-rich sulfamoyl nitrogen of the Br-DMS$^-$ anion can then undergo electrophilic attack by ozone (step j), resulting in the formation of the fragile intermediate (BrNO)-$SO_2N(CH_3)_2$ and in the concerted release of singlet oxygen. The intermediate (BrNO)-$SO_2N(CH_3)_2$ is stabilized by specific interactions with the solvent, according to our simulation.

Computational analysis of the charge distribution and bond lengths of this intermediate reveals that the BrNO group is loosely bound to the dimethylaminosulfinate $(SO_2)N(CH_3)_2^-$ group, and these fragments can cleave rapidly (step o). However, BrNO can also migrate from the sulfur atom to nitrogen atom of the dimethylaminosulfinate anion, following a mechanism that may be interpreted as an exothermic intramolecular nitrosation of the electron-rich dimethylaminosulfinate anion by the electrophile BrNO (step k). This intramolecular reaction leads to the formation of intermediate ($SO_2N(CH_3)_2$-BrNO),which can easily decompose to NDMA via a concerted and almost barrierless departure of both Br^- and SO_2 (step l).

4 CONCLUSION

Computational chemistry enabled us to estimated thermodynamic properties and kinetic constants of the elementary reactions involved in the NDMA formation mechanism. The results from our investigation provide new insights into the chemical equilibria, the reaction kinetics, and the detailed molecular events of this complex NDMA formation pathway.

References

(1) Schmidt, C. K.; Brauch, H.-J. *Environ. Sci. Technol.* **2008**, *42* (17), 6340–6346.

(2) Von Gunten, U.; Salhi, E.; Schmidt, C. K.; Arnold, W. A. *Environ. Sci. Technol.* **2010**, *44* (15), 5762–5768.

(3) Trogolo, D.; Mishra, B. K.; Heeb, M. B.; von Gunten, U.; Arey, J. S. *Environ. Sci. Technol.* **2015**, *49*, 4163–4175.

(4) Frisch, M. J.; Trucks, G. W.; Schlegel, H. B.; Scuseria, G. E.; Robb, M. A.; Cheeseman, J. R.; Scalmani, G.; Barone, V.; Mennucci, B.; Petersson, G. A.; Nakatsuji, H.; Caricato, M.; Li, X.; Hratchian, H. P.; Izmaylov, A. F.; Bloino, J.; Zheng, G.; Sonnenberg, J. L.; Hada, M.; Ehara, M.; Toyota, K.; Fukuda, R.; Hasegawa, J.; Ishida, M.; Nakajima, T.; Honda, Y.; Kitao, O.; Nakai, H.; Vreven, T.; Montgomery Jr., J. A.; Peralta, J. E.; Ogliaro, F.; Bearpark, M.; Heyd, J. J.; Brothers, E.; Kudin, K. N.; Staroverov, V. N.; Kobayashi, R.; Normand, J.; Raghavachari, K.; Rendell, A.; Burant, J. C.; Iyengar, S. S.; Tomasi, J.; Cossi, M.; Rega, N.; Millam, M. J.; Klene, M.; Knox, J. E.; Cross, J. B.; Bakken, V.; Adamo, C.; Jaramillo, J.; Gomperts, R.; Stratmann, R. E.; Yazyev, O.; Austin, A. J.; Cammi, R.; Pomelli, C.; Ochterski, J. W.; Martin, R. L.; Morokuma, K.; Zakrzewski, V. G.; Voth, G. A.; Salvador, P.; Dannenberg, J. J.; Dapprich, S.; Daniels, A. D.; Farkas, Ö.; Foresman, J. B.; Ortiz, J. V.; Cioslowski, J.; Fox, D. J. Gaussian Inc. Wallingford CT 2009.

(5) Zhao, Y.; Schultz, N. E.; Truhlar, D. G. *J. Chem. Phys.* **2005**, *123* (16), 161103.

(6) Grimme, S. *J. Chem. Phys.* **2006**, *124* (3), 034108.

(7) Schwabe, T.; Grimme, S. *Phys. Chem. Chem. Phys.* **2007**, *9* (26), 3397–3406.

(8) Davidson, E. R. *Chem. Phys. Lett.* **1996**, *260*, 514–518.

(9) Dunning Jr., T. H. *J. Chem. Phys.* **1989**, *90*, 1007–1023.

(10) Dunning Jr., T. H.; Peterson, K. A.; Wilson, A. K. *J. Chem. Phys.* **2001**, *114* (21), 9244.

(11) Kendall, R. A.; Dunning Jr., T. H.; Harrison, R. J. *J. Chem. Phys.* **1992**, *96*, 6796–6806.

(12) Zhao, Y.; Truhlar, D. G. *J. Chem. Phys.* **2006**, *125* (19), 194101.

(13) Zhao, Y.; Tishchenko, O.; Gour, J. R.; Li, W.; Lutz, J. J.; Piecuch, P.; Truhlar, D. G. *J. Phys. Chem. A* **2009**, *113* (19), 5786–5799.

(14) Marenich, A. V.; Cramer, C. J.; Truhlar, D. G. *J. Phys. Chem. B* **2009**, *113* (18), 6378–6396.

(15) Ben-Naim, A. *Solvation Thermodynamics*; Plenum Press, New York, 1987.

(16) Bryantsev, V. S.; Diallo, M. S.; Goddard III, W. A. *J. Phys. Chem. B* **2008**, *112* (32), 9709–9719.

(17) Hill, T. L. *An Introduction to Statistical Thermodynamics*; Dover Publications, Inc., New York, 1986.

(18) Steinfeld, J. I.; Francisco, J. S.; Hase, W. L. *Chemical Kinetics and Dynamics*, Second Ed.; 1999.

(19) Heeb, M. B.; Criquet, J.; Zimmermann-Steffens, S. G.; von Gunten, U. *Water Res.* **2014**, *48*, 15–42.

DETECTION METHODS TO MONITOR THE DEGRADATION OF ORGANIC CHLORAMINES

Zuo Tong How, Kathryn L. Linge, Francesco Busetti and Cynthia A. Joll

Curtin Water Quality Research Centre, Department of Chemistry, Curtin University, GPO Box U1987, Perth, Western Australia, Australia 6845

1 INTRODUCTION

Organic chloramines have been predicted to be potentially cytotoxic and genotoxic and some organic chloramines, like *N*-chloroglycine, were found to be cytotoxic and genotoxic even at micromolar concentrations.[1] Recently, it was recommended[2] to give high priority to the characterisation of toxicological properties of organic chloramines with varying physical-chemical properties, as they are known to form in drinking water. However, little information is currently available on the occurrence or stability of organic chloramines in drinking water.

Organic chloramines may form when dissolved organic nitrogen-containing compounds, such as amino acids, amides and amines, present in water systems react with free chlorine[3] or inorganic chloramines[4, 5] used for disinfection. However, the main pathway of formation of organic chloramines is considered to be involving reactions between amino acids and free chlorine.[6, 7] After formation, organic chloramines typically degrade to other disinfection by-products, including aldehydes and nitriles.[8] Depending on the chemical structure, the half-life of an organic chloramine may vary from 13 minutes to more than 48 hours.[9-11] Organic chloramine stability is affected by many factors, although, in general, more basic organic chloramines have been found to be more stable.[12]

A number of analytical methods have been used to detect the presence of organic chloramines in water, including direct UV analysis, use of *N,N*-diethyl-*p*-phenylenediamine (DPD) to measure organic chloramines as part of combined chlorine, and iodometric methods to measure organic chloramines as part of total chlorine. When formed, organic chloramines have absorbance lambda maxima (λ_{max}) between 250 and 280 nm, and molar absorptivities of around 350 M^{-1} cm^{-1}.[9] Many studies have utilised these spectroscopic properties to monitor the formation[13] and degradation[9, 10] of pure organic chloramines by direct UV analysis.

The DPD method measures the concentration of free chlorine and chloramines (commonly known as combined chlorine, and including both organic and inorganic chloramines) in a sample by the spontaneous reaction between free chlorine and DPD indicator to produce a red colour in the absence of iodide ion.[14] When iodide is subsequently added into samples containing chloramines, the iodide reacts with the chloramines releasing free chlorine to react with the DPD indicator to produce the red colour. A small amount of iodide is added to determine the monochloramines concentration, and excess iodide ion is added to determine the dichloramines concentration, as the excess iodide will release additional chlorine from the dichloramines which will react rapidly with the DPD indicator.[14] In fact, the measurement of free chlorine represents all species which can oxidise the DPD indicator reagent to the red coloured species, and is thus often referred to as free chlorine equivalents. Although the DPD method is not able to differentiate between inorganic and organic chloramines, it has previously been used to monitor the formation and degradation of organic chloramines,[1, 13] as well as confirm that all free chlorine has reacted.[10, 13]

Iodometric titration is a commonly used analytical method to determine the concentration of oxidizing agents.[15] A strong oxidant, such as free chlorine, oxidises the iodide into triiodide producing a pale yellow solution in an acidic environment. When an excess of iodide is present in the solution, an equivalent amount of triiodide is produced to the amount of oxidant present (Equation 1). The triiodide is then titrated against sodium thiosulfate (Equation 2) until it become colourless; to ensure a better colour contrast, starch indicator is normally used during the titration to determine the end point. Although the iodometric titration cannot differentiate between different oxidants (e.g. between free chlorine and chloramines), previous studies have utilised Reaction (1) from the iodometric titration method to improve the sensitivity of organic chloramines for UV detection by converting the organic chloramines with low absorptivity (ε= 330-450 M^{-1} cm^{-1} at 250 nm) to a compound (triiodide) with higher absorptivity (ε= 25750 M^{-1} cm^{-1} at 353 nm).[13, 16]

$$HOCl + 3I^- + H^+ \rightarrow I_3^- + H_2O + Cl^- \qquad (1)$$

$$I_2 + 2S_2O_3^{2-} \rightarrow 2I^- + S_4O_6^{2-} \qquad (2)$$

In the current study, a colorimetric measurement method was developed for the detection and measurement of the concentration of all organic chloramines which were not suitable for direct UV measurement due to interference in absorbance from the precursors or the by-products of organic chloramines at wavelengths between 250 and 280 nm. Both direct UV and the colorimetric measurement method were then used for the confirmation of the formation of organic chloramines and investigation of the degradation of selected organic chloramines. Four amino acids, asparagine, glycine, phenylalanine and serine, were selected for this study. This suite of amino acids included polar, non-polar and aromatic amino acids with a range of different amino group basicity (Table 1).

Table 1 *Molecular weight, classification, pK_a and structures of the amino acids selected in this study.*

Name	Molecular Weight (g mol^{-1})	Classification	pKa of amino group	Structure
Glycine	75.07	Non-Polar	9.58	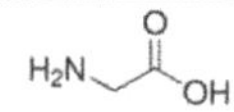
Phenylalanine	165.19	Non-Polar (Aromatic)	9.09	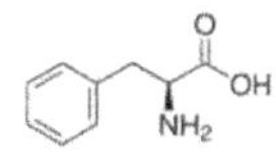
Serine	105.09	Polar	9.05	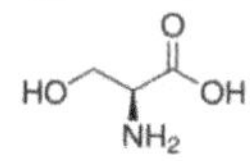
Asparagine	132.12	Polar	8.73	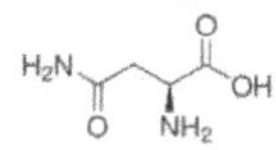

2 MATERIALS AND METHODS

2.1 Chemicals

Amino acids asparagine, glycine, phenylalanine and serine (purity ≥ 97%), aqueous sodium hypochlorite solution (10 – 15% chlorine), thioacetamide, and *N,N*-diethyl-*p*-phenylenediamine sulphate (purity ≥ 97%) were purchased from Sigma Aldrich (New South Wales, Australia). Potassium iodide, potassium dihydrogen phosphate and dipotassium hydrogen phosphate were purchased from Ajax FineChem (New South Wales, Australia); glacial acetic acid was purchased from Chem Supply (South Australia, Australia). Ethylenediaminetetraacetic acid was purchased from Asia Pacific Speciality Chemical (New South Wales, Australia), while sulfuric acid (purity 98%) was from Scharlab S.L. (Sentmenat, Spain). Monochlor F reagent was from Hach Pacific (Victoria, Australia). Ultrapure water was purified using an ion exchange system (IBIS Technology, Western Australia, Australia), followed by an Elga Purelab Ultra system with a 0.2 μm filter (Elga, High Wycombe, UK).

2.2 Formation of organic chloramines

Organic chloramines (0.06 or 0.6 mM) were prepared by adding sodium hypochlorite (HOCl) to solutions of the selected amino acids (Table 1) at a molar ratio of 1:5 (amino acids:HOCl). This molar ratio has been shown to minimise side reactions and ensure that the monochloramine is the major chloramine species formed.[13] The formation of organic chloramines from glycine and serine was confirmed by performing UV absorbance scans between 195 and 400 nm using an Agilent Technologies Cary 60 UV-Vis spectrometer (California, USA) with a deuterium arc lamp. As described in Section 3.1, a peak with λ_{max} between 250 and 280 nm is expected in the UV spectra if an organic chloramine is formed (Figure 1). For organic chloramines which have an interference peak at $\lambda = 250$ to 280 nm (e.g. phenylalanine), an alternative method to confirm organic chloramine formation was used. Such a method consists of first obtaining a UV spectrum of a chlorine solution followed by a UV spectrum after addition of the organic chloramine's precursor. The disappearance of the free chlorine peak at $\lambda = 292$ nm indicates that all chlorine has reacted (Figure 2), with the assumption that all chlorine reacted to form organic chloramines.

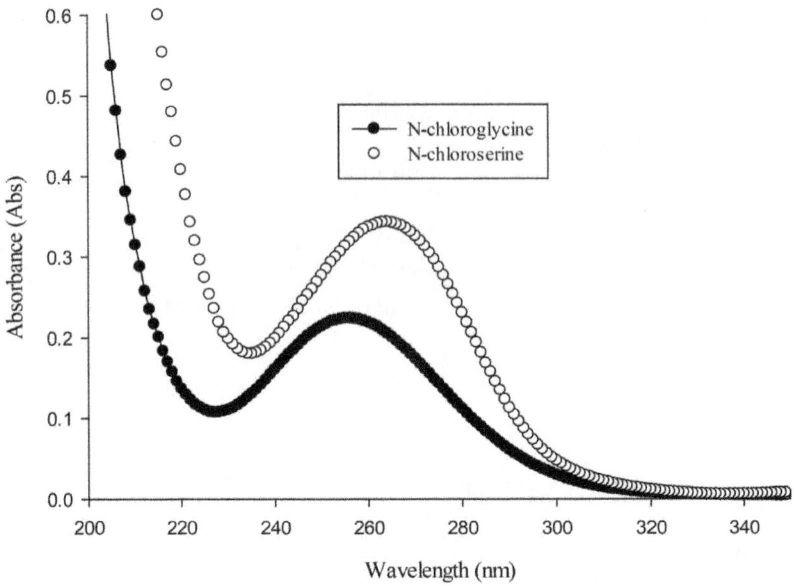

Figure 1 *UV spectra of organic chloramines (0.6mM) formed from chlorination of glycine and serine; a peak with λ_{max} between 250-280 nm confirms the formation of organic chloramines*

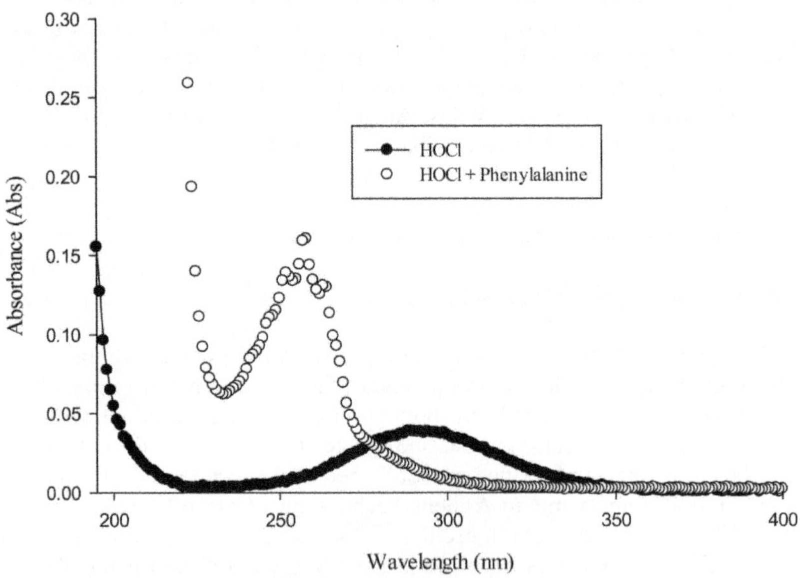

Figure 2 *UV spectra to confirm the reaction of free chlorine and phenylalanine. The disappearance of the peak at λ = 292 nm indicates that all free chlorine had reacted. The peak at λ = 254 nm is attributed to the excess phenylalanine present in solution.*

2.3 Measurement of organic chloramine degradation rates

The rates of degradation of organic chloramines were measured over 60 minutes at time intervals indicated in Table 2. Two measurement methods were used to determine the rate of formation and degradation of organic chloramines. The degradation of *N*-chloroglycine and *N*-chloroserine were monitored by direct UV measurement of the absorbance at $\lambda = 255$ nm. A colorimetric method, modified from a standard iodometric titration[17] (referred to as the triiodide colorimetric method), was used to measure the degradation of *N*-chloroasparagine and *N*-chlorophenylalanine. For the triiodide colorimetric method, glacial acetic acid (125 μL) and potassium iodide (125 μL of a 15 g L^{-1} solution) were added to 2.5 mL of sample and the absorbance was measured at $\lambda =$ 353 nm, assuming a molar absorptivity of triiodide of 25750 M^{-1} cm^{-1}.[18]

Table 2 *Time interval and duration used for the measurement of the degradation rates of selected organic chloramines*

Direct UV method	Time interval	1 min	5 min	10 min
	Duration	1-10 min	10-30 min	30-60 min
Triiodide method	Time interval	5 min		10 min
	Duration	1-30 min		30-60 min

3 RESULTS AND DISCUSSION

3.1 Monitoring formation and degradation of organic chloramines using direct UV measurement

A peak at $\lambda = 250$ to 280 nm is expected if an organic chloramine is formed, and this was found for both *N*-chloroglycine and *N*-chloroserine (Figure 1). However, interferences in the UV spectra for both phenylalanine and chlorinated asparagine meant that the direct UV method could not be used to monitor the formation of organic chloramines from these amino acids. Comparison of the UV spectra for all tested amino acids (Figure 3) shows that phenylalanine has a peak at $\lambda = 254$ nm which means that chloramines could not be monitored between $\lambda = 250$ and 280 nm. While asparagine did not have an interference peak in the region of $\lambda = 250$ to 280 nm in the UV scan of the original amino acid, the degradation of *N*-chloroasparagine resulted in a by-product that had an absorbance at $\lambda = 270$ nm (Figure 4), indicating that direct UV was not a suitable method for the measurement of the degradation of *N*-chloroasparagine.

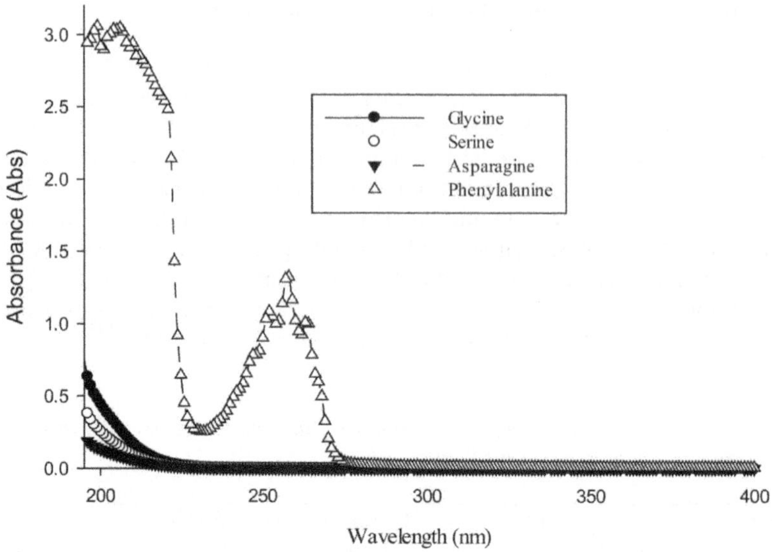

Figure 3 *UV spectra of selected amino acids. Phenylalanine has a peak at 254 nm and therefore the direct UV method was not suitable to measure the degradation of N-chlorophenylalanine.*

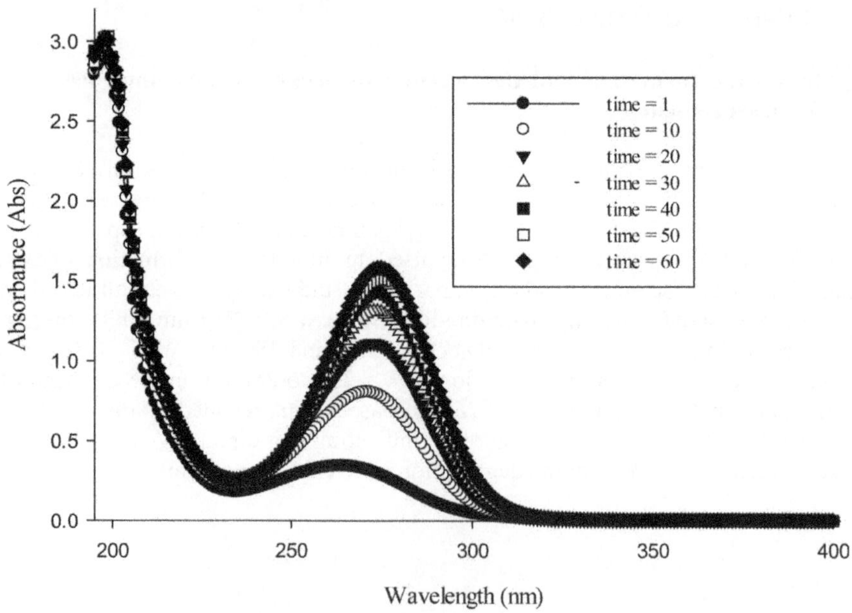

Figure 4 *UV spectra of chlorinated asparagine between 1 to 60 min of chlorination. The increased absorbance and the shift of λ_{max} indicated that the degraded product of* N*-chloroasparagine had similar UV absorbance to the* N*-chloroasparagine. The direct UV method was therefore not suitable for the measurement of the degradation of* N*-chloroasparagine*

3.2 Measurement method for the organic chloramines not suitable for direct UV measurement

Direct UV detection is a suitable method for the detection and measurement of a number of organic chloramines. However, as discussed in Section 3.1, not all organic chloramines are suitable for direct UV measurement due to interference peaks between $\lambda = 250$ and 280 nm.

For those amino acids that cannot be monitored by direct UV measurement, chloramine rates of degradation can be estimated by measuring the changes in combined chlorine (inorganic and organic chloramines) over time using the standard DPD colorimetric method (4500-Cl G),[19] with the assumption that no inorganic chloramine is present in solution. The DPD method has previously been used to monitor the formation and degradation of organic chloramines[1, 13], as well as confirm that all free chlorine has reacted[10, 13]. However, in this study it was found that the free chlorine concentration measured by the DPD method was extremely variable, increasing uncertainty in the combined chlorine measurements. It has previously been found that combined chlorine concentrations higher than 0.5 mg L^{-1}-Cl_2 can breakthrough into the free chlorine reading when using DPD methods.[14] The concentration of organic chloramines expected for this study was from 4 to 40 mg L^{-1}-Cl_2, therefore an interference in the free chlorine measurements was expected. While adding thioacetamide immediately after the DPD method is recommended to stop further reaction with combined chlorine,[14] this modification did not reduce the variability of the free chlorine concentration in this study. Therefore, the DPD method was not considered to be suitable for the measurement of the degradation of organic chloramines.

As an alternative, the standard iodometric titration method for total chlorine (free chlorine and combined chlorine) (4500-Cl B)[17] was modified for colorimetric analysis.[16] Instead of determining the concentration of triiodide by titrating against sodium thiosulfate titrant, the concentration of triiodide was measured at $\lambda = 353$ nm. The total chlorine measurement was used as a proxy for organic chloramine concentration, but assuming that all free chlorine reacted with the excess amino acid, and that only organic chloramines formed. The assumption that all chlorine reacted with the amino acids was confirmed by the disappearance of the free chlorine peak at $\lambda = 292$ nm, as described in Section 2.2. It was confirmed that inorganic monochloramine did not form by using analysis with Monochlor F reagent. As di/tri inorganic chloramines could not be detected, it was assumed that they were not formed. Free chlorine was used as an oxidant to obtain the calibration for the response of triiodide to oxidant concentration, and a linear response (R^2=1) was achieved from 0 to 12 mg L^{-1}-Cl_2. Since the concentration of organic chloramines used for this experiment was between 4 to 10 mg L^{-1}-Cl_2, this indicated the degradation of organic chloramines (i.e. change in response) will be linear in the working concentration. Therefore, the triidode method would be a suitable and convenient method for the monitoring of the degradation of organic chloramines. Finally, the degradation of *N*-chloroglycine was measured by both the direct UV and the triiodide methods. The degradation rate constant for *N*-chloroglycine measured by the direct UV method ($k = 0.02 \times 10^{-4}$ s^{-1}) was found to be comparable to that obtained by the triiodide method ($k = 0.03 \times 10^{-4}$ s^{-1}).

3.3 Measured organic chloramine degradation rates

The degradation of organic chloramines was monitored by either the direct UV method or triiodide method described in Section 2.3. The measurements were then used to plot the first-order linear kinetic plot, ln (At/Ao) vs time in seconds, where At was the absorbance at a specific time and Ao was the initial absorbance (Figure 5). Figure 5 shows that the degradation of *N*-chloroglycine, *N*-chloroserine, *N*-chloroasparagine and *N*-chlorophenylalanine could all be modelled as pseudo first-order reactions, as the correlation coefficients of their linear regression curves were more than 0.97 for their first-order linear kinetic plots. The rate constants of the degradation of these organic chloramines were determined from the gradient of the linear regression curves of their first-order linear kinetic plots for the 4 organic chloramines, and their half-lives were calculated using the half-life equation for first-order reactions (Equation 4). The half-life for the selected organic chloramines ranged from 15 to 5775 min (Table 3). Among the four organic chloramines, only the rate constant for the degradation of *N*-chloroglycine has been previously studied, and this study is the first to determine the rate constants for the degradation of *N*-chloroserine, *N*-chloroasparagine and *N*-chlorophenylalanine. The rate constant of 0.02×10^{-4} s^{-1} for the degradation of *N*-chloroglycine from this study agreed well with the rate constant of 0.04×10^{-4} s^{-1} previously reported by Hand *et. al.*.[9] Other studies have also found the degradation of organic chloramines to show first-order kinetics.[10, 11]

$$\text{Half-life} = \frac{\ln(2)}{k}, \text{ where } k \text{ is the rate constant} \qquad (3)$$

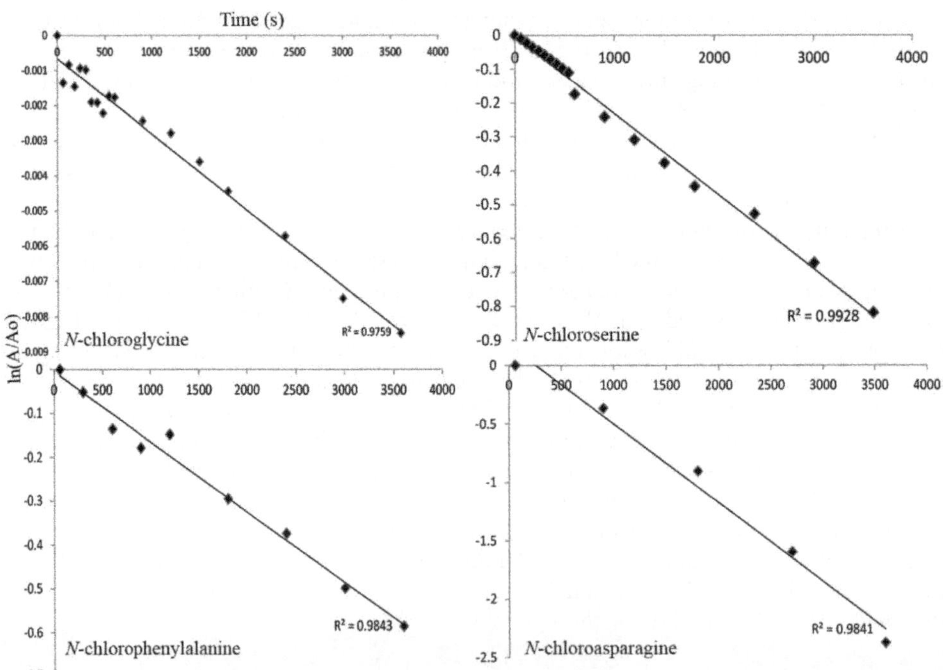

Figure 5 *Pseudo first-order plots for the degradation of the selected organic chloramines*

The stability of the organic chloramines (based on their half-life) varies greatly, with the non-polar amino acids (glycine and phenylalanine) producing more stable organic chloramines than those produced from the polar amino acids (serine and asparagine). In addition, the trend of the half-life of the selected organic chloramines is consistent with a previous report that more basic organic chloramines are more stable,[12] i.e. more basic precursors formed more stable organic chloramines (pK_a values in Table 1). However, further study with an extended number of amino acids is required to confirm the possible trend of stability with half-life.

As *N*-chloroglycine was found to be stable (i.e. half-life > 5700 min) and since *N*-chloroglycine is known to be toxic at micromolar concentrations[1], a need for a more in-depth investigation of the occurrence of this species in drinking water is required due to its potential risk to human health. In contrast, organic chloramines with shorter half-life are unlikely to be found in drinking water distribution systems. However, for these less stable organic chloramines, it is still necessary to consider the formation of additional by-products from these organic chloramines to understand their overall risk.

Table 3 *Organic chloramines, and their rate constant of degradation and half-life.*

Organic Chloramines	k ± SD ($x10^{-4}$)(s^{-1}) (n=4)	Half-life (min)
N-Chloroglycine	0.02 ± 0.01	5775
N-Chloroserine	2.1 ± 0.2	55
N-Chloroasparagine	6.7 ± 3	15
N-Chlorophenylalanine	1.6 ± 0.03	75

4 CONCLUSIONS

Two analytical methods, direct UV and an alternative triiodide colorimetric method, were successfully applied to the measurement of the degradation of organic chloramines.

The stability of organic chloramines was found to vary, with their half-life ranging from 15 to 5775 min, where more basic organic chloramines were found to be more stable. Although only *N*-chloroglycine was found to be the stable, more studies on the stability of different organic chloramines are required to better understand the risk associated with the presence of organic chloramines in drinking water.

References

1 S. Laingam, S. M. Froscio, R. J. Bull and A. R. Humpage, *Environ. Mol. Mutagen.*, 2012, **53**, 83.

2 R. J. Bull, D. A. Reckhow, X. F. Li, A. R. Humpage, C. Joll and S. E. Hrudey, *Toxicology*, 2011, **286**, 1.

3 J. V. Hunter and S. D. Faust, *Principles and applications of water chemistry* Wiley, New York, 1967.

4 M. Snyder, *Inorg. Chem.*, 1982, **21**, 2545.

5 R. A. Isaac and J. C. Morris, *Environ. Sci. Technol.*, 1983, **17**, 738.

6 A. J. Ellis and F. G. Soper, *J. Chem. Soc. (Resumed)*, 1954, **0**, 1750.

7 J. Yoon and J. N. Jensen, *Environ. Sci. Technol.*, 1993, **27**, 403.

8 A. Nweke and F. E. Scully, *Environ. Sci. Technol.*, 1989, **23**, 989.
9 V. C. Hand, M. P. Snyder and D. W. Margerum, *J. Am. Chem. Soc.*, 1983, **105**, 4022.
10 F. E. Scully, Jr. and M. A. Bempong, *Environ. Health Perspect.*, 1982, **46**, 111.
11 X. L. Armesto, M. Canle, M. V. García, M. Losada and J. A. Santaballa, *J. Phys. Org. Chem.*, 1996, **9**, 552.
12 I. H. Pitman, H. Dawn, T. Higuchi and A. A. Hussain, *J. Chem. Soc. B*, 1969, 1230.
13 D. A. Reckhow, Analytical Methods Development for Organic *N*-Chloramines, WRF, Denver,CO, 2011, pp. 69-89.
14 APHA, AWWA and WEF, 4500-Cl F. DPD Ferrous Titrimetric Method, ed. A. D. Eaton, APHA, AWWA, WEF, Maryland, United State of America, 2005.
15 G. D. Christian, Titrations Inolving Iodine: Iodimetry and Iodometry, ed. B. Deborah, Wiley, United States of America, 2004.
16 M. Bedner, W. A. MacCrehan and G. R. Helz, *Jour. Chromatogr. Sci.*, 2002, **40**, 447.
17 APHA, AWWA and WEF, 4500-Cl B. DPD Iodometric Method I, ed. A. D. Eaton, APHA, AWWA, WEF, Maryland, United State of America, 2005.
18 D. A. Plamer, R. W. Ramette and R. E. Mesmer, *Jour. Sol. Chem.*, 1984, **13**, 673.
19 APHA, AWWA and WEF, 4500-Cl G. DPD Colorimetric Method, ed. A. D. Eaton, APHA, AWWA, WEF, Maryland, United State of America, 2005.

REAL TIME MONITORING OF DISINFECTION BY-PRODUCTS IN CHLORINATED WATERS BY USING ABSORBANCE AND FLUORESCENCE INDICES

Paolo Roccaro[1], Federico G. A. Vagliasindi[1], Gregory V. Korshin[2]

[1]Department of Civil Engineering & Architecture, University of Catania, Viale A. Doria 6, 95125, Catania, Italy, proccaro@dica.unict.it
[2]Department of Civil and Environmental Engineering, Box 352700 University of Washington Seattle, WA 98195-2700 United States

1 INTRODUCTION

Chlorination is one of the most efficient public health measures ever undertaken. However, it has deficiencies such as its limited efficiency in deactivating *Giardia lamblia* and *Cryptosporidium* and the generation of harmful disinfection by-products (DBPs). DBPs are generated in the reactions between chlorine and natural organic matter (NOM) present in all source surface waters[1].

The concentration and reactivity of NOM play a major role in DBP formation[2,3], but other water quality parameters (e.g. pH, temperature, concentrations of ammonia, bromide and iodide) affect DBPs formation and speciation[4-6]. The nature of halogen atoms incorporated in DBPs also affects their toxicity which follows an I >> Br >> Cl order[7,8].

THMs and HAAs occur at the highest concentrations (up several hundred µg/L) in chlorinated water. The next tier of DBPs also includes nitrogenous compounds such as haloacetonitriles (HANs), halonitromethanes (HNMs)[7]. In general, nitrogenous DBPs tend to be more cytotoxic and genotoxic than carbonaceous DBPs[7,14].

NOM is a complex and heterogeneous entity, whose isolation and fractionation have been widely used to study its properties and reactivity[3,15,16]. Several surrogates such as TOC, DOC, UV absorbance indexes, have been used to quantify NOM reactivity in DBPs formation. Among those parameters, absorbance of ultraviolet light at 254 nm (A_{254}) and its specific value ($SUVA_{254}$), which is the ratio between A_{254} and DOC, are probably the most widely used[17,18]. However, questions have been raised about the strength of $SUVA_{254}$ as an indicator of NOM reactivity[19,20]. DBP formation potential tests are widely used to assess the reactivity of NOM in DBPs generation[15,20] but they require very stringent standard conditions[21-23].

Several models have been developed in order to understand the mechanism of DBPs formation, or to predict DBPs yields[24]. The modeling of DBP formation aims to establish relationships between DBPs levels in waters and the parameters of water quality that can be linked to their formation. Several studies have focused on the effects of chlorine dose and NOM characteristics, with particular interest to the reactivity sites. Other water quality parameters, such as pH, water temperature, bromide level, chlorine dose, and reaction time are also often taken into account[25]. Fundamentally, models for DBPs formation have been developed using statistical and limited mechanistic approaches[26,27].

Alternative approaches based on spectroscopic indices have been also developed[28-30] in order to monitor DBPs levels during chlorination at varying operating conditions or to measure the reactivity of a given NOM in DBPs formation. In this chapter these approaches are presented with a particular emphasis to the monitoring of individual DBPs species for real time application.

2 MATERIALS AND METHODS

2.1 Water sources, materials, and spectroscopic approach

In this study, the formation of DBPs in several water systems located in Italy and the United States was examined. Several classes of DBPs, including THMs, HAAs, HANs, halonitromethanes (HNMs), chloral hydrate, were considered. The relationships between the formation of these DBPs and the changes in the absorbance or fluorescence of NOM were explored in chlorinated raw and coagulated waters with different water quality parameters (e.g. DOC, SUVA, bromide, pH) and varying operating conditions (e.g. chlorine dose, reaction time, temperature). Specific attention was given to the formation of DBPs in treated water in order to evaluate the application of this approach for real time monitoring of DBPs in water distribution systems.

The investigated spectroscopic indices used to quantify changes in NOM during the halogenation process included differential absorbance, defined as $\Delta A_\lambda(t)=A_\lambda(t)-A_\lambda(t=0)$, measured at 272 nm ($\Delta A_{272}$) and changes in fluorescence quantified by the differential position of wavelength that corresponds to 50% of the maximum intensity of fluorescence, $\Delta\lambda^{em}{}_{0.5}(t) = \lambda^{em}{}_{0.5}(t) - \lambda^{em}{}_{0.5}(t=0)$, and the differential ratio of fluorescence intensities measured at 500 and 450 nm, defined as $\Delta(I_{500}/I_{450})=(I_{500}/I_{450})(t)-(I_{500}/I_{450})(t=0)$[28-30].

Selected raw and coagulated waters were also fractionated and the effluents from the XAD-8/XAD-4 columns were used for chlorination along with the unfractionated samples. The details on the fractionation of these waters were provided in Roccaro and Vagliasindi[31].

All chemicals were ACS reagent grade or better. Solvents used in extractions were high-purity grade. Reagent water was obtained from a Millipore Super-Q Plus water system. Chlorine stock solution was prepared by dilution of a reagent-grade sodium hypochlorite solution (5% available chlorine) with Milli-Q water.

2.2 Experimental methods

Chlorination experiments were conducted in dark conditions at widely varying reaction times (from 5 minutes to 7 days), Cl_2/DOC weight mass ratios (0.25 to 2 mg/mg), bromide concentrations (from its background levels to 2 mg/L), pH (7, 8 and 9) and temperatures (4, 20 and 34°C). The samples were chlorinated with sodium hypochlorite in the presence of 0.03 mol/L phosphate or borate buffer. Requisite amounts of Na_2SO_3 or NH_4Cl were added to quench the residual chlorine. Quenched chlorinated samples were stored at 4°C for no more than 10 days before being analysed for DBPs concentrations.

2.3 Analytical methods

Chlorine concentrations were determined using the DPD colorimetric method. DOC was analyzed using a Shimadzu VCSH organic carbon analyser. Concentrations of DBPs were determined using standard analytical procedures (EPA methods 551.1 and 552.2) and a Perkin-Elmer AutoSystem gas chromatograph equipped with an electron capture detector. Only chlorinated samples which had chlorine residual at the required reaction time were analyzed for HANs and residual chlorine concentrations were used to model DHAN formation as described in more detail in the sections that follow. Other aspects of these analyses are described in our previous publications[32,33].

3 RESULTS AND DISCUSSION

3.1 Formation and speciation of DBPs in chlorinated waters

In agreement with the data of prior research, the content and reactivity of NOM and the level of several water quality parameters such as chlorine dose, bromide concentration, reaction time, temperature, pH, significantly impacted the concentration of DBPs. For instance, higher chlorine doses, reaction times and temperatures resulted in higher DBPs concentration in chlorinated waters with a monotonic increase observed for many DBPs, such as THMs, HAAs, HANs (excluding trichloroacetonitrile, TCAN) and HNMs[13,33].

Bromide concentrations dramatically impacted the speciation of these DBPs classes. For instance, the concentrations of chloroform (trichloromethane, TCM), dichloroacetic acid (DCAA), trichloroacetic acid (TCAA) and dichloroacetonitrile (DCAN) decreased with increasing bromide concentration, while the reverse was observed for concentrations of bromoform (tribromomethane, TBM), dibromoacetic acid (DBAA), tribomoacetic acid (TBAA) and dibromoacetonitrile (DBAN). Dichlorobromomethane (DCM), dibromochloromethane (DBM), bromochloroacetic acid (BCAA), bromodichloroacetic acid (BDCAA), dibromochloroacetic acid (DBCAA) and chlorobromoacetonitrile (BCAN) changed non-monotonically as a function of bromide[26,27].

The effect of pH on the investigated DBPs was in agreement with prior research[6,34].

3.2 Spectroscopic changes in NOM during chlorination

In agreement to results observed for different surface raw waters, e.g. [28,29], the differential absorbance spectra, defined as $\Delta A_\lambda(t)=A_\lambda(t)-A_\lambda(t=0)$, of chlorinated raw, coagulated and fractionated waters, had a prominent peak near 272 nm and their intensity increased with increasing reaction time. For instance, Figure 1 shows the differential absorbance spectra obtained by the chlorination of a coagulated water at a chlorine dose of 3 mg/L.

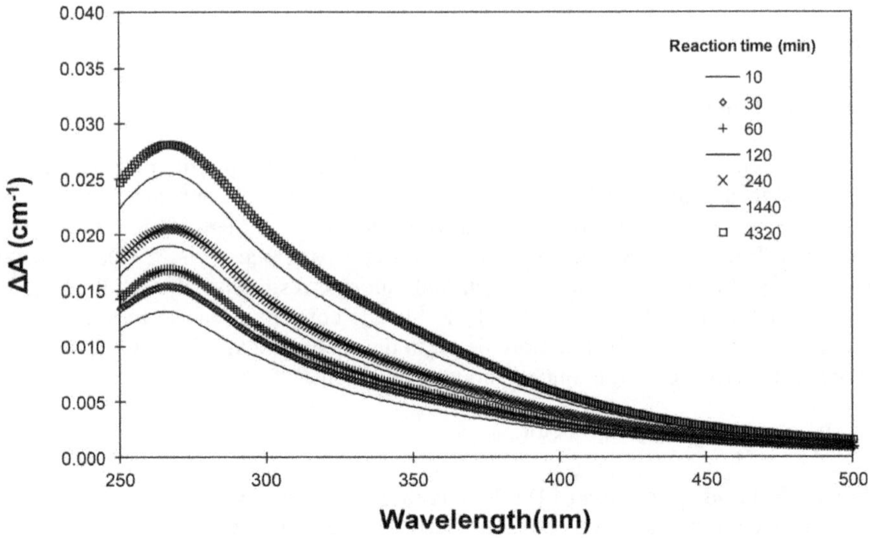

Figure 1 *Differential absorbance spectra obtained for chlorinated coagulated water at varying reaction time and chlorine to DOC ratio 1.0 mg/mg.*

The intensity of fluorescence emission spectra was non-monotonic, as observed in prior research for several surface raw waters[30], but the position of the normalized emission band at its half-intensity (for wavelengths > λmax), denoted as $\lambda^{em}{}_{0.5}$, exhibited a blue shift that became increasingly prominent at higher chlorine doses or reaction times. For instance, Figure 2 shows the normalized fluorescence spectra for a coagulated water with a chlorine dose of 1.5 mg/L.

3.3 Use of spectroscopic approaches to monitor DBPs formation and speciation

Prior studies showed that very strong correlations exist between differential absorbance (ΔA_{272}) and DBPs formation at varying chlorination parameters[29,32].
A sole correlation can model and therefore predict the formation of a target individual DBP or a class of DBPs at varying chlorine doses, reaction times and temperatures[29]. For instance, Figure 3 shows that very strong correlations exist between TCM concentrations and ΔA_{272} values obtained by the chlorination of raw or coagulated water at varying chlorine doses and reaction time. This result confirms that DBPs formation is strongly related with changes of differential absorbance at 272 nm caused by the halogenation processes, even in the case of treated water which contains less NOM and less reactive sites. However, chlorinated waters having different NOM resulted in different correlations between ΔA_{272} and DBPs, as reported in prior research[31]. The latter result, in turn, is also very useful because can help to quantify the NOM reactivity. Indeed, it was reported that time specific ΔA_{272} is a strong surrogate for quantifying the NOM reactivity in DBPs formation and can be employed alternatively to the widely used $SUVA_{254}$[31].

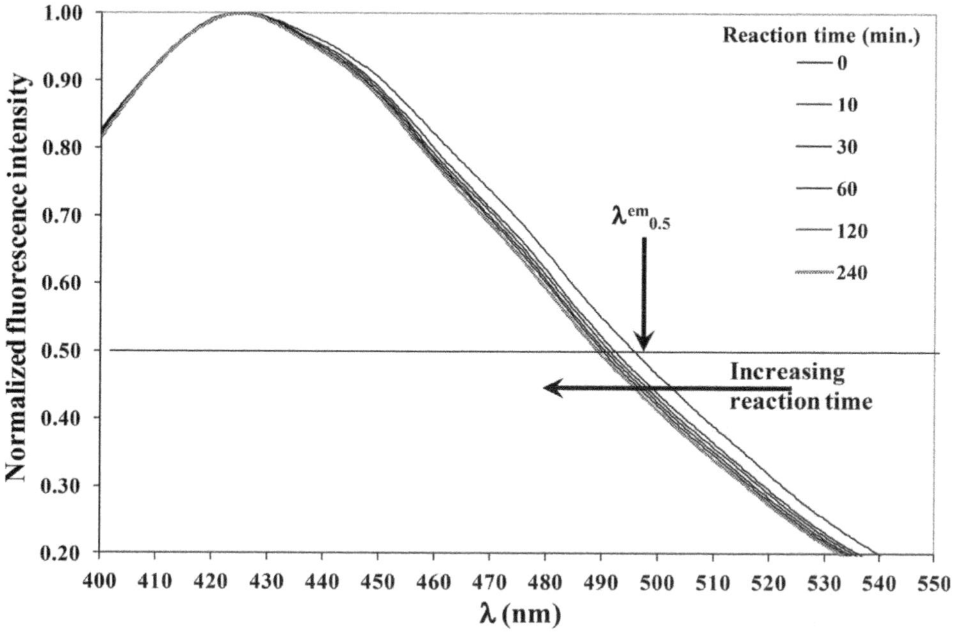

Figure 2 *Normalized fluorescence spectra obtained for chlorinated coagulated water at varying reaction time and chlorine to DOC ratio 0.5 mg/mg.*

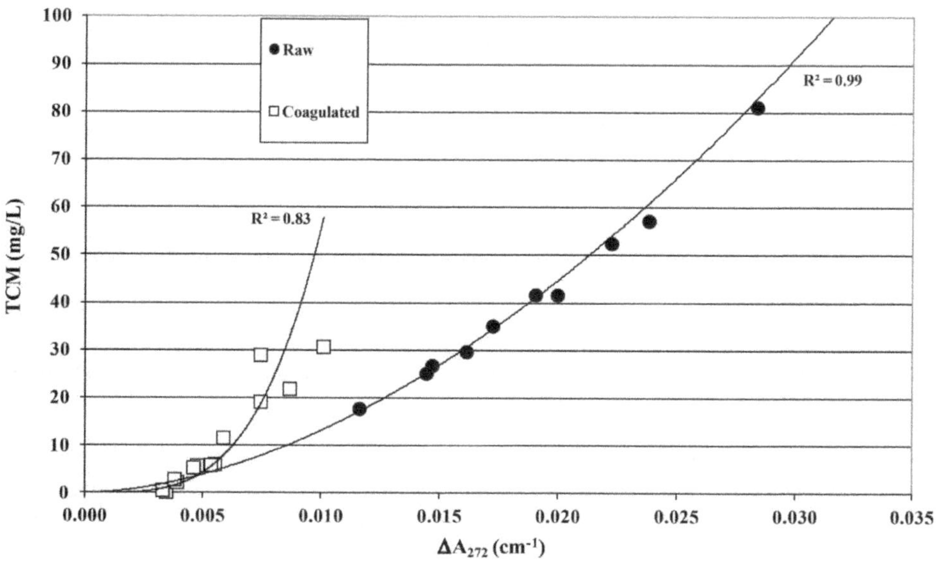

Figure 3 *Correlations between the concentration of TCM and ΔA_{272} values for chlorinated raw and coagulated water at chlorine to DOC ratio from 0.75 to 1.5, reaction time from 10 minutes to 3 days, temperatures 20°C.*

Similarly to results observed in previous studies[30], several fluorescence indexes were found to be strongly correlated with the concentration of individual DBP or classes of DBP. For instance, Figure 4 shows the relationship between the concentration of selected individual DBPs and $\Delta(I_{500}/I_{450})$ values for a chlorinated water at varying chlorine doses, reaction times and temperatures. Similar performance was obtained by using $\Delta\lambda^{em}_{0.5}$ index. The use of fluorescence based indexes can be envisioned to result in a practical advantage. Indeed, the correlations between fluorescence indexes and the individual or group of DBPs was the same regardless of the NOM content and reactivity[30], therefore can be used in systems where a change in NOM properties can take place.

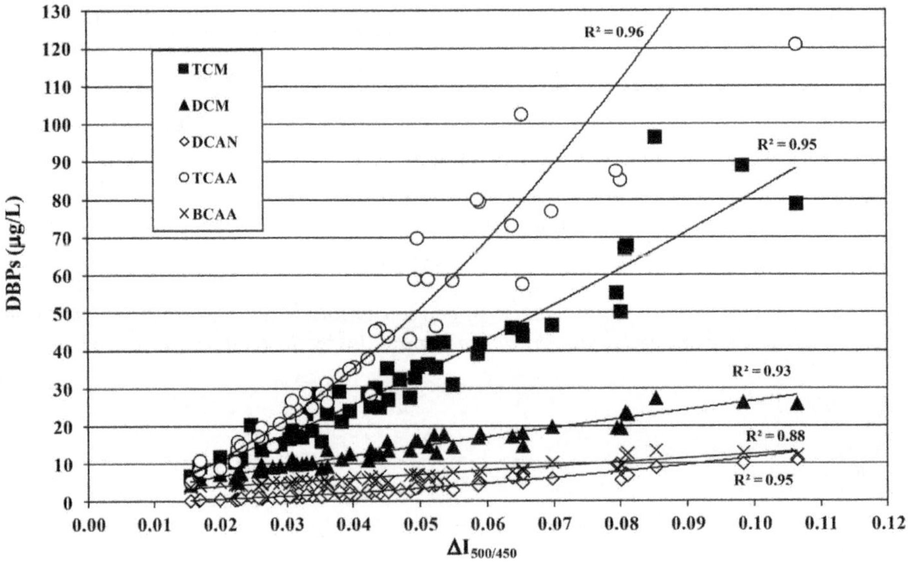

Figure 4 *Correlations between the concentration of selected individual DBPs and $\Delta I_{500/450}$ values for chlorinated raw water at chlorine to DOC ratio from 0.25 to 2.0, reaction time from 10 minutes to 7 days, temperatures from 3 to 34°C.*

3.4 Suitability of spectroscopic approaches for in situ and real time applications: the case of NOM reactivity and DBPs formation

Overall, the obtained results highlight that the selected spectroscopic indexes that are strongly correlated with the concentrations of individual or groups of DBPs in chlorinated raw waters[28-30], can be used also to predict and control the concentrations of DBPs in drinking waters (coagulated waters) whose NOM content and reactivity have been lowered at the water treatment plant. These surrogates can also be employed to monitor the formation of emerging DBPs species, such as HANs and HNMs present at trace levels in treated waters. Furthermore, the reactivity of NOM in DBPs formation can be predicted by using time specific differential absorbance.

Finally, considering that both absorbance and fluorescence based indexes are easy to measure with any temporary resolution and they can be measured *in situ* (some high sensitivity commercial devices are available and other are under development), this approach can be implemented in practically any drinking water treatment plant or distribution system to monitor DBPs levels in real time.

4 CONCLUSION

Changes in NOM during chlorination of drinking water can be quantified by spectroscopic indexes based on absorbance and fluorescence. These indexes are correlated with the concentration of individual or classes of DBPs, including trace level emerging DBPs, and are suitable for application of in situ and real time measurement.

References

1. D.L. Sedlak, U. von Gunten, 2011, *Science*, **331**, 42.
2. D.A. Reckhow, P.C. Singer, R.L. Malcolm, 1990, *Environ. Sci. Technol.,* **24**, 1655.
3. J.A. Leenheer, J.-P. Croue, 2003, *Environ. Sci. Technol.,* **37**, 18A.
4. G.A. Cowman, P.C. Singer, 1996, *Environ. Sci. Technol.,* **30**, 16.
5. G. Hua, D.A. Reckhow, J. Kim, 2006, *Environ. Sci. Technol.*, **40**, 3050.
6. A. Obolensky, P.C. Singer, 2008, *Environ. Sci. Technol.*, **42**, 5654.
7. S.D. Richardson, M.J. Plewa, E.D. Wagner, R. Schoeny, D.M. DeMarini, 2007, *Mutation Res.*, **636**, 178.
8. L. Zhang, L. Xu, Q. Zeng, S.-H. Zhang, H. Xie, A.-L. Liu, W.-Q. Lu, 2012, *Mutation Research*, **741**, 89.
9. T. Bond, J. Huang, M.R. Templeton, N. Graham, 2011, *Water Res.*, **45**, 4341.
10. B.Y. Chen, P. Westerhoff, 2010, *Water Res.*, **44**, 3755.
11. J. Hu, H. Song, J.W. Addison, T. Karanfil, 2010, *Water Res.*, **44**, 105.
12. H.H. Chang, C.Y. Chen, G.S. Wang, 2011, *Water Res.*, **45**, 3753.
13. W.H. Chu, N.Y. Gao, Y. Deng, M.R. Templeton, D.Q. Yin, 2011, *Chemosphere*, **85**, 1187.
14. M.G. Muellner, E.D. Wagner, K. Mccalla, S.D. Richardson, Y.-T. Woo, M.J. Plewa, 2007, *Environ. Sci. Technol.*, **41**, 645.
15. J.-P. Croué, G.V. Korshin, M.M. Benjamin, American Water Works Association Research Foundation: Denver, CO. 2000.
16. M. Kitis, T. Karanfil, A. Wigton, J.M. Kilduff, 2002, *Water Res.*, **36**, 3834.
17. J.P. Croué, J.F. DeBroux, G.L. Amy, G. Aiken, J. Leenheer, 1999, Singer, P.C., (Ed.), Formation and Control of Disinfection By-Products in Drinking Water, American Water Works Association: Denver.
18. M. Kitis, T. Karanfil, J. E. Kilduff, A. Wigton, 2001, *Water Sci. Technol.*, **43**, 9.
19. J.L. Weishaar, G. Aiken, B.A. Bergamaschi, M.S. Fram, R. Fujii, K. Mopper, 2003, *Environ. Sci. Technol.*, **37**, 4702.
20. N. Ates, M. Kitis, U. Yetis, 2007, *Water Res.*, **41**, 4139.
21. APHA, AWWA, and WEF, 1998, Standards Methods for the Examination of Water and Wastewater, 20th ed., Washington DC.
22. R.S. Summers, S.M. Hooper, H.M. Shukairy, G. Solarik, D. Owen, 1996, *J. Am. Water Works Assoc.*, **88**, 80.
23. B. Koch, S.W. Krasner, M.J. Sclimenti, and W.K. Schimpff, 1991, *J. Am. Water Works Assoc.,* **83**, 62.
24. R. Sadiq, M. Rodriguez, 2004, *Sci Total Environ*, **321**, 21.

25. S. Chowdhury, P. Champagne, P.J. McLellan, 2009, *Sci Total Environ,* **407**, 4189.
26. P. Roccaro, H.-s. Chang, F.G.A. Vagliasindi, G.V. Korshin, 2013, *Water Res,* **47**, 5995.
27. P. Roccaro, G.V. Korshin, D. Cook, C.W.K. Chow, M. Drikas, 2014, *Water Res,* **62**, 117.
28. G.V. Korshin, C.-W. Li, M.M. Benjamin, 1997, **Water Res.**, **31**, 946.
29. P. Roccaro, H.-S. Chang, F. G. A. Vagliasindi, and G. V. Korshin, *Water Res.*, 2008, **42**, 1879.
30. P. Roccaro, F. G. A. Vagliasindi, and G. V. Korshin, *Environ Sci. Technol.*, 2009, **43**, 724.
31. P. Roccaro, F. G. A. Vagliasindi, *Water Res.*, 2009, **43**, 744.
32. G.V. Korshin, W.W. Wu, M.M. Benjamin, O. Hemingway, *Water Res.*, 2002, **36**, 3273.
33. P. Roccaro, F. G. A. Vagliasindi and G.V. Korshin, *Water Science and Technology*, 2011, **63**, 40.
34. K. Ichihashi, K. Teranishi, A. Ichimura, *Water Res.*, 1999, **33**, 477.

IONIC DISINFECTION BY-PRODUCTS - HOW TO ANALYZE

Detlef Jensen, Christian Fröhling
Thermo Fisher Scientific GmbH, Im Steingrund 4, 63303 Dreieich, Germany
E-mail: detlef.jensen@thermofisher.com

1 INTRODUCTION

As a vital limited resource required for survival, water fit for human consumption is ideally produced from natural sources such as below ground aquifers and surface reservoirs. Because a clean water supply is the single most important determinant of public health, water from these ground and surface water sources must undergo a purification process to ensure that all undesirable chemicals, biological organisms, and suspended solids are removed. As part of the purification process, water is often treated with disinfectants such as chlorine, chlorine dioxide, chloramine, and ozone,[1,2] which react with naturally occurring organic and inorganic matter in the source water to form amongst others anionic disinfectant by-products (DBPs), oxo halides and haloacetic acids that are potentially harmful to humans. The World Health Organisation (WHO) states that efficient disinfection must never be compromised in attempt to meet the guidelines for disinfection by-products and that the microbiological quality of the water must always take precedence.[3]

One example disinfectant is ozone. Due to its high degree of oxidative reactivity, ozone effectively kills all microbial organisms that may be present, and also serves to decolorize and deodorize the treated water. However, it also reacts with bromide to form the carcinogenic compound bromate. The World Health Organization[3], European Commission,[4] U.S. EPA,[5] and Japanese Ministry of Health, Labour, and Welfare[6] have established a regulatory maximum contaminant level (MCL) of 10 µg/L for bromate in municipal drinking waters treated with ozone for disinfection. Furthermore, the European Commission established a MCL of 3 µg/L bromate in ozonated natural mineral waters and spring waters.[7] As a consequence many standardization organizations like AFNOR, DIN, ISO, US EPA - only to name a few - developed analytical strategies allowing the simplified determination of bromate not only at the MCL, but also ensuring the required European minimum detection limit of 2.5 µg/L for drinking water[4] and 0.75 µg/L for natural mineral and spring waters[7]. In addition to the parametric bromate value the European Communities[8] set a parametric value of 100 µg/L for total trihalomethanes (TTHMs) (i.e. a group of four disinfection by-products, namely chloroform, bromoform, dibromochloromethane and bromodichloromethane). While no parametric values have been set for disinfection by-products other than THMs and bromate in Europe, there is a requirement of the Drinking Water Regulations that "...any contamination from disinfection by-products is kept as low as possible without compromising the disinfection....".[4] While there may not be specific parametric values for DBPs, other than THMs or bromate, they must not be present in concentrations that constitute a potential danger to human health.[9]

U.S. EPA and ISO methods (U.S. EPA 552.1,[10] 552.2,[11] 552.3,[12] and ISO 23631[13]) published analytical methods for the determination of five haloacetic acids, referred to as HAA5, for which the US EPA have set an upper limit in drinking water of 60 µg/l. The HAA5 are monochloroacetic acid, dichloroacetic acid, trichloroacetic acid, monobromoacetic acid, and dibromoacetic acid. These methods involve liquid-liquid microextraction, derivatisation and GC with either Electron Capture Detection (ECD) or Mass Spectrometry Detection (MS).

As this work focuses on anionic DBPs, Ion Chromatography (IC) is the method we are going to discuss. Recently its application was expanded to the determination of haloacetic acids, despite already available GC-methods and standards. Reasons are the simplified overall method. Compared to GC-methods no sample preparation and derivatisation are needed, and eluent generation is integrated and automated within the setup.[14] IC is an instrumental technique used for the quantitative analysis of solutions. Since its introduction in 1975,[15] analytical chemists have rapidly moved to replace traditional wet chemical methods of analysis with IC. The basis of IC is separation followed by selective detection. First, the analyte ions are physically separated from each other on a chromatographic column filled with ion exchange resin. The ions are pumped through the separator in a solution called the eluent (i.e. mobile phase). Analyte ions that are charged at neutral pH are detected by measuring their electrical conductivity as they elute from the separator column and pass through an ion exchange membrane device called a suppressor[16]. The suppressor is used to increase sensitivity and selectivity, minimizing the conductivity of the eluent, and facilitating the use of gradient elution in IC at the same time. This approach is known as suppressed conductivity detection. Chromatographic separation combined with chemical eluent suppression and conductivity detection provides a sensitive, selective, accurate, and rapid method for detecting ionic species in solution. Continuously regenerated membrane-based suppressors facilitate the use of MS-detectors as in combination with the separation, the suppressor acts as a desalter, and in the process neutralizes the high pH eluents used for anion separations.

Because of the increased global interest in DBPs we will review standardized IC methods with respect to their analytical performance and application.

2 METHODS AND RESULTS

2.1 Bromate Determinations Using Suppressed Conductivity Detection

Early developments were focussed on the determination of trace bromate levels in drinking water. Initial methods developed by the U.S. EPA and ISO combine the determination of bromate and the so called standard anions[17, 18]. The use of high capacity columns, such as the Thermo Scientific Dionex IonPac AS9-HC or the Dionex IonPac AS23 column, using sodium bicarbonate/sodium carbonate-based eluents and suppressed conductivity detection, allow for the determination of bromate at the requested MDL and ensure the required limit of detection (Figure 1).

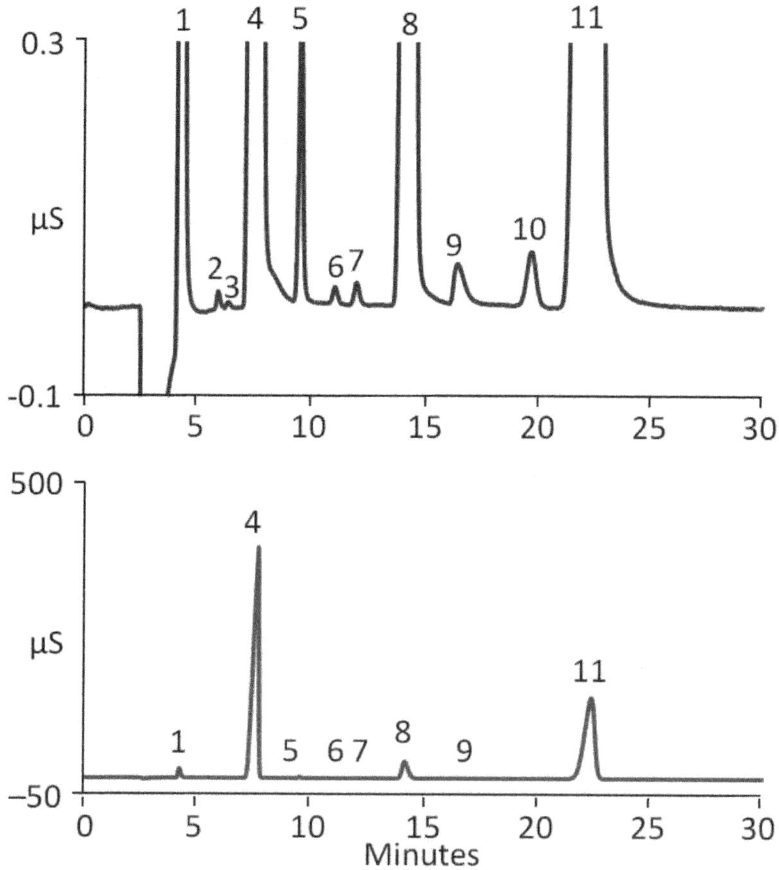

Figure 1 *Simultaneous separation of inorganic anions and bromate. Column: Thermo Scientific Dionex IonPac AS23, Eluent: 4.5 mmol/L Na_2CO_3 + 0.8 mmol/L $NaHCO_3$, Flow: 1 mL/min, Injection vol.: 200 µL, Detection: suppressed conductivity, Suppressor: Thermo Scientific Dionex ASRS Anion Self-Regenerating suppressor. Lower trace: Large scale chromatogram, upper trace: enlarged chromatogram. Peaks: 1. Fluoride 1.0 mg/L (ppm), 2. Chlorite (0.01), 3. Bromate (0.005), 4. Chloride (50), 5. Nitrite (0.1), 6. Chlorate (0.01), 7. Bromide (0.01), 8. Nitrate (10), 9. Carbonate (50), 10. Phosphate (0.1), 11. Sulfate (50).*

Figure 1 represents a typical separation of the so called seven common anions, sample carbonate, and the DBPs chlorite, chlorate, and bromate under isocratic conditions with suppressed conductivity detection. ISO 15061[18] specifically demands a chromatographic resolution (R) of neighbouring peaks of $R \geq 1.3$. For many samples, like the one shown in Figure 1, this can be achieved. Simultaneous determination of the common anions and bromate is possible.

The main challenge with this approach, however, is to ensure the resolution between chloride and bromate. This is especially true for chloride rich waters. Hence, ISO 15061 describes a manual sample preparation to remove chloride using silver loaded cation exchangers, an automated sample concentration step to ensure lowest possible limits of detection (LODs), and a step change elution of the analytes. At the time ISO 15061 was released this was the only - but labour intensive - approach available.

Modern IC uses a simplified approach to generate high purity eluents that employs a Reagent-Free-Ion-Chromatography (RFIC) system. This system allows IC to be conducted without the need for manual eluent preparation or the manual handling of chemicals. A RFIC system electrolytically generates, for example hydroxide based eluents simply by feeding deionized water into an eluent generator. Through controlling the electrolytic current, the hydroxide concentration can be adjusted and gradient elution is greatly simplified. As this electrolytic process takes place in the high pressure part of the IC instrument, eluent contamination, e.g. by carbon dioxide from the surrounding atmosphere, is prevented, and offline eluent degassing is no longer necessary. The practical benefits for the determination of DBPs result from the improved baseline performance and sensitivity as a result of high purity eluents, and from simplified gradient elution that improves the peak capacity of the analytical approach. A representative example is given in Figure 2, showing improved separation between chloride and bromate as well as the separation of the common anions in one run.

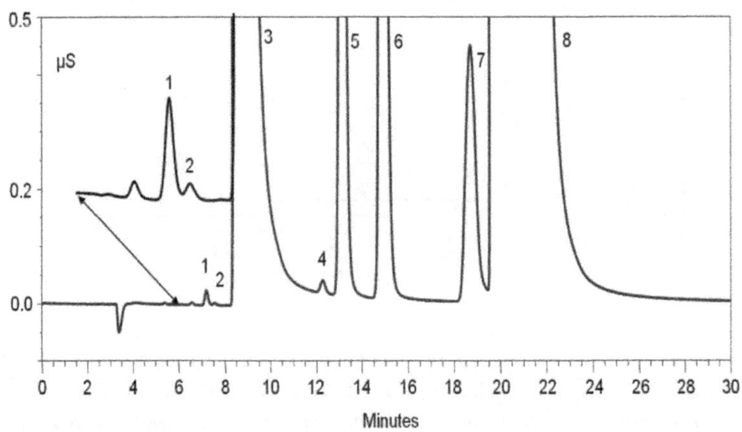

Figure 2 *Simultaneous separation of inorganic anions and bromate using gradient elution. Column: Dionex IonPac AS19, Eluent: 20 mmol/L KOH, Flow: 0.75 mL/min, Injection vol.: 100 µL, Detection: suppressed conductivity, Suppressor: Dionex ASRS Anion Self-Regenerating suppressor. Lower trace: Large scale chromatogram, upper trace: enlarged chromatogram. Peaks: 1. Chlorite, 2. Bromate (2 µg/L), 3. Chloride (100 mg/L), 4. Chlorate, 5. Bromide, 6. Nitrate, 7. Carbonate, 8. Sulfate (300 mg/L).*

To prevent conversion of residual hypochlorite or hypobromite to chlorate or bromate ethylendiamine (EDA) is added to the sample. Under certain conditions, however, carbonate from the sample can react with the excess EDA, and the reaction products can possibly interfere with the detection of certain analytes by suppressed conductivity detection. In such situations and if conductivity detection is to be applied the use of a column (e.g. Dionex IonPac AS27) with specifically designed chromatographic selectivity solves the issue. Otherwise alternative detection schemes can be used as described in the following.

2.1.1 Analytical Performance of Different Eluents. Both analytical approaches (manually prepared carbonate/bicarbonate eluents and electrolytically generated hydroxide eluents) were compared with respect to their analytical performance (Table 1).

The separations were performed under isocratic conditions. The statistical evaluation reveals that hydroxide-based eluents have a clear advantage over carbonate-based mobile phases with respect to the overall analytical performance. The limits of detection were improved five-fold for the listed oxo halides including bromate, when hydroxide eluents were used.

Table 1 *Comparison of Analytical Performance of Methods using Different Columns/Eluents*

Analyte	Range (µg/L)	Linearity (r^2)	Retention Time Precision (%RSD[a,b])	Peak Area Precision (%RSD)	MDL Standard (µg/L)	MDL Calculated (µg/L)
Dionex IonPac AS19 Column – Hydroxide Eluent						
Chlorite	2-50	0.9999	0.04	1.20	1.0	0.18
Bromate	1-25	0.9995	0.03	1.40	2.0	0.31
Chlorate	2-50	0.9999	0.01	0.54	1.0	0.28
Dionex IonPac AS23 Column – Carbonate/bicarbonate Eluent						
Chlorite	10-50	0.9999	0.07	2.20	5.0	1.02
Bromate	5-25	0.9998	0.07	2.63	5.0	1.63
Chlorate	10-50	0.9998	0.11	2.48	9.0	2.05

[a] RSD = relative standard deviation, $n = 7$

[b] Quality control standard contained 10 ppb each of chlorite, chlorate, and bromide and 5 ppb bromate

2.2 Bromate Determinations Using UV Detection

Due to the European requirements a different approach was developed addressing the need for lower LOD for bromate[7]. Based on the original work authored by Salhi et al.[19] post column derivatisation methods were evaluated and optimized. ISO 11206[20] describes the detection of bromate at levels ≥ 0.5 μg/L using an acidified solution of potassium iodide containing a catalytic amount of molybdenum(VI) as a post-column reagent. Bromate reacts with iodide to form the tri-iodide (I_3^-) ion in a post-column reaction (PCR) step, which is detected by its UV absorption at 352 nm. The standard presents two different approaches, which can easily be differentiated by the eluent's pH.

2.2.1 High pH-Eluents with Post-column-derivatisation for Bromate following ISO 11206. This is an approach that is technically equivalent to U.S.EPA Method 326.0. The oxo halide anions chlorite, chlorate, and bromate as well as bromide are separated on a Dionex IonPac AS9-HC column and measured by using suppressed conductivity detection (as in ISO 10304-1[21] and EPA Method 300.1[17]), followed by post-column reaction (PCR) to enhance detection of bromate. Sensitivity for bromate is improved by more than a factor of 10 through the use of a post-column reaction in which hydroiodic acid (HI) generated in situ from potassium iodide (KI) reacts with bromate in the column effluent to form the triiodide anion (I_3^-) that absorbs at 352 nm.

Each sample that contains excess chlorite must be treated to remove chlorite and then reanalyzed for bromate, because elevated levels of chlorite also reacts with I_3^- and it can interfere when quantifying low concentrations of bromate. The interference from chlorite can be minimized by reducing the chlorite with ferrous sulfate prior to the injection.

Figure 3 shows the chromatograms from a direct injection of bottled water. The top trace (A) was obtained with the conductivity detector, and the bottom trace (B) was obtained with the UV/Vis absorbance detector. The bottle label read: "Prepared using filtration, reverse osmosis, deionization, and ozonation". The DBP precursor bromide and the DBP bromate were both observed in the bottled water sample.

UV-detection adds additional detection selectivity to the analytical methods that determine DBPs like bromate and chlorite. Due to the experimental conditions (pH of the reaction solution > 1) this approach requires heating of the reaction coil to 80 °C.

2.2.2 Low pH-Eluents with Post-column Derivatisation for Bromate following ISO 11206. The second approach uses a low pH eluent and is tailored for the selective trace determination of only bromate in presence of chlorite. The respective evaluations of the ISO committee showed that using an acidic eluent and maintaining the pH of the derivatisation solution below or equal to pH 1 eliminated the need to heat the reaction coil. Also, chlorite does not interfere, even if present in a large excess. The original method described in ISO 11206 uses sulphuric acid as the eluent, limiting the application to a small set of commercially available columns. During our studies we found that if methanesulphonic acid (MSA) is used as the eluent, more columns of different chemical and chromatographic properties can be used, offering the benefit of greater application flexibility. Bromate was separated on a Thermo Scientific Dionex CarboPac PA1 column in less than 18 minutes using isocratic conditions and detected by UV absorbance at 352 nm after derivatisation. These conditions prevent interference from chlorite, as described in the previous paragraph. The limit of detection of this approach was determined to 0.07 μg/L and the limit of quantification was calculated to be 0.22 μg/L.[22]

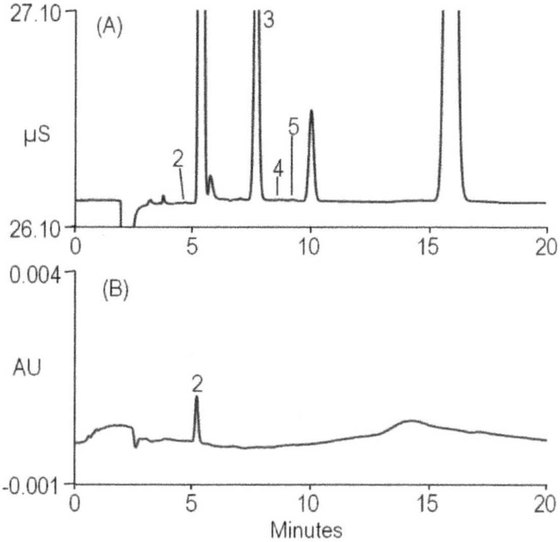

Figure 3 *Determination of DBP anions in bottled water. (A) Suppressed conductivity detection and (B) UV absorbance detection after PCR with acidified KI (352 nm). Column: Dionex IonPac AS9-HC, Eluent: 9 mmol/L* Na_2CO_3*, Flow: 1.3 mL/min, Injection vol.: 225 µL, Post-column reagent: Acidified KI*[20]*, PCR flow rate: 0.4 mL/min, Post-column heater: 80 °C, Peaks A: 1. Chlorite n.d. µg/L (ppb), 2. Bromate (1.52), 3. DCA*, 4. Bromide (1.12), 5. Chlorate (1.08), Peak B: 6. Bromate-UV (1.84), * DCA = Dichloroacetate, quality control surrogate*

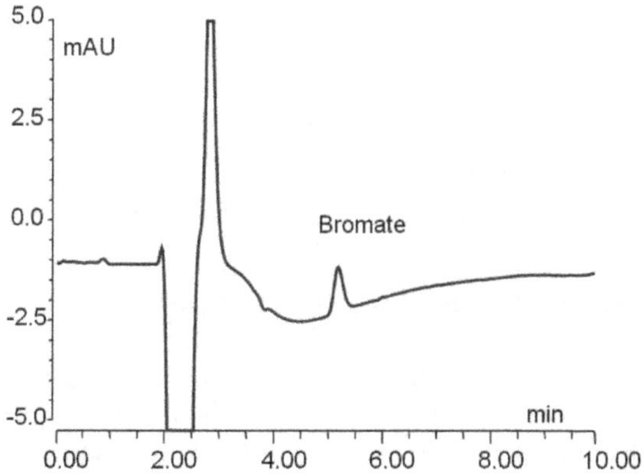

Figure 4 *Bromate determination in a drinking water sample using an acid eluent. Column: Dionex CarboPac PA1 (4 × 250 mm), Eluent: 200 mmol/L MSA, Flow: 1 mL/min, Injection vol.: 500 µL, Temperature: 30 °C, Detection: UV 352 nm (after PCR), Bromate 1.2 µg/L. The trace ends at 10 min, although the run time is extended to 18 min due to a later eluting component.*

2.3 Determination of Chlorite and Chlorate in Swimming Pool Water

With the release of DIN 19643[23] in 2012 on the treatment of water of swimming pools and public baths a parametric value for chlorate in these waters was set for the first time. Currently, many swimming pool operators that use hypochlorite for disinfection may not be in compliance with this value.[24] A method to determine oxo halides like chlorite and chlorate in the presence of chloride, bromide, and nitrate is described in ISO 10304-4.[25]

Special attention must be paid to the separation of chlorate from both nitrate and bromide. As recommended in ISO 10304-4 the resolution of neighbouring peaks must be equal or exceed 1.3, so that an IC column with suitable chromatographic selectivity is needed. As the nitrate concentration in swimming pool waters can significantly increase over time the column should separate bromide, chlorate, and nitrate. A representative example is shown in Figure 5. This method can be easily applied and used together with the determination of standard anions in a variety of aqueous samples, as described in ISO 10304-1.[21]

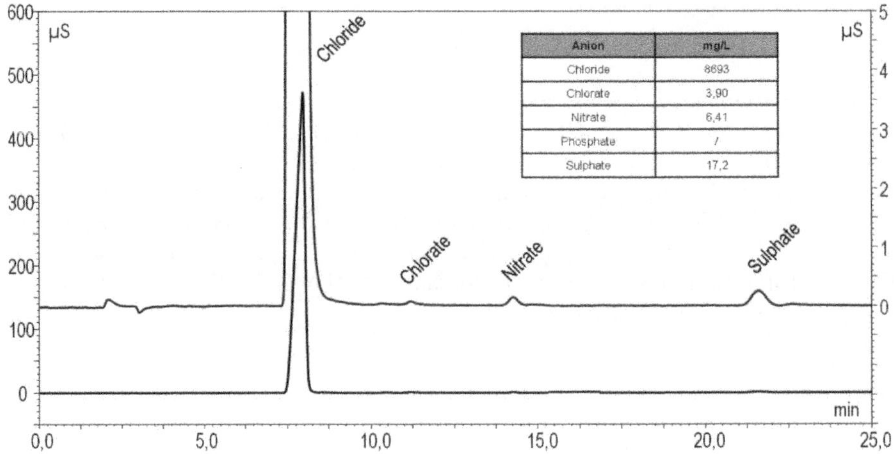

Figure 5 *Anion Determination in Disinfected Swimming Pool Water. Column: Dionex IonPac AS23, Eluent: 4.5 mmol/L Na_2CO_3 + 1.5 mmol/L $NaHCO_3$, Flow: 0.25 mL/min, Injection vol.: 25 µL, Detection: Suppressed conductivity. Suppressor: Dionex ASRS Anion Self-Regenerating suppressor. Concentrations provided in the figure.*

2.4 Haloacetic Acid Determination using IC-ESI-MS/MS

Despite already available GC methods and standards, the use of IC was recently expanded to the determination of haloacetic acids. The IC method is a simplified overall method. Compared to GC methods no sample preparation and derivatisation is needed, and eluent generation is part of the setup thus eliminating the labour and possible errors associated with eluent preparation.[14]

This method[14] is a direct-injection IC method using negative-ion electrospray ionization and tandem mass spectrometry (IC-ESI-MS/MS) for the determination of haloacetic acids, bromate, and Dalapon (2,2-dichloropropionic acid) in finished drinking water. In order to address matrix effects in MS detection, known as ion suppression effects, a real time matrix elimination step removing the major common anions is a key feature of this method. Anions like chloride, nitrate, carbonate, and sulfate, present in drinking water samples, which would cause matrix suppression, are diverted from the MS. The method describes the use of a Reagent-Free IC system, gradient elution, and suppressed conductivity detection. If manually prepared hydroxide eluents are used, care must be taken to prevent formation of carbonate in the mobile phase from exposure to the atmosphere, which could cause analyte retention times to drift. The use of a continuously regenerated, membrane based suppressor is essential to desalt and neutralize the high ionic strength hydroxide eluent prior to the MS interface. U.S. EPA Method 551.7 reports IC-ESI-MS/MS detection limits in the mid to low ng/L-range.

In cases where the special features of MS/MS are not needed, single quadrupole MS systems are being used for the determination of DBPs in combination with IC. A typical example is shown in Figure 6. Please note that higher specificity, selectivity, and sensitivity can be obtained with MS instruments allowing Multiple Reaction Monitoring (MRM).

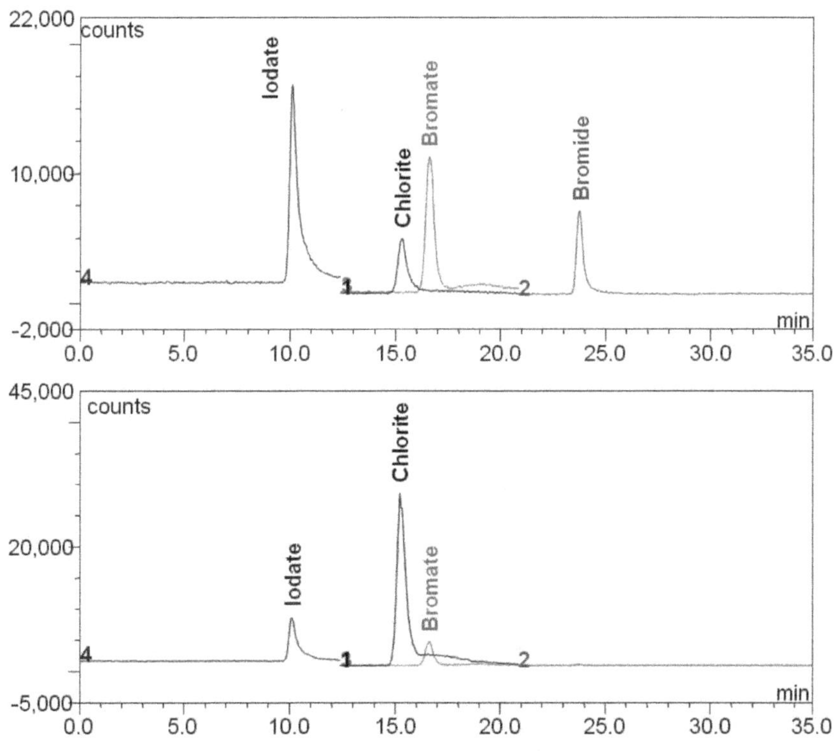

Figure 6 *Determination of DBPs using a Single Quadrupole MS; Column: Dionex IonPac AS19, Flow: 0,25 mL/min, Detection: Injection vol.: 250 µL, Selected Ion Monitoring (SIM.); Desalting: Dionex ASRS Anion Self-Regenerating suppressor (external water mode), Ionisation: ESI in negative mode, Needle Voltage: 3 kV, Cone Voltage: 50 V (chlorite) and 75 V, (iodate, bromate, and bromide), Probe temperature: 400°C, S.I.M: 67, 79, 129, 175 m/z. Upper trace: Standard solution, lower trace: fortified drinking water.*

2.5 Additional Analytical Approaches

2.5.1 Ion Chromatography with Inductively Coupled Plasma (ICP) Detection. Today's IC systems can be easily connected to ICP instruments. Though originally designed for the detection of metals in various samples, an ICP - equipped with either optical or MS detection - in combination with IC opens an easy access to speciation analysis in various samples. Early examples for DBPs like bromate and haloacetic acids can be found in the literature,[26,27,28] and US EPA has released a method for IC-ICP/MS for bromate.[29]

For hyphenation with an ICP, the column's eluent is pneumatically nebulised into a high frequency plasma, where energy transfer processes are responsible for the desolvation, atomization, and ionization of the sample components. As this occurs after the chromatographic separation, the differentiation of species (speciation) is possible based on their different chromatographic behaviour. The ions from the plasma are then introduced in the MS and separated on the basis of their mass-to-charge ratio.

This method can not only be applied to a variety of brominated species. The chromatographic selectivity of modern liquid chromatographic column technologies allows for the separation of a large variety of different DBPs. While Figure 7 shows a standard separation of iodine species in about 65 seconds, similar IC approaches are applied to the determination of DBPs in more complex environmental samples, and in demanding food and beverage samples such as milk.[30]

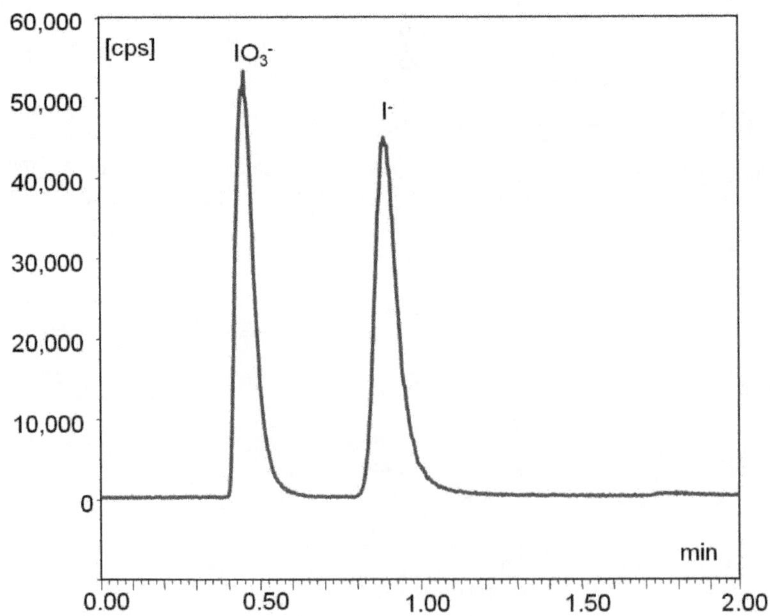

Figure 7 *Fast Separation of Iodine Species. Column: Dionex IonPac AG19 (2×50 mm), Eluent: 10 mmol/L KOH, Flow: 0.5 mL/min, Injection vol.: 100 μL, Detection: ICP-MS (m/z= 129), Concentration: 1 μg/L each.*

2.5.2 Two-Dimensional IC (2-D IC) for the determination of DBP. Trace determinations of ionic DBP are often difficult because of the high ionic strength that can be experienced in different drinking water samples. A number of approaches to overcome matrix effects have been discussed in the previous sections of this article. A different approach is to integrate matrix elimination steps into the analytical setup.

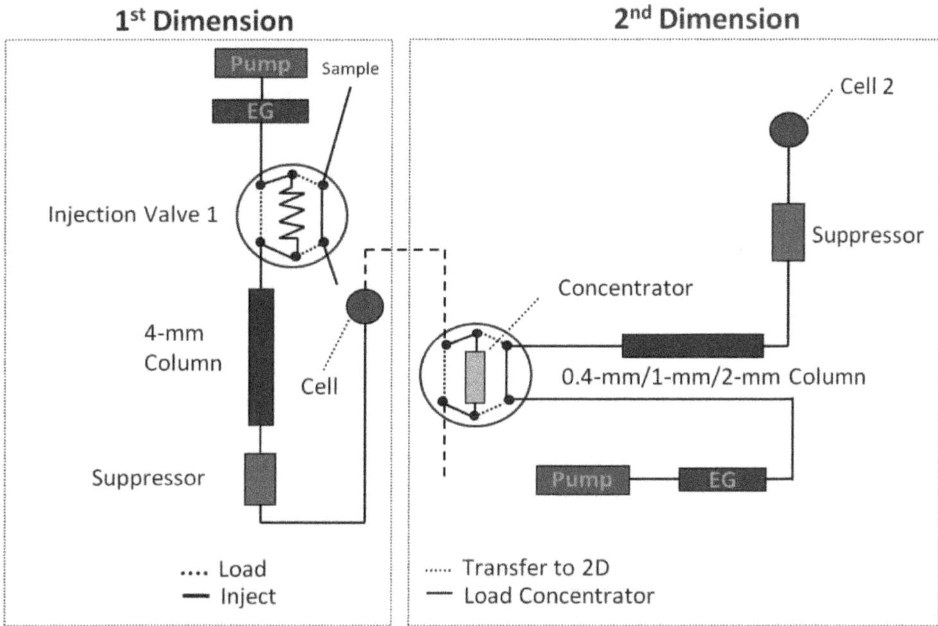

Figure 8 *Schematic setup for a 2-D IC approach. The analysis in the first dimension is performed using 4-mm column chemistry and in the second dimension using 0.4-mm, 1-mm, or 2-mm column chemistry leading to improved sensitivity proportional to the flow rate ratio of the 1st dimension to the 2nd dimension.*

The schematic setup for a two-dimensional analysis with integrated matrix elimination is shown in Figure 8.[31,32] It is applicable to the determination of trace concentrations of bromate in municipal and natural mineral waters with high ionic strength matrices (Figure 9).

The first dimension uses a high capacity 4-mm Dionex IonPac AS19 column to resolve the bromate from the matrix ions. The matrix ions are diverted to waste while a 2 mL plug (cut volume) containing the bromate is transferred to the second dimension for analysis. Bromate is well resolved in the second dimension using a 2-mm Dionex IonPac AS24 column. This method is fully automated and achieves bromate detection limits equivalent to or better than post column addition methods. The 2-D IC method avoids the cost and disposal of the chemicals required for post-column configurations and simplifies the experimental setup. Additionally, it avoids potential column overload during analysis of high ionic strength matrices.

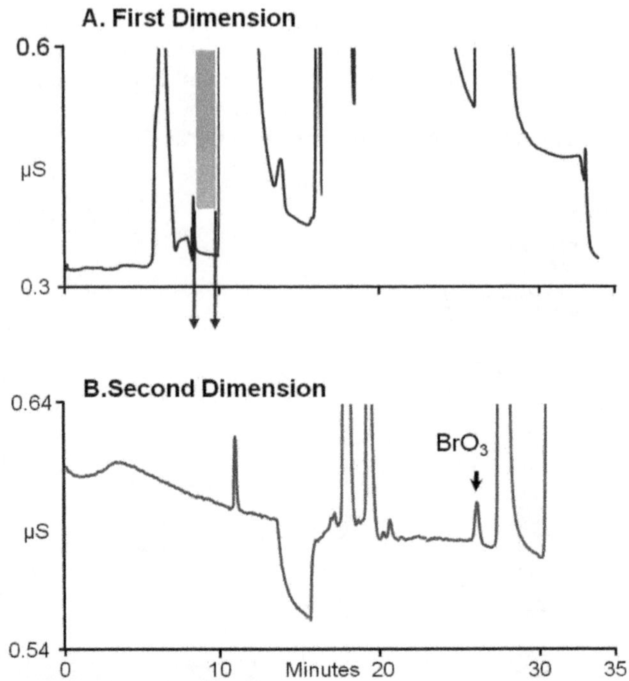

Figure 9 *Trace Analysis of Bromate (0.5 µg/L) in Bottled Water by 2-D IC*[31] *1st Dimension: Column: Dionex IonPac AS19 (4 mm), Flow: 1 mL/min, Eluent:10-65 mmol/L KOH, Suppressor: Dionex ASRS 300 Anion Self-Regenerating suppressor (4 mm), Injection vol.: 1000 µL, Temperature: 30 °C, 2nd Dimension: Column: Dionex IonPac AS24 (2 mm), Flow: 0.25 mL/min, Eluent: 35 mmol/L KOH, Suppressor: ASRS 300 Anion Self-Regenerating suppressor (2 mm), Temperature: 30 °C, Injection vol.: 2000 µL of the suppressed effluent from the 1st dimension (8-10 min) on Dionex IonPac UTAC Ultratrace Anion Concentrator column.*

3. CONCLUSION

Ion chromatography allows the trace determination of various ionic DBPs in waters of different ionic strength, composition, and origin. Multiple analytical approaches are available that incorporate different elution profiles, detection schemes, and matrix elimination approaches, thus addressing the different analytical needs. Most of these methods have already been adapted and converted to national or international standards providing validated methods to the final user. In addition, in most of the methods time-consuming off-line sample preparation is eliminated, improving consistency and reliability of the analytical data between different analysts and laboratories. IC is suited for the determination of inorganic DBPs like bromate, chlorite, and chlorate and for the determination of haloacetic acids. Though not specifically addressed in this work, many of the methods involving suppressed conductivity and MS detection can also be applied to the trace determination of perchlorate.

References

1 Drinking Water Treatment; EPA 810-F-99-013; U.S. Environmental Protection Agency, 1999.
2 World Health Organization. Disinfectants and Disinfection By-Products; International Programme on Chemical Safety-Environmental Health Criteria 216; Geneva, Switzerland, 2000.
3 WHO Library Cataloguing-in-Publication Data, Guidelines for drinking-water quality - 4th ed, ISBN 978 92 4 154815 1.
4 European Parliament and Council Directive No. 98/83/EC, Quality of Water Intended for Human Consumption, 1998.
5 US EPA. National Primary Drinking Water Regulations: Disinfectants and Disinfection Byproducts. Fed. Regist. 1998, 63 (241), 69389-69476.
6 Ministry Directive No. 101, Ministry of Health, Labour, and Welfare, Tokyo, Japan, 2002.
7 European Parliament and Council Directive No. 2003/40/EC, 2003.
8 European Communities (Drinking Water)(No. 2) Regulations, 2007 (S.I. 278 of 2007)
9 European Union (Drinking Water) Regulations, 2014 (S.I. No. 122 of 2014)
10 Method 551.1, U.S. Environmental Protection Agency. Cincinnati, OH, 1995.
11 Method 551.2, U.S. Environmental Protection Agency. Cincinnati, OH, 1995.
12 Method 551.3, U.S. Environmental Protection Agency. Cincinnati, OH, 2003.
13 ISO 23631, 2006, www.beuth.de.
14 Method 551.7, U.S. Environmental Protection Agency. Cincinnati, OH, 2009.
15 H. Small, T. S. Stevens und W.C. Baumann, Anal. Chem. 47 (1975) 1801.
16 Eluent Suppressors for Ion Chromatography, Download at: www.thermoscientific.com.
17 Method 300.1, U.S. Environmental Protection Agency. Cincinnati, OH, 1997.
18 ISO 15061, 2001, www.beuth.de.
19 E. Salhi, U. Von Gunten, *Water Research*.1999, 33, p 3239-3244.
20 ISO 11206, 2011, www.beuth.de.
21 ISO 10304-1, 2007, www.beuth.de.
22 J. Djoukeng, D. Jensen, Determination of Bromate by ISO Method 11206, Technical Note 116, http://www.thermoscientific.com/content/dam/tfs/ATG/CMD/CMD%20Documents/Application%20&%20Technical%20Notes/113626-TN116-IC-Bromate-ISO-Method-07Jun2012-TN70067E.pdf.
23 DIN 19643 part 1 to 3, 2012, www.beuth.de.
24 A. Nahrstedt, presentation "Chlorit und Chlorat im Schwimmbeckenwasser", Seminar on Disinfection By-products, June 11, 2013, Langenau, Germany.
25 ISO 10304-4, 1997, www.beuth.de.
26 D.T. Heitkemper, L.A. Kaine, D.S. Jackson, K.A. Wolnik. *J. Chromatogr. A.*, 1994, 671, p 101-108
27 J.T. Creed, M.L. Magnuson, J.D. Pfaff, C.A. Brockhoff. *J. Chromatogr. A.*, 1996, 753 p 261-267.
28 J.T. Creed, M.L. Magnuson, C.A. Brockhoff, *Environ. Sci. Technol*, 1997, 31, p 2059-2063.
29 Method 321.8, U.S. Environmental Protection Agency. Cincinnati, OH, 1997.

30 S. McSheehy-Ducos, J. D. Willis, L. Rottmann, The Determination of Major and Trace Elements in Milk using ICP-Q-MS, Poster Presentation, Recent Advances in Food Analysis (RAFA), Prague, 2013.
31 Determination of Sub-µg/L Bromate in Municipal and Natural Mineral Waters Using Preconcentration with Two-Dimensional Ion Chromatography and Suppressed Conductivity Detection, B. DeBorba, J. Rohrer, Application Note 187, 2007,
www.thermoscientific.com.
32 Method 302.0, U.S. Environmental Protection Agency. Cincinnati, OH, 2009.

Health and Toxicology

HEALTH AND TOXICOLOGICAL ASPECTS OF DISINFECTION BY-PRODUCTS IN DRINKING WATER – A UK REVIEW

G. Lau

Centre for Radiation, Chemical and Environmental Hazards, Public Health England, 151 Buckingham Palace Road, London SW1W 9SZ, UK

1 INTRODUCTION

Water companies are required to disinfect water at water treatment works (WTWs) before supplying wholesome[1] drinking water to consumers through their distribution networks.[2] Disinfection of drinking water is fundamental to preventing the spread of waterborne diseases, such as cholera.[3] Chemical processes such as chlorination or chloramination for microbial disinfection of drinking water are practised widely and proven effective. Water companies must design, operate and maintain the disinfection process in order to keep the presence of disinfection by-products (DBPs) as low as possible without compromising the effectiveness of the disinfection. This is a Council Directive 98/83/EC requirement that became part of the Water Supply (Water Quality) Regulations 2010 in the UK.

Since the discovery of DBP formation in the mid-1970s,[4] there has been a debate on the potential toxicity of DBPs in drinking water.[5] The main focus of concern with regard to the health effects of DBPs in drinking water has been potential carcinogenicity and adverse developmental or reproductive effects.[6] However, studies to date have been inconsistent and may not be directly applicable to the UK due to different disinfection practices employed in different countries and constituents of the raw water.[3] Information regarding DBPs in drinking water and a summary of statements published by two independent scientific advisory committees to the UK Government on potential carcinogenicity and toxicity of chemicals are provided in this concise review.

2 CHLORINATION AND CHLORAMINATION

Chlorination can be achieved by using different techniques at WTWs and is one of the most commonly used primary disinfection processes for drinking water in the UK.[7] The effectiveness of disinfection depends on the form of chlorine employed, its concentration and contact time. Typically, low levels of chlorine disinfectant residual i.e. 0.1 to 0.5 mg/l, in the UK,[2] are deliberately maintained within distribution networks, as a secondary disinfection measure, to protect against recontamination and limit microbial growth.[8] This compares favourably to the health based guideline value of 5 mg/l set by the World Health Organization (WHO)[2] as much higher levels of chlorine disinfectant residual are common in other countries.[9]

Chloramination is essentially a secondary disinfection process where ammonia is mixed with chlorine to form chloramine disinfectant residual in drinking water leaving a WTW.[2] Monochloramine, dichloramine and trichloramine (nitrogen trichloride) are typical chloramines with monochloramine being the most abundant and used primarily as a residual for drinking water disinfection. In 2009, nearly 50% of all water companies in the UK practised chloramination in some of their water supplies.[10] Although chloramines are less potent disinfectants than chlorine and can give rise to taste and odour complaints e.g. trichloramine,[8] they provide longer-lasting protection as they do not break down quickly in distribution networks. In recent years, chloramination has become the preferred secondary disinfection process in order for water companies to meet regulations on DBPs in the US.[6,11]

3 DISINFECTION BY-PRODUCTS

Typically, DBPs are formed as a result of chemical interactions between disinfectants e.g. chlorine or chloramines, and trace organic matter e.g. fulvic or humic substances, in source waters.[6] The formation of any particular DBPs can be predicted from the source water quality and disinfectant used in a given water supply.[2] Treated drinking water from surface waters generally contains higher concentrations of DBPs than groundwater due to the higher concentrations of organic material in surface waters.[12] Trihalomethanes (THMs), halogenated acetic acids (HAAs), haloketones and haloacetonitriles are the main DBPs formed as a result of chlorination. Although monochloramine produces THMs in lower concentrations, other DBPs may result,[8] which are yet to be characterised.[3]

Based on current global research, THMs and HAAs have typically served as surrogates reflecting the concentrations of other DBPs, due to their dominance in chlorinated and chloraminated drinking water.[2,8] HAAs and other DBPs are not directly regulated in the UK.[7] Individual WHO guideline values for THMs (bromoform, bromodichloromethane, chloroform, dibromochloromethane) and the UK prescribed concentration for total THMs in drinking water are shown in Table 1.

THMs	UK (ug/l)	WHO (μg/l)
Bromoform	100	100
Bromodichloromethane		60
Chloroform		300
Dibromochloromethane		100

Table 1 *Individual WHO guideline values for THMs and UK prescribed concentration for total THMs in drinking water.*

Where disinfection is practised, measures to minimise DBP formation should be taken into consideration.[8] Whilst removal of any organic precursors from source waters prior to disinfection remains the most effective means of controlling DBPs formation,[2] other strategies that can be adopted to reduce DBP concentrations are:

- Using a different chemical disinfectant with a lower tendency to produce DBPs with the source water;
- Using non-chemical disinfection e.g. ultraviolet; and
- Removing DBPs prior to distribution.[6,7]

4. HEATH AND TOXICOLOGICAL ASPECTS OF DISINFECTION BY-PRODUCTS

There is widespread potential for human exposure to DBPs in drinking water through day-to-day activities e.g. cooking and bathing.[13] Concern about the potential health effects of DBPs has driven the investigation of the possible association between exposure to DBPs and incidence of cancer, and more recently, with adverse reproductive outcomes.[12] In the past, toxicological studies have shown that certain DBPs in drinking water may cause an increased incidence of cancer and reproductive effects in laboratory animals.[13] In addition, epidemiological studies have suggested that chronic exposure to chlorinated drinking water may be associated with an increased risk for bladder, colon and rectal cancers[14] and adverse reproductive or developmental health effects, such as spontaneous abortion or foetal anomalies in individuals.[12]

In 2008, the UK Committee on Carcinogenicity of Chemicals in Food, Consumer Products and the Environment (COC) evaluated data from several epidemiological studies published since 1999 on chlorinated drinking water and a range of cancers as shown below:

- Adult leukaemia
- Bladder cancer
- Brain cancer
- Childhood acute lymphoblastic leukaemia (ALL)
- Colorectal cancer
- Pancreatic cancer[3]

Although four of the studies were from Canada, with others from Australia, France, Italy, Spain and the US, the aim of the evaluation was to determine whether or not there was a recognised carcinogenic hazard from DBPs in drinking water. In these studies, the following exposure parameters were compared:

- Chlorinated vs. non-chlorinated drinking water
- Duration of exposure to chlorinated drinking water
- Source of drinking water
- THM levels (usually total THMs)[3]

COC commented in their 2008 statement that whilst the studies on bladder,[14-19] brain,[20] and colon[21] cancers and exposure to DBP i.e. THMs, provided some evidence for an association between these cancers and exposure to THMs in men, the evidence for an association in women was conflicting for bladder cancer and non-existent for the others. In addition, results of three studies on exposure to THMs and adult leukaemia[22] and childhood ALL[23, 24] were found to be inconclusive and statistically insignificant and no association was found between pancreatic cancer and increasing THMs levels in another study.[3,25]

Considering the number of interpretation issues associated with these studies such as incomparable exposure parameters due to their wide ranges and possible residual confounding factors e.g. smoking status, the small relative risks found and that disinfection practices and constituents of the raw water may be different in other countries, the study results may not be directly applicable to the UK. Subsequently, COC concluded in their statement that evidence for a causal association between cancer and exposure to THMs was limited and that any such association was unlikely to be strong.[3]

Furthermore, a major two-phase study carried out by the UK Small Area Health Statistics Unit (SAHSU), part of the Medical Research Council-Public Health England (MRC-PHE) Centre for Environment and Health, which investigated potential associations between DBPs i.e. THMs, and the risks of adverse reproductive outcomes (first phase)[26] and congenital anomalies (second phase)[27] were published in 2004 and 2008 respectively. In the first phase of the study, SAHSU in collaboration with Northumbrian Water (North East region), Severn Trent Water (Mid Wales and East Midlands region) and United Utilities (North West region), utilised routinely collected THM measurements in drinking water and records extracted from national birth and stillbirth registers held at SAHSU to assess the risks involved.[26]

The UK Committee on Toxicity of Chemicals in Food, Consumer Products and the Environment (COT) evaluated the first phase of the study and commented in their 2004 statement that in the North West region, exposure to THMs showed an inverse association with mean birth weight and a direct association with low (<2,500 g) and very low (<1,500 g) birth weights and stillbirth. However, there was evidence of residual confounding by social deprivation in the region, which was not adequately adjusted for in the study. In the Mid Wales and East Midlands region, the prevalence of very low birth weight decreased with increasing exposure to THMs and no association was found for low birth weight or stillbirth. In addition, no association was found between exposure to THMs and birth weight and stillbirth in the North East region, although this may be attributed to a relatively small number of births in the region as shown in Table 2.[28]

Water Company	Region covered in the UK	Estimated population (millions)
Northumbrian Water	North East	2.6
Severn Trent Water	Mid Wales and East Midlands	7.4
United Utilities	North West	6.8

Table 2 *Profiles of water companies participated in the first phase of the study by SAHSU.*[26]

COT evaluated additional data from several epidemiological studies[29-41] and findings from animal[42-48] and DBP exposure/uptake[49-51] studies and reviews[12,52,53] published since 1998. Considering the contrasting findings in the three regions by the SAHSU study and inconsistent and inconclusive results from the epidemiological studies, COT concluded overall in their statement that a causal relationship was unfounded between exposure to THMs and the following adverse reproductive outcomes:

Low Apgar score;
Infant death;
Infant's body length and head circumference at birth;
Length of gestation;
Low and very low birth weights;
Neonatal hypothyroidism and jaundice;
Perinatal death;
Pre-term delivery;
Spontaneous abortion; and
Stillbirth.[28]

In 2008, COT evaluated the second phase of the study, where routinely collected THM measurements from 12 water companies in the UK (estimated total population of 44.2 millions) and records extracted from the National Congenital Anomalies System held at SHASU and regional registries via the British Isles Network of Congenital Anomaly Registers. These data were used for SAHSU's main analyses covering the following broad categories of congenital anomalies:

Cleft lip/palate;
Diaphragmatic hernia and abdominal defects;
Major cardiac defects;
Neural tube defects;
Respiratory defects; and
Urinary tract defects.[7]

COT commented in their 2008 statement that a higher prevalence of anomalies was observed in the most deprived areas. Cleft lip/palate and urinary tract defects were 50 to 100% higher in males. A U-shaped relationship was found between prevalence of congenital anomalies and maternal age except for neural tube defects, where the prevalence decreased with increasing age. Residual confounding factors identified for the study i.e. sex, maternal age and socioeconomic status, were found not to have a significant effect on risk estimates in this study. Although further statistical analyses showed a number of associations between exposure to THMs and major cardiac defects and a restricted set of isolated anomalies i.e. congenital anomalies of the oesophagus, gastroschisis and ventricular septal defects, no statistically significant trends were established and little evidence was found for a relationship between exposure to THMs in drinking water and the risk of congenital anomalies.[7]

Since the SAHSU study only focused on THMs in drinking water and did not consider variations such as individual movements and water consumption during pregnancy and other sources of exposure e.g. bathing, showering and swimming, possible associations might have been obscured by a lack of individual exposure data for interpretation.[7] COT carried out further evaluation including data from a preliminary UK study on HAAs in drinking water,[54] epidemiological studies on congenital anomalies[30,34,55-64] and adverse reproductive outcomes[39,65-68] published since 2004 and a review[69] that considered developmental and reproductive animal studies on individual DBPs and toxicological studies on congenital anomalies. Subsequently, COT concluded in their statement that overall no consistent relationship was found between exposure to THMs and adverse reproductive outcomes and congenital anomalies in human studies. In animal studies, effects were mainly seen at high doses associated with maternal toxicity and thus were not considered to be predictive of effects in human exposure at lower levels of DBPs.[7]

5. CONCLUSIONS

COC and COT concluded in their 2008 statements that there was limited evidence to suggest a causal association between cancer and exposure to THMs[3] and that no consistent relationship was found between exposure to THMs and adverse reproductive outcomes and congenital anomalies in human studies.[7,28] Effects reported in animal studies were largely associated with high doses, which are not considered predictive of effects in humans exposed to much lower levels of DBPs.[7] In addition, studies in other countries on THMs exposure evaluated by COC and COT, may not be directly applicable in the UK, due to different disinfection practices employed and varying constituents of the raw water.[3]

UK specific studies on potential associations between THMs and the risks of adverse reproductive outcomes and congenital anomalies have been limited since COC and COT's statements in 2008. Although lately the MRC-PHE Centre for Environment and Health collaborated in a large UK-wide study aiming to provide evidence for an association between THMs and adverse semen outcomes in humans, little evidence was found for an association between various THMs i.e. chloroform, total brominated THMs and total THMs, and a number of semen quality outcomes.[70]

THMs and HAAs have typically served as surrogates of other DBPs,[2,8] with THMs being the most widely measured individual DBPs in epidemiological studies.[54] However, THMs may not necessarily be a good marker of other DBPs or the most significant from a health point of view.[12,54,71,72] Prospective study designs should consider individual and seasonal variations and include impact analyses of potential causative agents and residual confounding factors.[28] Whilst research to determine the potential health effects of DBPs continues, effective microbial disinfection should not be compromised in order to minimise DBPs in chlorinated/chloraminated drinking water[3,7,14,28] or meet regulatory guidelines for DBPs as emphasised by WHO.[8]

References

1 Drinking Water Inspectorate, 2010. *What are the drinking water standards?* [Online] London: Drinking Water Inspectorate. Available at: http://dwi.defra.gov.uk/consumers/advice-leaflets/standards.pdf [Accessed 2nd January 2015].

2 Drinking Water Inspectorate, 2013. *Drinking water 2012*. [Online] London: Drinking Water Inspectorate. Available at: http://dwi.defra.gov.uk/about/annual-report/2012/central-eastern.pdf [Accessed 2nd January 2015].

3 Committee on Carcinogenicity of Chemicals in Food, Consumer Products and the Environment, 2008. *Second statement on chlorinated drinking water and cancer*. Available at: http://www.pwtag.org/researchdocs/Used%20Ref%20docs/16%20HPA%20Secondstatement%20on%20CBPs%20and%20cancer%20COC08.S1.pdf [Accessed 2nd January 2015].

4 J. Rook, *Water Treat. Exam.*, 1974, **23**, 234.

5 Health Protection Agency, 2007. *Review of chemical toxicity to the reproductive system, with particular reference to development toxicity.* Available at: http://webarchive.nationalarchives.gov.uk/20140722091854/http://www.hpa.org.uk/webc/HPAwebFile/HPAweb_C/1194947320972 [Accessed 2nd January 2015].

6 S. Krasner, *Philos. Trans. R. Soc., A.* 2009, **367**, 4077.

7 Committee on Toxicity of Chemicals in Food, Consumer Products and the Environment, 2008. *COT statement on a SAHSU study on chlorination disinfection by-products and risk of congenital anomalies in England and Wales.* [Online] London: Committee on Toxicity of Chemicals in Food, Consumer Products and the Environment. Available at: http://cot.food.gov.uk/sites/default/files/cot/cotstatementdbp200802.pdf [Accessed 2nd January 2015].

8 World Health Organization, 2011. *Guidelines for drinking-water quality.* 4th ed. [Online] Geneva: World Health Organization. Available at: http://whqlibdoc.who.int/publications/2011/9789241548151_eng.pdf [Accessed 2nd January 2015].

9 Drinking Water Inspectorate, 2010. *Chlorine.* [Online] London: Drinking Water Inspectorate. Available at: http://dwi.defra.gov.uk/consumers/advice-leaflets/chlorine.pdf [Accessed 2nd January 2015].

10 Drinking Water Inspectorate, 2009. *Request for information: chloramination of UK tap water.* [Online] London: Drinking Water Inspectorate. Available at: https://www.whatdotheyknow.com/request/19663/response/48321/attach/2/acknowledgement%20letter.pdf [Accessed 2nd January 2015].

11 Environmental Protection Agency, 2009. *Chloramines Q & A's.* [Online] Washington: Environmental Protection Agency. Available at: http://www.epa.gov/ogwdw/disinfection/chloramine/pdfs/all29_q.pdf [Accessed 2nd January 2015].

12 M. Nieuwenhuijsen, M. Toledano, N. Eaton, J. Fawell and P. Elliott, *J. Occup. Environ. Med.*, 2000, **57**, 73.

13 G. Boorman, V. Dellarco, J. Dunnick, R. Chapin, S. Hunter, F. Hauchman, H. Gardner, M. Cox and R. Sills, *Environ. Health Perspect.*, 1999, **107**, 207-217.

14 C. Villanueva, K. Cantor, S. Cordier, J. Jaakkola, W. King, C. Lynch, S. Porru and M. Kogevinas, *Epidemiol.*, 2004, **15**, 357.

15 C. Chevrier, B. Junod and S. Cordier, *Epidemiol.*, 2004, **15**, 605.

16 G. Ranmuthugala, L. Pilotto, W. Smith, T. Vimalasiri, K. Dear and R. Douglas, *Epidemiol.,* 2003, **14**, 617.

17 C. Villanueva, F. Fernandez, N. Malats, J. Grimalt and M. Kogevinas, *J. Epidemiol. Community Health*, 2003, **57,** 166.

18 C. Villanueva, K. Cantor, W. King, J. Jaakkola, S. Cordier, C. Lynch, S. Porru and M. Kogevinas, *Int. J. Cancer*, 2006, **118**, 2040.

19 M. Villanueva, K. Cantor, J. Grimalt, N. Malats, D. Silverman, A. Tardon, R. Garcia-Closas, C. Serra, A. Carrato, G. Castano-Vinyals, R. Marcos, N. Rothman, F. Real, M. Dosemeci and M. Kogevinas, *Am. J. Epidemiol.*, 2007, **165**, 148.

20 K. Cantor, C. Lynch, M. Hildesheim, M. Dosemeci, J. Lubin, M. Alavanja and G. Craun, *Am. J. Epidemiol.*, 1999, **150**, 552.

21 W. King, L. Marrett and C. Woolcott, *Cancer Epidemiol., Biomarkers Prev.*, 2000, **9**, 813.

22 K. Kasim, P. Levallois, K. Johnson, B. Abdous, P. Auger and Canadian Cancer Registries Epidemiology Research Group, *Am. J. Epidemiol.*, 2006, **163**, 116.

23 C. Infante-Rivard, E. Olson, L. Jacques and P. Ayotte, *Epidemiol.*, 2001, **12**, 13.

24 C. Infante-Rivard, D. Amre and D. Sinnett, *Environ. Health Perspect.*, 2002, **110**, 591.

25 M. Do, N. Birkett, K. Johnson, D. Krewski, P. Villeneuve and Canadian Cancer Registries Epidemiology Research Group, *Environ. Health Perspect.*, 2005, **113**, 418.

26 M. Toledano, M. Nieuwenhuijsen, N. Best, H. Whitaker, P. Hambly, C. de Hoogh, J. Fawell, L. Jarup and P. Elliott, *Environ. Health Perspect.*, 2004, **113**, 225.

27 M. Nieuwenhuijsen, M. Toledano, J. Bennett, N. Best, P. Hambly, C. de Hoogh, D. Wellesley, P. Boyd, L. Abramsky, N. Dattani, J. Fawell, D. Briggs, L. Jarup and P. Elliott, *Environ. Health Perspect.*, 2008, **116** , 216.

28 Committee on Toxicity of Chemicals in Food, Consumer Products and the Environment, 2004. *Statement on chlorinated drinking water and reproductive outcomes.* Available at: http://cot.food.gov.uk/sites/default/files/cot/cotstatementchlorwater0408.pdf [Accessed 2nd January 2015].

29 L. Dodds, W. King, A. Allen, B. Armson, D. Fell and C. Nimrod, *Epidemiol.*, 2004, **15**, 179.

30 L. Dodds, W. King, C. Woolcott and J. Pole, *Epidemiol.*, 1999, **10**, 233.

31 L. Fabiani, F. Materazzo, F. Ensabella, A. Giuliani, F. Patacchiola, V. Oleandri and V. Leoni, *Annali di lgiene: Medicina Preventiva e di Comunità*, 2003, **15**, 933.

32 M. Gallagher, J. Nuckols, L. Stallones and D. Savitz, *Epidemiol.*, 1998, **9**, 484.

33 J. Jaakkola, P. Magnus, A. Skrondal, B-F. Hwang, G. Becher and E. Dybing, *Occup. Environ. Med.*, 2001, **58**, 437.

34 B. Kallen and E. Robert, *Reprod. Toxicol.*, 2000, **14**, 303.

35 W. King, L. Dodds and A. Allen, *Environ. Health Perspect.*, 2000, **108**, 883-886.

36 E. Righi, G. Fantuzzi, M. Montanari, A. Bargellini, G. Predieri and G. Aggazzotti, *Annali di lgiene: Medicina Preventiva e di Comunità*, 2003, **15**, 649.

37 K. Waller, S. Swan, G. Windham and L. Fenster, *J. Exposure Anal. Environ. Epidemiol.*, 2001, **11**, 522.

38 J. Wright, J. Schwartz and D. Dockery, *Occup. Environ. Med.*, 2003, **60**, 173.

39 J. Wright, J. Schwartz and D. Dockery, *Environ. Health Perspect.*, 2004, **112**, 920.

40 C-Y., Yang, *Toxicol.*, 2004, **198**, 249.

41 C-Y. Yang, B-H. Cheng, S-S. Tsai, T-N. Wu, M-C. Lin and K-C. Lin, *Environ. Health Perspect.*, 2000, **108**, 765.

42 S. Bielmeier, D. Best, D. Guidici and M. Narotsky, *J. Toxicol. Sci.*, 2001, **59**, 309.

43 S. Bielmeier, D. Best and M. Narotsky, *J. Toxicol. Sci.*, 2004, **77**, 101.

44 J. Chen, G. Douglas, T. Thirkill, P. Lohstroh, S. Bielmeier, M. Narotsky, D. Best, R. Harrison, K. Natarajan, R. Pegram, J. Overstreet and B. Lasley, *J. Toxicol. Sci.*, 2003, **76**, 75.

45 J. Chen, T. Thirkill, P. Lohstroh, S. Bielmeier, M. Narotsky, D. Best, R. Harrison, K. Natarajan, R. Pegram, J. Overstreet, B. Lasley and G. Douglas, *J. Toxicol. Sci.*, 2004, **78**, 166.

46 M. Christian, R. York, A. Hoberman, R. Diener, L. Fisher and G. Gates, *Int. J. Toxicol.*, 2001, **20**, 239.

47 M. Christian, R. York, A. Hoberman, R. Diener and L. Fisher, *Int. J. Toxicol.*, 2001, **20**, 225.

48 M. Christian, R. York, A. Hoberman, L. Fisher and W. Brown, *Int. J. Toxicol.*, 2002, **21**, 115.

49 L. Backer, D. Ashley, M. Bonin, F. Cardinall, S. Kieszak and J. Wooten, *J. Exposure Anal. Environ. Epidemiol.*, 2000, **10**, 321.

50 M. Lynberg, J. Nuckols, P. Langlois, D. Ashley, P. Singer, P. Mendola, C. Wilkes, H. Krapfl, E. Miles, V. Speight, B. Lin, L. Small, A. Miles, M. Bonin, P. Zeitz, A. Tadkod, J. Henry and M. Forrester, *Environ. Health Perspect.*, 2001, **109**, 597.

51 A. Miles, P. Singer, D. Ashley, M. Lynberg, P. Mendola, P. Langlois and J. Nuckols, *Environ. Sci. Technol.*, 2002, **36**, 1692.

52 F. Bove, Y. Shim and P. Zeitz, *Environ. Health Perspect.*, 2002, **110** , 61.

53 C. Graves, G. Matanoski and R. Tardiff, *Regul. Toxicol. Pharmacol.*, 2001, **34**, 103.

54 E. Malliarou, C. Collins, N. Graham and M. Nieuwenhuijsen, *Water Res.*, 2005, **39**, 2722.

55 A. Aschengrau, S. Zierler and A. Cohen, *Arch. Environ. Health*, 1993, **48**, 105.

56 F. Bove, M. Fulcomer, J. Klotz, J. Esmart, E. Dufficy and J. Savrin, *Am. J. Epidemiol.*, 1995, **141**, 850.

57 M. Cedergren, A. Selbing, O. Löfman and B. Källen, *Environ. Res.*, 2002, **89**, 124.

58 L. Dodds and W. King, *Occup. Environ. Med.*, 2001, **58**, 443.

59 B-F. Hwang, P. Magnus and J. Jaakkola, *Am. J. Epidemiol.*, 2002, **156**, 374.

60 J. Klotz, and L. Pyrch, *Epidemiol.*, 1999, **10** , 383.

61 P. Magnus, J. Jaakkola, A. Skrondal, J. Alexander, G. Becher, T. Krogh and E. Dybing, *Epidemiol.*, 1999, **10**, 513.

62 G. Shaw, D. Ranatunga, T. Quach, E. Neri, A. Correa and R. Neutra, *Epidemiol.*, 2003, **14**, 191.

63 B-F. Hwang, J. Jaakkola and H-R. Guo, *Environ. Health*, 2008, **7**.

64 T. Luben, J. Nuckols, B. Mosley, C. Hobbs and J. Reif, *Occup. Environ. Med.*, 2008, **65**, 420.

65 A. Hinckley, A. Bachand and J. Reif, *Environ. Health Perspect.*, 2005, **113** , 1808.

66 C. Infante-Rivard, *Environ. Health Perspect.*, 2004, **112**, 1213.

67 C. Porter, S. Putnam, K. Hunting and M. Riddle, *Am. J. Epidemiol.*, 2005, **162**, 334.

68 Awwa Reserach Foundation, 2005. *Drinking water disinfection by-products and pregnancy outcome*. [Online] Washington: AWWA Research Foundation. Available at: http://www.waterrf.org/PublicReportLibrary/91088F.pdf [Accessed 2nd January 2015].

69 R. Tardiff, M. Carson and M. Ginevan, *Regul. Toxicol. Pharmacol.*, 2006, **45**, 185.

70 N. Iszatt, M. Nieuwenhuijsen, J. Bennet, N. Best, A. Povey, A. Pacey, H. Moore, N. Cherry and M. Toledano, *Occup. Environ. Med.*, 2013, **70**, 754.

71 X. Xu, M. Mariano, D. Laskin and P. Weisel, *Toxicol. Appl. Pharmacol.*, 2002, **184**, 19.

72 X. Xu and P. Weisel, *Environ. Sci. Technol.*, 2003, **37**, 569.

INDEXES OF DISINFECTION BY-PRODUCTS (DBPS) EXPOSURE AND IRRITANT SYMPTOMS IN SWIMMING POOL WORKERS

E. Righi, G. Fantuzzi, G. Predieri, G. Aggazzotti

Department of Diagnostic, Clinical and Public Health Medicine, University of Modena and Reggio Emilia, Via Campi 287, 41125 Modena, Italy

1 INTRODUCTION

The onset of respiratory and other irritant symptoms in pool workers is of growing concern as, recently, different epidemiological studies have reported an unusually high prevalence of ocular and respiratory symptoms, as well as asthmatic conditions, in occupationally exposed subjects as a consequence of their exposure to disinfection by-products (DBPs).[1,2]

Among the hundreds of DBPs identified, the most commonly investigated are trihalomethanes (THMs), which are considered representative of volatile DBPs in swimming pools environments.[3-6] Exposure for pool workers occurs mainly through inhalation, however, in specific situations, especially for trainers, uptake of DBPs via dermal absorption and ingestion of small quantities of pool water can occur as well. Moreover, lifeguards and trainers are usually exposed to levels of THMs significantly higher than employees involved in other tasks, such as bartenders, receptionists or office managers.[7] Nevertheless, epidemiological studies exploring the potential association between THMs exposure and irritant health outcomes are sparse.[8]

Indeed, the interest of the scientific community has been mainly focused on occupational exposure to chloramines, and in particular to trichloramine (NCl_3), the most volatile compound among chloramines.[1,2] NCl_3 is a strong mucous membrane irritant and, therefore, is considered as the putative causal agent of ocular and respiratory symptoms, as well as asthma, in subjects working in swimming facilities. Different studies have been carried out: most reported a significantly high frequency of acute ocular and respiratory symptoms in typical indoor swimming pools working conditions, while for asthma-related endpoints evidence is still inconclusive.[9-17]

Overall, the determination of airborne NCl_3 levels poses some methodological issues. Different sampling and analytical methods have been proposed, however, none of them is considered specific enough to differentiate between inorganic and organic chloramines.[18-20] Further, no analytical method for personal biological monitoring is available. Finally, recently, some authors have suggested that there may be other volatile DBPs with potential irritant properties in indoor pool environments which could be responsible or co-responsible of the high prevalence of irritant health outcomes.[14,21]

In two previous studies we evaluated the frequency of different ocular and respiratory symptoms in a group of pool workers in relation to their exposure to

DBPs.[8,16] Recently, we were able to extend the monitoring to a further group of workers. The aim of this study was to re-evaluate associations between different irritant health effects and occupational exposure to both THMs and NCl_3 in this enlarged cohort of subjects in order to assess the role played by different index of exposure to irritant chemicals. THMs exposure was evaluated by both environmental and biological monitoring, while NCl_3 levels were investigated in environmental air.

2 METHODS

2.1 Study design

In this cross sectional study 23 public indoor swimming pools in Northern Italy were investigated between 2008 and 2013. Each swimming pool was monitored once and, in order to operate in similar environmental and ventilation conditions, sampling sessions were carried out during the winter period, always in the morning and at the beginning of the working week. During each sampling session environmental monitoring was carried out and information about pool water treatments and disinfection procedures was collected. Pool workers (170 subjects) were asked to participate in the study. All subjects agreed to participate and provided an informed consent. Each subject was asked to fill in a questionnaire and to provide an alveolar air sample in order to perform personal biological monitoring as well.

2.2 Pool workers health status assessment

A self-administrated questionnaire was drawn up to record information about personal data (age, sex, smoking habits and medical history), specific working activities inside the swimming pool (tasks, number of working hours per day and week, occupational history) and previous jobs. The frequency of self-reported ocular symptoms (red eyes, itchy eyes, eye infections, watery eyes), respiratory symptoms (nasal itch, sneezing, cold, runny nose, blocked nose, voice loss) and asthma-related symptoms (wheezing, shortness of breath, coughing and dyspnea/asthma diagnosed by physician) was recorded as well. Possible answers to questions about symptoms were never, sometimes and often. Five subjects filled the questionnaire partially and were excluded from the study.

2.3 Environmental and biological monitoring

Water temperature, pH, free and combined chlorine and THMs levels were evaluated in each swimming pool. Temperature, pH, free and combined chlorine in pool water were analyzed at three different points near the edge of the pool: the mean value was considered representative of these parameters in the pool water during the sampling session. To assess THMs levels in water, three samples of pool water were collected during each session from a depth of 20 cm at three different points near the edge of the pool; the mean value was considered representative of THMs concentration in the water.

In environmental air both NCl_3 and THMs levels were measured. NCl_3 levels were evaluated in samples collected at a height of 100 cm, 3 m distant from the edge of the pool by an active air sampling (Buck Pump, Supelco) at a flow rate of 1000 cm^3/min for 100 min. THMs environmental air levels were assessed by collecting air samples at a height of 150 cm above the edge of the pool. Samples were collected over a period of 2h in Tedlar bags (2l) using a pump (Buck Pump, Supelco) calibrated at a flow rate of 15 cm^3/min.

Finally, each pool worker was asked to provide an alveolar air sample. Samples were collected in a 34 cm^3 one-way glass tubes with two valves. The sampling method has been already described in previous studies.[7,22,23] Briefly, subjects were asked to breathe normally into the tube with open valves. At the end of expiration, the valves were closed. Fifteen subjects refused to provide an alveolar air sample and therefore were excluded from the biological monitoring evaluation part of the study.

2.4 Analytical Methods

Water temperature and pH were detected on site with a thermometer and a portable pH meter; free and combined residual chlorine levels were evaluated by the colorimetric N,N-diethyl-p-phenylenediamine/potassium iodide (DPD/KI) method, according to EN ISO 7393-2:2000, using a photometer (PC Compact-Aqualytic).

Airborne NCl_3 levels were determined according to the simple, non-toxic colorimetric analytical method recently suggested by Predieri and Giacobazzi.[20] The method is based on the entrapment of airborne NCl_3 in two impingers through an active ambient air sampling. The impingers (one for analysis and one as a blank) are filled with a DPD/KI water solution (15 cm^3). The solution is then analysed as combined chlorine, according to the DPD/KI colorimetric method. Airborne NCl_3 values were estimated as the difference between the reading values from the two impinger solutions and, after stoichiometric conversion, the observed values were expressed as $\mu g/m^3$.

THMs in water samples were analysed using a static head space gas chromatographic technique (SHS-GC). A Varian 3380 gas chromatograph equipped with a ^{63}Ni electron-capture detector (ECD) and a Vocol capillary column (30 m × 0.53 mm I.D., film thickness 3.0 μm (Supelco)). Quantitative and qualitative analyses of THMs were managed using Chromeleon Dionex, 6.0 software. Calibration was performed according to the external standard method. Standards were prepared with pure analysis reagents (Supelco) in specific concentrations of methanol. A working standards solution (concentration 1-8 μg/l) was prepared every day in synthetic water (chloride 100 ppm, nitrate 20 ppm and sulfate 150 ppm). Aliquots of 5 ml from the water sample, external standard and control blank were placed into 10 ml glass vials, sealed, and heated to a specific temperature (1 hour at 37 °C). After incubation, 100 μl of the head-space sample was injected directly into the GC using a gas-tight syringe (Hamilton). The detection limit of the method was 0.1 μg/l.

Levels of THMs in ambient and alveolar air were analysed, after heating to 37°C, by direct injection of samples into the GC using a gas-tight syringe (Hamilton). Calibration was performed as mentioned above, and the limit of detection, both in environmental and alveolar air samples, was 1 $\mu g/m^3$.

2.5 Statistical Methods

Statistical analyses were performed using IBM Statistical Package for Social Sciences (SPSS) version 21 for Windows. Environmental data were summarized by calculating arithmetic mean, standard deviation (SD), median and range. The Kolmogorov-Smirnov test for normality was used to analyze variable distributions. As data were not normally distributed, non-parametric statistical tests were applied. Correlations between chemical-physical parameter and/or DBP levels were evaluated by Spearman's rank correlation coefficient. Subjects' main characteristics were compared according to the main working tasks (lifeguards and trainers vs other employees) by performing X^2 test, Mann-Whitney U test or Kruskal-Wallis test. Different categories of exposure to THMs or NCl_3 were then identified according to levels detected respectively in alveolar air or in ambient air. Associations between self-reported health symptoms and DBPs exposure were evaluated using a multivariate logistic regression, in order to adjust results for potential confounding variables, such as sex, age and smoking habits. Odds ratio (OR) and 95% confidence interval (95% CI) were computed. The level of statistical significance for all statistical tests was set at $p<0.05$.

3 RESULTS

Chemical-physical parameters and DBP levels observed in the 23 examined swimming pools are reported in Table 1. In most pools, water temperature and pH levels appear in agreement with the values suggested respectively for temperature (24-30°C) and pH (6.4-7.5) by Italian guidelines.[24] On the contrary, free and, above all, combined residual chlorine levels often resulted higher than Italian guideline values (for free chlorine: 0.7–1.5 mg/l; for combined chlorine <0.4 mg/l).[24] The most common agent involved in pool water disinfection was sodium hypochlorite (9 pools), followed by calcium hypochlorite (6 pools), trichloroisocyanuric acid (5 pools) and sodium dichloroisocyanurate (3 pools). When disinfection treatments were taken into account, no significant differences in water physical-chemical parameter were observed with the exclusion of pH which tended to be significantly lower in pool water treated with calcium hypochlorite (mean value: 7.0±0.2).

The levels of THMs in pool water ranged from 6.8 to 134 μg/l with a mean value of 45.5±32.7 μg/l. Only one swimming pool water exceeded the value of 100 μg/l recommended for total THMs in drinking water by the European Directive 98/83/CE, however 10 water pools (42%) exceeded the more restrictive value adopted by the Italian normative for drinking water (30 μg/l).[25,26] Both THMs and NCl_3 levels in ambient air showed a wide range of values (12.6-257 μg/m^3 and 204-1107 μg/m^3, respectively) and in 18 swimming pools (78%) NCl_3 levels were higher than 500 μg/m^3, the recommended value suggested by WHO guidelines.[6] No significant differences were observed in DBPs levels in water and air according to disinfection treatment. Airborne levels of NCl_3 and THMs were significantly correlated (r_s=0.498, p=0.016) and airborne NCl_3 levels appeared significantly correlated to combined chlorine in pool water as well (r_s=0.450; p=0.031). On the contrary, water temperature, pH and free chlorine levels did not result correlate to THMs or NCl_3 concentrations.

Table 1 *Chemical-physical parameters and environmental DBPs levels in 23 Italian Indoor swimming pools*

Parameter	*N*	*Mean±SD*	*Median*	*Min-Max*
Pool water:				
Temperature (°C)	23	28.5±0.9	28.6	26.4-31.2
pH	23	7.2±0.2	7.3	6.8-7.6
Free chlorine (mg/l)	23	1.3±0.3	1.3	0.7-2.0
Combined chlorine (mg/l)	23	0.5±0.2	1.8	0.1-0.9
THMs (μg/l)	23	45.5±32.7	31.3	6.8-134
Ambient air:				
THMs (μg/m^3)	23	98.6±59.9	83.1	12.6-257
NCl_3 (μg/m^3)	23	660±217	650	204-1107

Table 2 *Main characteristics of swimming pool workers.*

Variables	*Total (n.165)*	*Lifeguards and trainers (n. 110-67%)*	*Other employees (n. 55-33%)*	*p*
Gender:				
Male	79-47.9%	54-49.1%	25-45.5%	
Female	86-52.1%	56-50.9%	30-54.5%	0.392
Age (years)[a]:				
<30	57-34.5%	41-37.6%	16-29.6%	
30–39	63-38.2%	46-42.2%	17-31.5%	
≥40	43-26.1%	22-20.2%	21-38.9%	0.038
Smoking habits:				
Smoker	58-35.2%	36-32.7%	22-40.0%	
Ex-smoker	16-9.7%	63-57.3%	28-50.9%	
Never smoked	91-55.2%	11-10.0%	5-9.1%	0.653
Occupational activity at indoor swimming pools (years):				
1–10	108-65.5%	72-65.5%	36-65.5%	
>10	57-34.5%	38-34.5%	19-34.5%	0.567

[a]age is missing for 2 subjects

The main characteristics of pool workers are reported in Table 2. Subjects' mean age was 34±9 years, 86 (52%) workers were female and 107 (65%) were no-smoker or ex- smoker. 110 subjects (67%) spent most of the working shift near the pool as lifeguards or trainers, while 55 (33%) were usually involved in different tasks far from the poolside. No statistical differences were observed between males and females or according to main tasks with the exclusion of age.

A total THMs mean value of 28.9 ±20.3 μg/m^3 was observed in alveolar air samples collected from the 150 workers who agreed to participate in the biological monitoring (Table 3). Lifeguards and trainers showed significantly much higher levels of THMs in alveolar air than workers involved in different tasks (Mann-Whitney U test $p<0.0001$). THMs levels in alveolar air appeared significantly correlated to THMs levels in water (r_s=0.294, $p<0.001$), THMs in ambient air (r_s=0.624, $p<0.001$) and NCl_3 in ambient air (r_s=0.450, $p<0.001$).

Among self-reported ocular symptoms, swimming pool workers frequently declared having red eyes (47.9%) and itchy eyes (43.6%). Cold and sneezing were the most frequent self-declared respiratory symptoms (69.7% and 54.5%, respectively), followed by blocked nose (53.9%) and voice loss (44.8%), while only 13 subjects (7.9%) reported suffering from an asthmatic condition.

Table 3 *THMs levels in alveolar air in swimming pool workers*

subjects	*THMs in Alveolar air (μg/m^3)*				
	n	*mean±SD*	*Median*	*Range*	*p*
All subjects	150	28.9±20.4	24.1	1.0-123	
Trainers/lifeguards	101	35.0±21.5	30.2	1.0-123	
Other tasks	49	16.4±9.6	14.9	1.7-37	<0.0001

Regarding NCl_3 (Table 4), three exposure categories were set up: subjects working outside the water poolside, although within the swimming pool facilities, were considered as the referent group, while lifeguards and trainers were classified in two different levels of exposure according to NCl_3 value suggested by WHO guidelines (500 $\mu g/m^3$).[6] The risk of suffering from red and itchy eyes, runny nose, cold and voice loss resulted closely associated with NCl_3 levels higher than 500 $\mu g/m^3$. On the contrary, no significant increased risks were observed in workers exposed to high NCl_3 levels for asthma and other related symptoms.

Table 4 *Risk of suffering from self-reported irritant symptoms and NCl_3 airborne levels for 165 swimming pool workers.*

Symptoms		*NCl_3 in ambient air ($\mu g/m^3$)*	
	Ref[a]	*≤500*[b] *OR*[c] *(95% CI)*	*>500*[b] *OR*[c] *(95% CI)*
Respiratory symptoms:			
Runny nose	1	3.04 (0.84-11.01)	3.75 (1.58-8.90)
Blocked nose	1	2.46 (0.74-8.22)	1.72 (0.86-3.46)
Cold	1	1.21 (0.36-4.12)	2.22 (1.05-4.69)
Voice loss	1	3.13 (0.95-10.34)	3.67 (1.72-7.85)
Asthma-related symptoms:			
Coughing	1	2.24 (0.70-7.23)	1.45 (0.71-2.97)
Wheezing	1	0.24 (0.03-2.02)	0.69 (0.29-1.64)
Shortness of breath	1	-	1.15 (0.36-3.67)
Dyspnea/asthma	1	-	0.69 (0.29-1.64)
Ocular symptoms:			
Red eyes	1	0.98 (0.29-3.25)	3.41 (1.62-7.18)
Itchy eyes	1	0.75 (0.22-2.60)	2.69 (1.28-5.64)
Eye infections	1	-	2.99 (0.81-11.07)
Watery eyes	1	1.70 (0.43-6.75)	1.72 (0.71-4.16)

[a]Referent group: subjects working usually outside the poolside. [b]Lifeguards and trainers were classified in two groups according to NCl_3 airborne levels measured at the pool edge. [c]ORs are adjusted for sex, age and smoking habits.

When evaluating THMs exposure (Table 5), alveolar air levels were considered as the main exposure index and subjects were divided into 3 groups according to distribution terziles (1st terzile cutoff value: 18 $\mu g/m^3$; 2nd terzile cutoff value: 33 $\mu g/m^3$). Subjects classified in the highest exposure group appeared at increased risk of suffering from red eyes and itchy eyes, blocked nose and runny nose. Regarding asthma-related symptoms, the risk of suffering from coughing appeared significantly higher in subjects with high levels of THMs in alveolar air, and an increased risk, although not statistically significant, was observed both for wheezing and for dyspnea/asthma symptoms as well.

Table 5 *Risk of suffering from self-reported irritant symptoms and THMs levels in alveolar air samples for 150 swimming pool workers.*

Symptoms	*THMs in alveolar air*[a]		
	<18 μg/m³[b]	*18-33 μg/m³ OR*[c] *(95% CI)*	*>33 μg/m³ OR*[c] *(95% CI)*
Respiratory symptoms:			
Runny nose	1	1.46 (0.58-3.71)	2.41 (0.98-5.98)
Blocked nose	1	1.63 (0.72-3.72)	4.38 (1.84-10.42)
Cold	1	1.41 (0.60-3.34)	1.67 (0.69-4.04)
Voice loss	1	0.71 (0.31-1.63)	1.71 (0.75-3.88)
Asthma-related symptoms:			
Coughing	1	3.12 (1.28-7.62)	3.15 (1.29-7.68)
Wheezing	1	2.43 (0.80-7.38)	1.29 (0.40-4.10)
Shortness of breath	1	2.40 (0.56-10.25)	0.94 (0.18-5.05)
Dyspnea/asthma	1	3.21 (0.62-16.62)	2.25 (0.38-13.47)
Ocular symptoms:			
Red eyes	1	3.74 (1.57-8.92)	3.20 (1.36-7.54)
Itchy eyes	1	2.07 (0.89-4.83)	2.35 (1.01-5.48)
Eye infections	1	4.06 (0.76-21.65)	3.32 (0.64-17.18)
Watery eyes	1	1.25 (0.45-3.44)	1.44 (0.53-3.90)

[a]Three exposure categories were built up according to distribution terziles; [b]Referent group; [c] ORs are adjusted for sex, age and smoking habits.

4 DISCUSSION

Most epidemiological studies address work-related exposure in indoor swimming pool facilities via environmental monitoring. This method, however, shows some limitations. For instance, it does not take into account factors that can influence individual intake of DBPs such as tasks performed inside swimming pool facilities. Biological monitoring is a good procedure for investigating occupational exposure and, among human biomarkers, alveolar air analysis is considered a simple and non-invasive method for evaluating workers' exposure to volatile substances in indoor swimming pools.[22,23,27-29] For all these reasons, in this study, due to the well-known sampling and methodological constrains, NCl_3 exposure was limited to environmental monitoring, while THMs exposure was evaluated in a more detailed way by both environmental and personal biological monitoring.

NCl_3 levels in ambient air appeared high and many indoor swimming pools showed NCl_3 levels exceeding the recommended WHO guideline value.[6] Overall, NCl_3 values appeared slightly higher than those reported in some previous studies.[9,10,13,15,18] High levels of combined residual chlorine in pool waters were measured as well: the overuse of chlorine in pool water disinfection treatments may be responsible for the high NCl_3 concentrations, as suggested by the observed significant correlation.

Environmental levels of THMs, on the contrary and in agreement with other studies carried out in Italian swimming pools, appeared generally low.[7,23,30] Nevertheless, the possibility of carrying out personal biological monitoring allowed us to highlight significantly different individual intakes in occupationally exposed subjects: the highest levels in exhaled air were in fact measured in trainers or lifeguards working mainly at the poolside. The significant correlations observed between THMs in alveolar air and both THMs and NCl_3 levels in ambient air confirm that this biomarker is a good index of occupational exposure to different DBPs categories in swimming pool environments.

Both airborne NCl_3 levels and workers' alveolar air THMs levels appeared significantly associated to many acute irritant ocular and respiratory symptoms. The most consistently associated symptoms resulted red and itchy eyes and runny nose. These results are in agreement with other previous epidemiological studies that have linked acute symptoms in people attending swimming pools to DBPs, mainly to NCl_3.[8-18] Regarding chronic respiratory symptoms, pool workers with higher THMs levels in alveolar air appeared at increased risk of suffering from some asthma-related endpoints. A similar association, on the contrary, did not emerge so clearly when NCl_3 exposure was taken into account although the high levels of NCl_3 detected in the investigated swimming pools.

Some methodological issues related to health status assessment may exist and could have altered the observed results, especially for asthma-related symptoms. Firstly, information was collected using a self-administered questionnaire, with no medical control, recall bias may therefore have affected data and misclassifications may have occurred. Further, the low number of workers suffering from asthma-related symptoms may have significantly reduced the power of statistical analyses. Moreover, no questions about the onset of symptoms were included in the questionnaire and thus reverse causation or the healthy worker effect cannot be excluded.

Indeed, results of most recent studies aimed at investigating the risk of developing chronic respiratory symptoms in pool workers by measuring objective endpoints, including airway hyperresponsiveness or other specific biomarkers inflammation, are not conclusive as well.[14,17] Fornander and colleagues, for instance, observed a high occurrence of airway irritation, expressed both as fractional exhaled nitric oxide and as changes in protein profiling of the nasal lavage fluid, among indoor swimming pool personnel.[14] However, the high prevalence could not be explained by NCl_3 exposure levels.[14] Further, Schmalz and colleagues have conducted an in vitro toxicological study exposing human alveolar epithelial lung cells both to indoor pool air samples and to NCl_3, as the only contaminant.[21] Their results suggest that a mixture of NCl_3 and other DBPs could be responsible of the inflammatory effects to the lung cells in pool environments as in in vitro significant toxicological effects can be observed only when cells were exposed to levels of NCl_3, as the only contaminant, much higher than those detected in pool air samples.[21]

5 CONCLUSIONS

Overall, which compound is primary involved in developing and/or exacerbating irritant symptoms in pool workers is still not clear and, therefore, which compound should be selected as an exposure index or as a proxy measure is uncertain. In this study, in accordance with previous ones, significant associations between high NCl_3 airborne levels and different irritant health outcomes, but not with asthma related symptoms, have been observed; however, many methodological limitations do not allow to investigate in detail workers' personal exposure. THMs biological monitoring, on the contrary, is easy to perform, is able to identify different levels of exposure related to different tasks and appears significantly correlated both to environmental DBPs and to irritant health symptoms, including some asthma-related endpoints. For all these reasons, breath analysis should be considered a suitable biomarker of personal occupational exposure to volatile compounds in epidemiological studies involving swimming pool workers.

References

1 A. Florentin, A. Hautemanière and P. Hartemann, *Int. J. Hyg. Environ. Health,* 2011, **214**, 461.
2 C.M. Villanueva and L. Font-Ribera, *Ann. Ist. Super. Sanità,* 2012, **48**, 387.
3 S. Chowdhury, K. Alhooshani and T. Karanfil, *Water Res.*, 2014, **53**, 68.
4 C. Zwiener, S.D. Richardson, D.M. De Marini, T. Grummt, T. Glauner and F.H. Frimmel, *Environ. Sci. Technol.,* 2007, **41**, 363.
5 S.D. Richardson, D.M. DeMarini, M. Kogevinas, P. Fernandez, E. Marco, C. Lourencetti, C. Ballesté, D. Heederik, K. Meliefste, A.B. McKague, R. Marcos, L. Font-Ribera, J.O. Grimalt and C.M. Villanueva, *Environ. Health Perspect.*, 2010, **118**, 1523.
6 World Health Organization, *Guidelines for Safe Recreational-Water Environments, Vol. 2, Swimming Pools, Spas and Similar Recreational-Water Environments,* World Health Organization, Geneva, 2006.
7 G. Fantuzzi, E. Righi, G. Predieri, G. Ceppelli, F. Gobba and G. Aggazzotti, *Sci. Total Environ.* 2001, **264**, 257.
8 G. Fantuzzi, E. Righi, G. Predieri, P. Giacobazzi, K. Mastroianni and G. Aggazzotti, *Int. J. Environ. Res. Public Health,* 2010, **7**, 1379.
9 N. Massin, A.B. Bohadana, P. Wild, M. Héry, J.P. Toamain and G. Hubert, *Occup. Environ. Med.*, 1998, **55**, 258.
10 K.M. Thickett, J.S. McCoach, J.M. Gerber, S. Sadhra and P.S. Burge, *Eur. Respir. J.,* 2002, **19**, 827.
11 J.H. Jacobs, S. Spaan, G.B. van Rooy, C. Meliefste, V.A. Zaat, J.M. Rooyackers and D. Heederik, *Eur. Respir. J.,* 2007, **29**, 690.
12 Centers for Disease Control and Prevention, *MMWR Morb. Mortal. Weekly. Rep.*, 2009, **58**, 81.
13 J. Parrat, G. Donze′, C. Iseli, D. Perret, C. Tomicic and O. Schenk, *Ann. Occup. Hyg.,* 2012, **56,** 264.
14 L. Fornander, B. Ghafouri, M. Lindahl and P. Graff, *Int. Arch. Occup. Environ. Health*, 2013, **86**, 571.
15 T Chu, S. Cheng, G. Wang and S. Tsai, *Sci. Total Environ.,* 2013, **461–462**, 317.
16 G. Fantuzzi, E. Righi, G. Predieri, P. Giacobazzi, P. Bechtold and G.Aggazzotti, *J. Expo. Sci. Environ. Epidemiol.,* 2013, **23**, 88

17 G.F. Nordberg, N-G. Lundstrom, B. Forsberg, A. Hagenbjork-Gustafsson, B.J. Lagerkvist, J. Nilsson, M. Svensson, A. Blomberg, L. Nilsson, A. Bernard, X. Dumont, H. Bertilsson and K. Eriksson, *BMJ Open*, 2012; **2,** e000973.

18 M. Hery, G. Hecht, J.C. Gendre, G. Hubert and I. Rebuffaud, *Ann. Occup. Hyg.*, 1995, **39**, 427.

19 S.-C. Weng, W.A. Weaver, M.Z. Afifi, T.N. Blatchley, J.S. Cramer, J. Chen and E.R. Blatchley III, *Indoor Air*, 2011, **21**, 391.

20 G. Predieri and P. Giacobazzi, *Int. J. Env. Anal. Chem.*, 2012, **92**, 645.

21 C. Schmalz, H.G. Wunderlich, R. Heinze, F.H. Frimmel, C. Zwiener and T. Grummt, *J. Water Health.*, 2011, **9**, 586.

22 G. Aggazzotti, G. Fantuzzi, E. Righi and G. Predieri, *J. Chromatogr. A,* 1995, **710**, 181.

23 G. Aggazzotti, G. Fantuzzi, E. Righi and G. Predieri, *Sci. Total Environ.,* 1998, **217**, 155.

24 Conferenza Permanente per i Rapporti tra lo Stato, le Province Autonome di Trento e di Bolzano, *Accordo del 16 Gennaio 2003*, Gazzetta Ufficiale della Repubblica Italiana. n. 51, marzo 3 2003.

25 European Union, *The Drinking Water Directive. Council Directive 98/83/EC of 3 November 1998 on the quality of water intended for human consumption,* Off. J. Eur. Communities L330 (05/12/1998).

26 Decreto Legislativo n.31 del 2 febbraio 2001, *Attuazione della direttiva 98/83/CE relativa alla qualità delle acque destinate al consumo umano,* Gazzetta Ufficiale della Repubblica Italiana. n. 52, 3/3/2001.

27 L. Erdinger, K.P. Kühn, F. Kirsch, R. Feldhues, T. Fröbel, B. Nohynek and T. Gabrio, *Int. J. Hyg. Environ. Health,* 2004, **207**, 571.

28 J. Caro, M. Gallego, *Environ. Sci. Technol.* 2008, **42**, 5002.

29 C. Lourencetti, J.O. Grimalt, E. Marco, P. Fernandez, L. Font-Ribera, C.M. Villanueva and M. Kogevinas, *Environ. Int.,* 2012 **45** 59.

30 E. Righi, G. Fantuzzi, G. Predieri and G. Aggazzotti, *Microchim. J.,* 2014, **113**, 23.

REACTION KINETICS OF CHLORINE WITH HUMAN BODY FLUIDS PRESENT IN SWIMMING POOL WATER

Pei Hua[a], Xiao Chen[a], Ekaterina Vasyukova[a] and Wolfgang Uhl[a,b]

[a]Chair of Water Supply Engineering, Institute of Urban Water Management, Technische Universität Dresden, 01062 Dresden, Germany. E-mail: Pei.Hua@tu-dresden.de
[b]Norwegian Institute for Water Research (NIVA), 0349 Oslo, Norway. E-mail: Wolfgang.Uhl@tu-dresden.de

1 INTRODUCTION

Chlorine is widely used as disinfectant and oxidant to control the microbial activity in swimming pool (SP) water. However, it reacts with the organic matter remaining in SP water to form potentially harmful disinfection by-products (DBPs), including inorganic chloramines and organohalogenated DBPs. Thus, the organic matter is usually referred to as DBPs precursors. There are mainly two types of DBPs precursors in SP water: natural organic matter (NOM) that comes from the filling water, and human body fluids (BF) that are excreted by swimmers.[1]

The filling water is usually tap water obtained from a water distribution system. Due to the disinfection of the water treatment plant finished water, a certain amount of DBPs is formed in the filling water before mixing with SP water, whereas the DBPs precursors in the filling water are not completely exhausted.[2]

Human BF are composed of urine, sweat, dirt, saliva, body cells, and some synthetic chemicals.[2] Previous studies reported the introduction rate of sweat ranging from 200 to 1760 mL/swimmer/event and it of urine ranging from 25 to 117 mL/swimmer/event.[3] To simulate the human BF, several recipes containing synthetic compounds were proposed in literature, assuming 50 mL urine and 200 mL sweat were released to SP water during an average swim for one person.[4, 5] The synthetic mixture of these compounds is referred to as body fluid analogue (BFA).

The NOM in the filling water and the BFA have very different characteristics, i.e. exhibit different reactivity towards chlorine consumption and DBPs formation. The characteristics of individual BFA components may also be quite different. Although some efforts have been made to understand the mechanisms for chlorination of individual BFA components, such as urea[6, 7] and citric acid,[3] the characteristics of other BFA components are still not well known. The information regarding the kinetics of DBPs formation under SP conditions, which can be helpful for developing DBPs control strategies, is limited as well. Furthermore, although some models are available for simulating residual chlorine and DBP concentrations in drinking water distribution systems,[8-10] there is still very little information regarding the models for describing residual chlorine and DBP concentrations in SP water. This study was therefore conducted to better understand the reactivity of chlorine with NOM in the filling water and BFA from the reaction kinetics standpoint. Specifically, the objectives were to: (i) compare the specific chlorine demand and trichloromethane formation potential (TCM

FP) of DBP precursors typically present in SP water, and (ii) investigate the chlorine consumption and TCM formation as a function of time for DBP precursors.

2 MATERIALS AND METHODS

2.1 Experimental design

The BFA components were chosen according to a composition suggested by Judd and Bullock.[4] It contained six organic precursors and ammonium chloride as shown in Table 1. The dissolved organic carbon (DOC) percentage contributions, which were calculated as the DOC of individual BFA components divided by the total DOC, and the nitrogen percentage contributions of individual BFA components to the total nitrogen of BFA are also given in Table 1.

Table 1 *DOC and total nitrogen concentrations of individual BFA components, and their corresponding percentage contributions to the total BFA*

BFA components	*DOC* mg/L	*DOC percentages contribution* %	*Total nitrogen* mg/L	*Nitrogen percentages contribution* %
Urea	2960	51.48	6900	85.17
L-Histidine	560	9.74	320	3.95
Hippuric acid	1040	18.09	134	1.65
Uric acid	180	3.13	160	1.98
Citric acid	240	4.17	-*	-
Creatinine	770	13.39	67	0.83
Ammonium Chloride	-	-	520	6.42
Total BFA	5750	100	8101	100

* -: no contribution

Two experiment series were conducted, and were summarized in Table 2. Experiment series *I* was conducted to determine the chlorine demand and TCM FP of all DBPs precursors, including individual BFA organic component, BFA, humic acid (HA) and a mixture of BFA+HA. For Experiment series *I*, each individual BFA organic component was prepared firstly to match the identical initial DOC concentration of 2 mg *C*/L. The initial DOC of BFA was 3 mg *C*/L. The concentration of HA, chosen as a surrogate of NOM in the filling water, was set to match 1 mg *C*/L. The BFA+HA had a total DOC of 4 mg *C* /L, as 3 mg *C*/L contributed by BFA and 1 mg *C*/L by HA. This total DOC concentration falls into the DOC range of a typical SP water. The theoretical total nitrogen concentrations of DBPs precursors are also shown in Table 2.

The Experiment series *II* was conducted to investigate the kinetics of chlorine decay and THM formation with respect to individual BFA components, BFA, HA and BFA+HA. As shown in Table 2 Experiment series *II*, the BFA concentration was set firstly to the same initial DOC of 3 mg *C*/L as it is in Experiment series *I*. Then, in the experiments with individual BFA components, their DOC concentrations were calculated according to the defined percentage contributions shown in Table 1. In other words, the sum of DOC in all single-component experiments was equal to the DOC of BFA, i e. 3 mg *C*/L. The initial DOC values of HA and BFA+HA were the same as those in Experiment series *I*, i.e. 1 mg *C* /L and 4 mg *C*/L, respectively. Table 3 shows the DOC percentage contributions of all individual organic precursors to the mixture solution of BFA+HA.

Table 2 *Experimental design*

Precursors	*Experiment series I*		*Experiment series II*	
	DOC mg/L	*Total Nitrogen* mg/L	*DOC* mg/L	*Total Nitrogen* mg/L
Urea	2.00	4.66	1.54	3.58
L-Histidine	2.00	1.14	0.29	0.17
Hippuric acid	2.00	0.26	0.54	0.07
Uric acid	2.00	1.77	0.09	0.08
Citric acid	2.00	-	0.13	-
Creatinine	2.00	0.74	0.40	0.35
Ammonium Chloride	-	0.27	-	0.27
BFA	3.00		3.00	
Humic acid	1.00		1.00	
BFA+HA	4.00		4.00	

Table 3 *DOC percentage contribution of individual precursor to BFA+HA mixture*

Precursor	*Citric acid*	*Histidine*	*Creatinine*	*Urea*	*Uric acid*	*Hippuric acid*	*HA*
Percentages	3%	7%	10%	39%	2%	14%	25%

To make sure the chlorine concentration is not a limitation for reaction, i.e. enough chlorine remains in the water sample after the maximum reaction time of 168 hours (h), and to set all the experiments with the same initial chlorine concentration (ICC), 50 mg Cl_2/L was chosen as ICC. According to the typical SP operation condition, pH was fixed at 7.2 by the addition of 0.1 M phosphate buffer and temperature was kept constant at 28 ± 1 °C.

2.2 Preparation of solutions

The BFA organic stock solutions were prepared by mixing individual BFA stock solutions with ultrapure water at pH = 12, while the ammonia chloride solution was prepared separately at pH = 2. pH was adjusted by using HCl (1 mol/L) or NaOH (2 mol/L). The BFA stock solution was prepared by mixing the BFA organic stock solution and ammonia chloride solution with a mixing ratio of 1:1 (volume:volume).

The humic acid stock solution at a concentration of 28 mg/L was prepared by dissolving the commercial humic acid sodium salt (Sigma-Aldrich, US) with ultrapure water, which was stirred firstly at least 3 h and then filtered through a 0.45-μm polycarbonate filter (Nuclepore Track-Etch Membrane, Whatman, Germany) to remove suspended solids. The chlorine stock solution with a concentration of 15.5 mM was prepared with ultrapure water using commercial concentrated sodium hypochlorite solution.

2.3 Analytical methods

The DOC concentration was determined by the catalytic combustion method using a LiquiTOC II organic carbon analyzer (Elementar Analysensystem GmbH, Germany). Free chlorine concentration was determined according to the N, N-diethyl-p-Phenylenediamine (DPD) Standard Colorimetric Method 4500-Cl G (APHA [11]). pH was measured by a pH meter 340i (Wissenschaftlich-Technische-Werkstätten, Germany).

DBP concentrations were determined by membrane inlet mass spectrometry (MIMS). Before each measurement, the trihalomethane standards without purification, which were purchased from Sigma-Aldrich, were diluted by ultrapure water to the target aqueous phase concentration and measured for the MIMS calibration and quality control. Water samples were first analysed in the mass spectrum scan mode (49-m/z-215) and then the trihalomethane concentrations were quantified with a selected ion monitoring. As BFA and HA were prepared with ultrapure water, which does not contain bromide, of all trihalomethane species, only TCM was detected. Ions at m/z 83 was chosen to quantify TCM with a quantification limit of 2.5 *u*g/L. Detailed operation conditions for MIMS can be found in previous studies.[12, 13]

3 RESULTS AND DISCUSSION

3.1 Specific chlorine demand and specific TCM formation potential

Figure 1 presents the specific chlorine demand (chlorine demand divided by the initial DOC concentration) and specific TCM FP (TCM FP divided by the initial DOC concentration) for six individual organic BFA components, HA and precursor mixtures (BFA and BFA+HA). The experimental results of chlorine demand and TCM FP were obtained from Experiment series *I*.

As shown in Figure 1, urea exhibited the highest specific chlorine demand of 19.5 mg Cl_2/mg *C*, followed by L-histidine with 13.5 mg Cl_2/mg *C* and uric acid with 12.0 mg Cl_2/mg *C*. These three BFA components show higher nitrogen concentrations than the other BFA components (Table 1), which is reasonably explained by a higher specific chlorine demand. As for precursor mixtures, the BFA showed the highest specific chlorine demand of 14 mg Cl_2/ mg *C*. and to a lesser extent, for BFA+HA, whereas that of the pure HA solution was nearly five times lower and exhibited 3 mg Cl_2/mg *C*. This suggests that BFA is more reactive towards chlorine than HA. The N-containing organic BFA components are responsible for the high specific chlorine demand of BFA.

Furthermore, citric acid exhibited significantly higher specific TCM FP (11.7 *u*mol/mg C) than all other individual BFA components. After citric acid, HA showed the second highest specific TCM FP (0.82 *u*mol/mg C) followed by BFA+HA (0.49 *u*mol/mg *C*) and BFA (0.45 *u*mol/mg *C*). It is noted that the specific TCM FP of HA was about 2 times higher than that of the BFA, which indicates HA was more reactive toward TCM formation than BFA. This finding, which together with the fact that HA is continuously injected with the fresh filling water into a real swimming pool, emphasizes the importance of an effective removal of HA in order to minimize TCM formation in SP water.

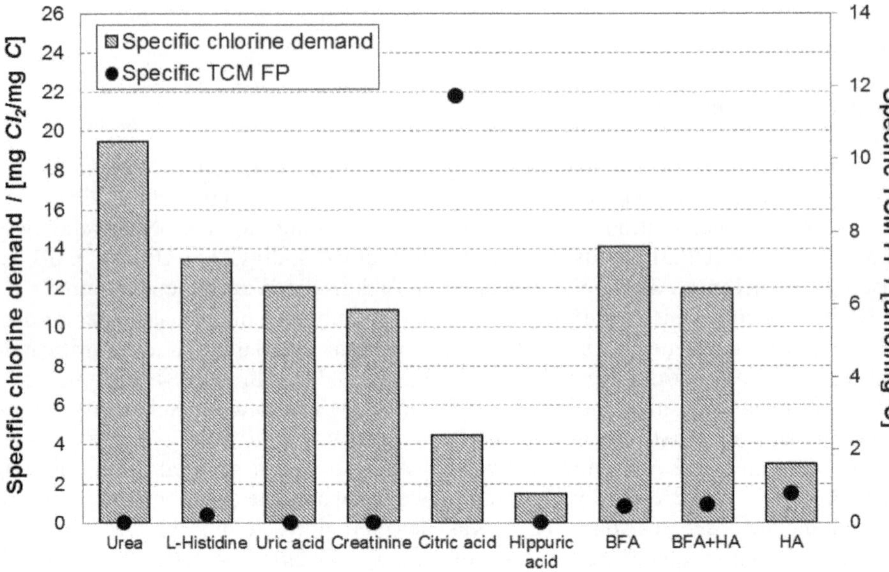

Figure 1 *Specific chlorine demand and specific TCM FP of organic precursors including individual organic BFA components, HA and precursor mixtures.*

3.2 Chlorine consumption of precursors as a function of time

Several parallel and independent chlorination experiments were conducted to investigate the chlorine consumption of individual DBP precursors as a function of time, which as shown in Table 2 (Experiment series *II*). The total DOC concentration of all individual components, including BFA components and HA, was equal to the DOC concentration of the BFA+HA (4 mg *C*/L). It should be illustrated that the DOC concentrations of each BFA components in BFA and BFA+HA were the same as those used for independent chlorination experiments.

It can be seen from Figure 2(a) that the total chlorine consumption of the BFA+HA increased with increasing incubation time and reached a plateau after 72 h. Urea contributed mostly to the chlorine consumption over the whole reaction period, which is due to the fact that urea is the main component of the BFA+HA (39% of total DOC of BFA+HA solution). However, the chlorine consumption of urea already reached a plateau at 24 h, which indicates that urea may belong to the fast reacting precursor. The chlorine consumptions of other BFA components were nearly one to two magnitudes lower than those of urea and precursor mixtures, which are shown in Figure 2(b) and (c). It can be seen that the chlorine consumptions of these BFA components nearly reached a plateau after 120 h, which implies that the reaction rate constants of those BFA components may be slower than that of urea. The chlorine consumption of citric acid and hippuric acid were the lowest compared with other BFA components as given in Figure 2(c).

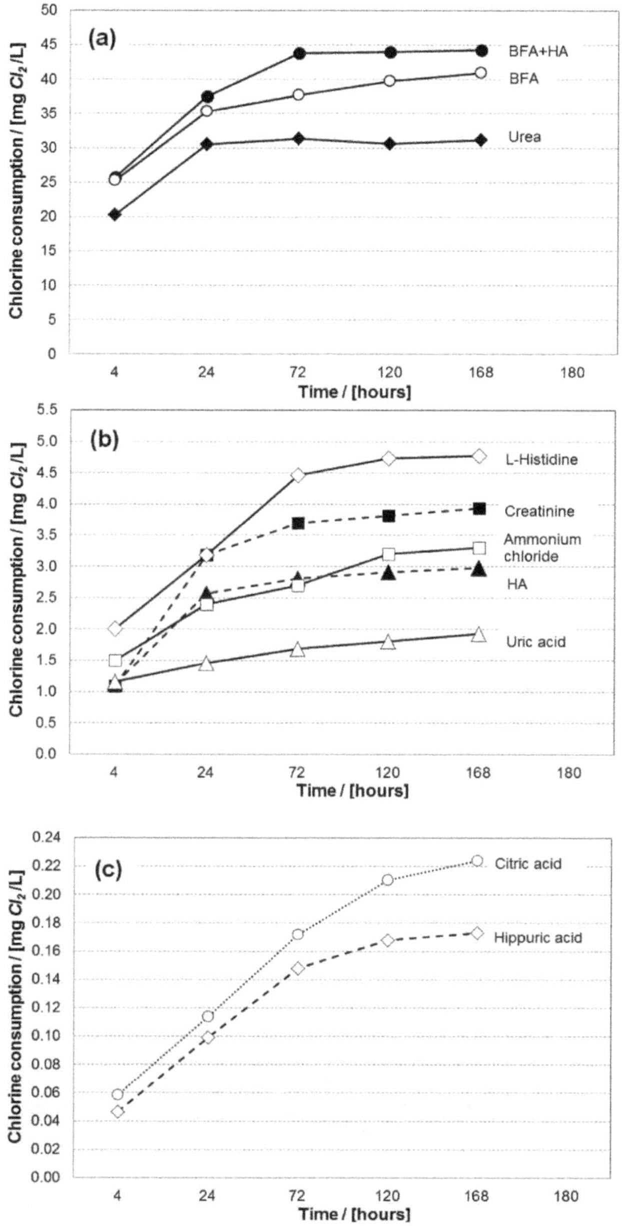

Figure 2 *Chlorine consumption of all DBP precursors as a function of time*

3.3 Trichloromethane formation of precursors as a function of time

Figure 3 shows the TCM formation of all organic BFA precursors (total DOC = 3 mg *C*/L), HA (DOC = 1 mg *C*/L), BFA and BFA+HA as a function of reaction time. It can be seen that although citric acid contributed less to the chlorine consumption as given in Figure 2(c), over 78% of the total TCM of BFA+HA was formed due to citric acid. Furthermore, the TCM concentration of HA nearly reached a plateau after 72 h, whereas that of citric acid reached a plateau at 120 h (Figure 3(a)). This suggests that HA has a higher reactivity with respect to the TCM formation than citric acid.

As shown in Figure 3(b), the TCM formations trends of the rest BFA components are not as clear as that of citric acid, which might be attributed to the quantification difficulty for such low TCM concentrations. Using the DOC percentage contribution regarding the individual BFA components to the BFA+HA given in Table 3, it can be demonstrated that urea, hippuric acid and creatinine are responsible for a total of 63% of DOC in the BFA+HA mixture. However, limited total TCM formation was caused by these components (less than 10% to TCM formation of BFA+HA). Therefore, efficient removal of citric acid and HA is considerably important to control and minimise TCM concentrations in SP water.

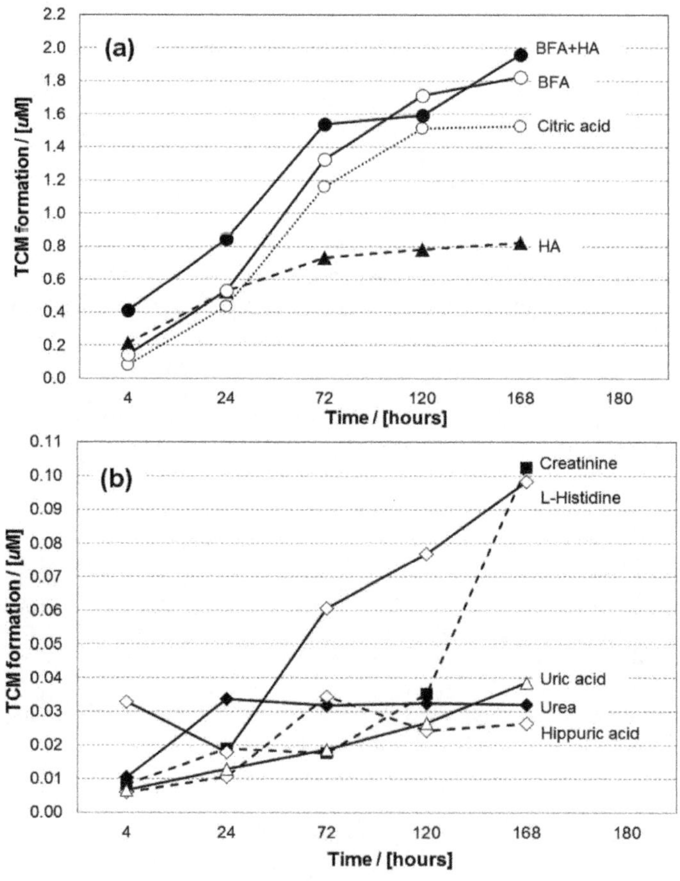

Figure 3 *TCM formations of all DBP precursors as a function of time*

4 CONCLUSIONS

In this study, the specific chlorine demand and specific trichloromethane formation potential (TCM FP) for organic precursors, such as individual body fluid analogue (BFA) components, humic acid (HA) as well as precursors mixtures (BFA and

BFA+HA), were determined experimentally. Considering the specific chlorine demand, urea exhibited the highest specific chlorine demand of 19.5 mg Cl_2/mg C, followed by L-histidine and uric acid. The higher nitrogen concentrations of these BFA components compared with others can be an explanation for a higher specific chlorine demand. The specific chlorine demand of the BFA was much higher than that of HA, indicating that BFA were more reactive towards chlorine than HA. However, the specific TCM FP of HA was about 2 times higher than that of the BFA, which suggests HA to be more reactive towards TCM formation than BFA.

The chlorine consumption and TCM formation as a function of time were also investigated for investigated DBP precursors. It was found that urea may belong to the fast reacting precursor towards chlorine consumption compared to other BFA components. As for TCM formation, citric acid and HA were the main precursors for TCM formation. However, the results suggested that HA had a higher reactivity than citric acid towards TCM formation.

The experimental data obtained in this study represent a basis for determining the reaction rate constants of chlorine decay and TCM formation for individual DBP precursors, which will be performed in the near future.

References

1. M. G. A. Keuten, F. M. Schets, J. F. Schijven, J. Q. J. C. Verberk and J. C. van Dijk, *Water Research*, 2012, **46**, 3682-3692.
2. A. Kanan and T. Karanfil, *Water Research*, 2011, **45**, 926-932.
3. L. Lian, Y. E, J. Li and E. R. Blatchley, *Environmental Science & Technology*, 2014, **48**, 3210-3217.
4. S. J. Judd and G. Bullock, *Chemosphere*, 2003, **51**, 869-879.
5. D. M. Goeres, T. Palys, B. B. Sandel and J. Geiger, *Water Research*, 2004, **38**, 3103-3109.
6. E. R. Blatchley and M. Cheng, *Environmental Science & Technology*, 2010, **44**, 8529-8534.
7. J. De Laat, W. Feng, D. A. Freyfer and F. Dossier-Berne, *Water Research*, 2011, **45**, 1139-1146.
8. S. Chowdhury, P. Champagne and P. J. McLellan, *Science of The Total Environment*, 2009, **407**, 4189-4206.
9. R. Sadiq and M. J. Rodriguez, *Science of The Total Environment*, 2004, **321**, 21-46.
10. I. Fisher, G. Kastl and A. Sathasivan, *Water Research*, 2011, **45**, 4896-4908.
11. APHA, *4500-Cl G: Standard method for determination of chlorine (residual)*, American Water Works Association and Water Environment Federation, Washington, D.C, 2005.
12. C. Shang and E. R. Blatchley, *Environmental Science & Technology*, 1999, **33**, 2218-2223.
13. X. Yang, J. Peng, B. Chen, W. Guo, Y. Liang, W. Liu and L. Liu, *Journal of Hazardous Materials*, 2012, **239–240**, 348-354.

Risk Assessment and Regulation of Disinfection By-products

MANAGING UNCERTAINTY IN THE PROVISION OF SAFE DRINKING WATER

John Fawell[1] and Steve Hrudey[2]

[1] Visiting Professor, Water Science Institute, Cranfield University Bedfordshire, UK.
[2]Professor Emeritus, Analytical and Environmental Toxicology Division, Faculty of Medicine and Dentistry, University of Alberta, Edmonton, Canada.

Microbiological threats remain the most urgent problem in delivering safe drinking water all over the world with a stubbornly substantial number of deaths that can be directly attributed to this cause. According to WHO, almost 1 billion people lack access to an improved water supply. There are 2 million annual deaths attributable to unsafe water, sanitation and hygiene and more than 50 countries still report cholera to WHO. In addition millions are exposed to unsafe levels of naturally occurring arsenic and fluoride. Many of these problems relate to developing countries but not exclusively since the greatest problems tend to be associated with small resource limited supplies, often in rural areas. Changes to water supply are not always easy to make and can be prohibitively expensive for many supplies where resources are very limited. Frequently, changes lead to more sophisticated systems of treatment that also require a greater level of expertise to manage them. It is not so long ago when the conditions in the cities of Europe and The United States resulted in endemic typhoid and frequent outbreaks of cholera. The answer was to change water supply and sanitation practice; changes which included filtration and eventually the addition of chlorine as a disinfectant. Subsequently chlorination became the most widespread disinfectant used in water supply, including more frequently in maintaining the safety of drinking water at a household level in developing countries where delivery of water may not be to the house or where rainwater collection is practiced.

Meanwhile, there is a limited range of options for dealing with the problem of waterborne microbial disease, particularly in many parts of the developing world. One is to manage sewage to prevent contamination of water sources but this is not sufficient on its own to ensure safe drinking water. The second is to prevent the pathogens in source waters reaching drinking water and this has proved highly successful. Chlorination has been one of the main components of a multi-barrier approach to treating drinking water for over 100 years and for some small supplies it is still the primary barrier. It is relatively cheap, easy to use and robust in its effectiveness, easy to measure, effective against most organisms; it is also a good indicator of the effectiveness of disinfection and has a residual action. It therefore remains an important tool in the fight against waterborne disease. Even in developed countries failures in disinfection can lead to outbreaks of disease and even deaths.

In the mid 1970s it was discovered that chlorination resulted in unwanted by-products, one of which, chloroform, was shown to cause tumours in laboratory animals and since then research in the field has increased almost exponentially with increasing numbers of by-products being identified at extremely low concentrations. In epidemiological studies a number of associations with cancers were identified in some studies but the only consistent finding has been an association with bladder cancer. However, many of the questions remain unresolved and although there is considerable evidence that many of these by-products can cause toxicity in laboratory animals and *in vitro* systems at high doses there is still no clear health effect for which there is sufficient evidence to confidently conclude causality. If we consider our overall confidence in the evidence that we have we can conclude that for pathogens the risk to health is high and we have considerable confidence that this is the case. For disinfection by-products we can conclude that the potential risk is high because of the widespread exposure but our confidence about the probability that they do cause significant effects is low.

All disinfectants produce some by-products and the database on chlorination in many respects reflects the much greater amount of research compared to other disinfectants. This is a dilemma for regulators and water suppliers because other means of disinfection are generally much more expensive and difficult to manage, particularly in resource limited environments. Although there was significant uncertainty regarding the actual health effects of THMs and HAAs, regulation was introduced to encourage control of their formation by removal of the natural precursors and adoption of practices that would discourage by-product formation. This was a cost-effective strategy that brought a range of additional benefits, mostly in developed countries where finance was available and there was a suitable skill base. While further potential problems continue to be identified these are unhelpful unless there is sufficient information to identify solutions that are also cost-effective. It is possible to introduce precautionary regulations but these always entail cost which will eventually fall on consumers. That means that the poorest in society who are already at higher risk of many diseases will bear a disproportionate burden. For regulators and water suppliers who have to make long-term decisions regarding infrastructure investment, cost and benefit are very important considerations. When balancing the certainty of waterborne microbial disease against what is still the small theoretical risk of long-term adverse health effects for which water is potentially only one of the causes of the diseases being considered, means that there need to be reliable cost effective solutions available to justify major change. In resource limited environments cost and skills are very important. It must also be remembered that it is not just surface water that is at risk of pathogen contamination, it is also true of many ground waters, household storage, and rainwater collection and through flooding and emergencies.

While research is important and legitimate, continuing to identify possible problems without providing information to allow suitable solutions to be identified is unhelpful. These data needs include identifying the scale of the risks in a variety of environments along with means of reducing disinfection by-products in a cost-effective, sustainable and achievable way. It is vital that we recognise that changing water infrastructure is not something that can be considered lightly. Of particular importance is not overstating the possible risks because this may lead to misunderstanding of the actual risks with tragic consequences. This has already happened in South America and has been only narrowly averted elsewhere.

Water supply is used to uncertainty and continues to develop and change with new developments. Many of these developments will lead to changes in treatment but there will be a significant cost attached which is usually not a possibility for most developing countries and small supplies. In attempting to balance the risks associated with waterborne pathogens and disinfection by-products it is not viable to consider cost and technology as the fulcrum balancing the two sides. Instead we must consider microbial safety to be the fulcrum, not to be compromised, and the balance is between the possible risks of disinfection by-products and the costs of dealing with them. The solutions also need to be as comprehensive as possible be it is not possible top constantly change water supply infrastructure.

Of course it is essential that chlorine, and other disinfectants, be used carefully and that has been the purpose of regulation to date. However, the certainty is that it remains an important weapon in the fight against pathogens in drinking water. It is also important that confidence in the possible risks from by-products should not be overstated and uncertainty needs to be fully acknowledged.

RISK-BASED APPROACH TO THE FORMATION OF DBPs OF CONCERN IN UK DRINKING WATER

G. Dillon[1], D. Shepherd[1], T. Hall[1], R. Gee[1], J. Parker[2] and P. Rumsby[1]

[1]WRc plc, Frankland Road, Swindon, Wiltshire SN5 8YF, UK
[2]UKWIR, UK Water Industry Research Ltd., 8th Floor, 50 Broadway, London SW1H 0RG, UK

1 INTRODUCTION

The disinfection by-product (DBP) rule was introduced in the UK Water Supply Regulations 2010 and requires water companies to "design, operate and maintain the disinfection process so as to keep disinfection by-products as low as possible without compromising the effectiveness of disinfection; and to verify the effectiveness of the disinfection process".[1]

To-date the emphasis has been on the regulated DBPs, mainly the trihalomethanes (THMs). The World Health Organization (WHO) advises that "maintaining THM and haloacetic acid (HAA) concentrations below the guideline values by controlling precursor compounds will provide adequate control over other chlorination by-products"; advice reiterated in the UK Chief Inspector of Drinking Water's annual report for 2011.[2] The same report stresses the duty of water companies to maintain DBPs below concentrations that constitute a potential danger to human health.

The increasing emphasis on a risk-based approach is likely to make consideration of other DBPs a regulatory priority. The very wide range of DBPs known to exist would make routine monitoring for the compounds themselves impractical, if not impossible. However, a knowledge of the factors which lead to formation of specific types of DBPs, particularly water quality and precursor concentrations, and information on removal of precursors by different treatment types and conditions, would allow other DBPs to be taken into account in a risk-based approach.

2 PRIORITISATION OF DBPS

The categorisation and prioritisation of DBPs by Weinberg *et al.*[3] for a USEPA survey provide a good basis for prioritising DBPs in relation to UK conditions. To allow relative risks to UK water supplies to be identified, each category of DBP has been considered in relation to mechanism of formation, known occurrence and toxicology. Based on formation mechanisms and likelihood of occurrence reported by Weinberg *et al.*[3] and in other studies in the USA and UK, the majority of the DBPs identified in the USEPA survey have been considered in this work. The main categories of DBPs considered are:

iodinated DBPs, nitrogen-containing DBPs, nitrosamines, trihalomethanes, haloacetic acids, and inorganic DBPs.

2.1 Occurrence

2.1.1 Iodinated DBPs. The only iodinated DBPs in the prioritised list identified for the USEPA survey[3] were the iodinated THMs: bromochloroiodomethane, dichloroiodomethane, dibromoiodomethane, chlorodiiodomethane, bromodiiodomethane and iodoform. Iodinated-THMs were highest where chloramination was used without free chlorine, i.e. chloramine used as the primary disinfectant, which is uncommon in the UK. Concentrations found were in the range 0.2-15 μg/l. A second US survey[4] included works using chloramination, and selected to have iodide (and bromide) in the source waters. Five iodo-acids were investigated: iodoacetic acid, bromoiodoacetic acid, (Z)-3-bromo-3-iodo-propenoic acid, (E)-3-bromo-3-iodo-propenoic acid and (E)-2-iodo-3-methylbutenoic acid; and two iodinated THMs: dichloroiodomethane and bromochloroiodomethane. Concentrations of the iodo-acids were in the tens of ng/l levels, with the total iodinated THMs in the low μg/l levels. Total iodinated DBPs were in the range <1 to 8 μg/l for waters containing around 1 to 100 μg/l iodide, with a trend of higher iodinated DBPs for the waters with higher iodide. All the works used chloramination, often with little or no free chlorine contact time.

A survey of seven treatment works was carried out in Scotland.[5] Four works used chloramination but included a free chlorine residual for around 30 minutes before ammonia addition; results indicated iodinated THMs to be 1-2% of the concentration of the total regulated THMs (chloroform, bromoform, dichlorobromomethane and dibromochloromethane).

Reaction between hypochlorous acid (HOCl) and iodide forms hypoiodous acid (HOI) which reacts with natural organics to form iodinated THMs or HAAs, but also reacts with HOCl to form iodate. This latter reaction means that iodinated DBP formation decreases with increasing free chlorine concentration and so greater concentrations of iodinated DBPs may be formed by chloramination after primary chlorination than by chlorination alone.

For chlorination at typical UK doses of around 1-3 mg/l, iodinated DBPs would be around 20% of the iodide concentration in the water at below pH 8, but can be more than double this at pH 9. Iodinated DBPs would be higher for chloramination but would depend on whether a free chlorine residual was used for primary disinfection and on the contact time. In the UK, chloramination is rarely (if ever) used for primary disinfection, so the 20% estimate is likely to apply for iodinated DBP formation at the works. However, when the water enters distribution after ammonia addition, some HOI may persist so iodinated DBP formation may continue in distribution. Ozone rapidly converts iodide to iodate and therefore prevents formation of organic iodinated DBPs.

2.1.2 Nitrogen-containing DBPs. The study carried out by Templeton *et al.*[6] looked at the occurrence and toxicology of a range of nitrogen-containing DBPs (N-DBPs). Very little occurrence data were available for the UK other than from a survey of seven water treatment works in Scotland,[5] which measured haloacetonitriles and one halonitromethane (chlorpicrin). Concentrations were typically less than 1 μg/l - lower than in the USEPA survey[3] - even though THM levels were higher than in the US survey, possibly because of lower levels of dissolved organic nitrogen (DON) in the raw waters in Scotland (although data on this were not available).

The main risk factors were identified as:

- DON >0.5 mg/l (or total Kjeldahl nitrogen >0.75 mg/l). Very little DON information was available for source water, so an algal-rich water or the presence of sewage effluent were used as surrogate indicators.
- Bromide >100 μg/l and/or iodide >5 μg/l in treated water before disinfection.
- pH 5.0-7.5.
- Use of chloramination and/or the use of chlorine dioxide prior to final disinfection (currently uncommon or possibly non-existent in UK water treatment).
- High HAA formation (>60 μg/l average).

A survey of seven UK water companies[6] was carried out to identify numbers of treatment works where these risk factors occurred. Around 13% of the treatment works included in the survey were defined as being at potential risk of N-DBP formation based on combinations of these risk factors.

A subsequent study by Templeton *et al.*[7] carried out a sampling survey of 20 water supply systems in the UK, selected to include treatment works with suspected risk factors for N-DBPs, but including some without these risk factors. Groups of N-DBPs investigated were the haloacetonitriles, haloacetamides and halonitromethanes. Concentrations of individual DBPs were always less than 2 μg/l, and mean total concentrations for groups of N-DBPs were: haloacetonitriles, 3.2 μg/l; haloacetamides, 1.5 μg/l; halonitromethanes, 0.4 μg/l, and cyanogen chloride with an average concentration of 5.8 μg/l.

The highest levels of N-DBPs were found for lowland surface water sources and for works using ozone (only used for the lowland sources in the survey). There were no clear relationships between N-DBP formation and raw water total organic carbon (TOC) or total THM/HAA formation, and N-DBP formation was no higher for works using chloramination than for those using chlorine. There were no consistent trends between N-DBPs and distance from the treatment works in distribution.

A USEPA report[8] identified chloramination as the principal cause of cyanogen halide formation, but the presence of DON could increase formation through reaction with chloramines. There may also be a mechanism through reaction of chlorine with DON, but this is unlikely to be significant because cyanogen halides are degraded by chlorine. Pre-oxidation with chlorine dioxide or ozone may increase cyanogen halide formation.

2.1.3 Nitrosamines. Two mechanisms have been identified for the formation of nitrosamines:

- Nitrosation reactions where nitrite under acid conditions forms nitrous acid, which reacts with organic nitrogen compounds (particularly amines) to form nitrosamines. This is unlikely to be significant for practical UK water treatment conditions.
- Monochloramine or dichloramine reaction with aliphatic amines to form nitrosamines.

Occurrence of each nitrosamine depends on the availability of the appropriate amine precursor. Dimethylamine (DMA) is likely to be the most common precursor, so N-nitrosodimethylamine (NDMA) is likely to be the most prevalent nitrosamine.

In a survey of 43 water treatment works, NDMA was detected in final water at three works at concentrations less than 10 ng/l[9]. The source of the NDMA was confirmed subsequently as a contaminated ferric coagulant used at the works[10].

Risk factors for the presence of nitrosamines include:

- Use of a contaminated ferric coagulant which was identified as the main source of NDMA.[10] This source has been reduced as a result of changes in the coagulant manufacturing process.
- The presence of precursors, particularly DMA, in the raw water would be increased by sewage effluent discharges upstream of the intake. Higher DON (>0.5 mg/l) could be indicative of increased risk.
- Nitrite in the raw water is unlikely to be a risk factor as this would be readily oxidised by chlorine.
- Use of polyamines and polyDADMAC in treatment which could increase DMA levels.
- The use of chloramination (or poorly operated breakpoint chlorination to remove ammonia) with production in distribution likely to be the main mechanism. This could be exacerbated where there are known nitrite issues in distribution from the use of chloramination.

Nitrosamines can be removed by activated carbon but this is unlikely to be of practical significance as formation will be during final disinfection and in distribution.

2.1.4 Trihalomethanes. Concentrations of THMs produced by chlorination are a function of pH, temperature, free chlorine concentration, contact time, and concentration and the nature of oxidisable organic material in the water. Generally, to restrict THM production:

- Avoid chlorinating raw surface water and treat water to remove organic precursors (measured by colour, TOC, UV_{254} absorbance).
- Limit free chlorine concentrations and contact times to the minimum required for the process (and distribution systems).
- Dechlorinate as soon as possible after breakpoint chlorination.
- Use chloramination to provide a residual in distribution.
- Keep pH low as THM formation increases at pH >7 (although the scope for keeping pH low might be limited by considerations of corrosion in distribution).
- Consider the use of an alternative oxidant or UV for primary disinfection.

For many situations where a chlorine residual is maintained in distribution (rather than a chloramine residual), the majority of THMs are formed in the distribution system. It is generally found that for any particular water source THM formation correlates quite well with such quality determinands as DOC, UV_{254} absorbance and chlorine demand. However, such correlations tend to be source specific, primarily because all-embracing determinands such as DOC provide no indication of the reactivity of the organic material that contributes the precursors for THM formation.

2.1.5 Haloacetic Acids. A survey of HAA occurrence at 20 water treatment works and distribution systems in England and Wales was carried out by Parsons and Goslan (2009).[11] This survey found generally low levels of HAA9 concentrations leaving the works (maximum 28.9 μg/l, median 12.5 μg/l) and in distribution (range 8-74.8 μg/l, median 20.8 μg/l). Levels increased in most of the distribution systems investigated, except where chloramination was used.

A separate study[12] of HAAs at 7 water treatment works in Scotland found generally higher levels (11-134 μg/l), particularly for 2 upland reservoir works, potentially due to non-optimised coagulation control at these works. Systems with chlorine in distribution produced higher HAA concentrations than those with chloramination (after primary chlorination).

Graham *et al.*[13] surveyed HAA9 in three supply systems. For the groundwater system, the total HAAs were found at very low levels in all four seasons, with average and maximum concentrations of 0.6 μg/l and 1.9 μg/l, respectively. For the lowland surface water system, the average and maximum total HAA were 11.9 μg/l and 20.0 μg/l, respectively. The upland surface water exhibited the highest total HAAs with an average concentration of 21.3 μg/l and a maximum of 41 μg/l. Some seasonal effect was seen for this system, with higher levels in summer and winter. This may have arisen from water temperature or from seasonal differences in TOC levels in the water. A good correlation between HAA and THM formation was seen for the combined data across the three systems.

2.1.6 Inorganic DBPs. Chlorate and bromate can be produced during electrolytic generation of hypochlorite, either on site or during commercial production. Chlorate and chlorite can also be produced from disproportionation of hypochlorite solution during storage.

Bromate is produced from bromide in salt used for electrolysis to produce hypochlorite. Bromate is associated with ozonation and less commonly with UV/chlorination. The formation of bromate has been reported in bench-scale UV/chlorination studies (with up to 32% conversion of bromide). There have also been reported incidences in the UK of excessive bromate formation from the use of chlorine before UV at relatively high disinfection doses (up to about 100 mJ/cm^2) where bromide has been present in source waters at 100-150 μg/l.

Nitrite is an inorganic DBP associated with chloramination. It can also be formed from high dose UV using medium pressure lamps on high nitrate water. This is usually controlled by using shrouds on the lamps to filter out the lower wavelength UV which causes the reduction of nitrate.

2.2 Processes potentially producing DBPs

2.2.1 Ozonation. Ozonation does not generate THMs or other chlorinated by-products. However, there are two principal concerns:

- Bromate which is produced by oxidation of bromide. This can be reduced by ozonation at lower pH.
- Increased levels of biodegradable organics, measured as assimilable organic carbon (AOC) or biodegradable dissolved organic carbon (BDOC). These are not regulated, but high levels may cause biological growth in distribution. This can be avoided by ensuring that ozonation is not the final treatment stage and is followed by GAC which can remove AOC.

2.2.2 Chloramination. For effective chloramination, it is necessary to ensure that a balance is struck between maximising the production of monochloramine and minimising the concentration of excess ammonia. Free ammonia can be converted to nitrite through biological action (nitrification) in distribution. The conditions under which the ammonia residual entering the distribution system becomes available to support nitrification are not fully established. The degree of nitrification that occurs in distribution will depend on the presence of a specialist group of autotrophic bacteria (known as ammonia-oxidising bacteria or nitrifying bacteria). The survival and growth of these bacteria depends on temperature, pH, availability of ammonia and the concentration of any disinfecting residual (monochloramine). Growth is restricted by low temperatures and it is common to find that nitrite formation is a bigger problem during the summer than the winter.

Organic nitrogen compounds can be oxidised by microorganisms as a source of energy and the excess nitrogen released as ammonia. There is unlikely to be sufficient organic nitrogen of natural origin in a water supply for this transformation to act as a source of ammonia for nitrite formation. However, there is anecdotal evidence to suggest that chloramine residuals can be converted to ammonia by this transformation, but it would be difficult to distinguish this from nitrification driving the hydrolysis of chloramine.

2.3 Toxicological Considerations

While several hundred DBPs and potential DBPs have been identified[3] - and reduced to a priority list of 70 based on toxicological considerations - the amount and quality of toxicological studies remain limited. For a few DBPs, standards and/or guideline values have been set by authoritative bodies (such as WHO, USEPA and the European Union) and for these a full set of regulatory tests have been conducted. These DBPs are: THMs, HAAs, dichloroacetonitrile (and dibromoacetonitrile), nitrosamines, cyanogen chloride, 2,4,6-trichlorophenol, bromate, chlorate, chlorite, nitrite. For these chemicals, the UK standard or the WHO Guideline Values[15] have been used as Guidance Values (GVs) in this project.

For other nitrogenated, iodinated and non-halogenated DBPs, toxicological data are much more limited and the recent interest in the possible toxicity of, for example, iodinated DBPs, is based mainly on a screening exercise using *in vitro* cell assays. Some members of DBP groups, for example, the nitrogenated DBPs, halonitromethanes and haloacetonitriles, and nitrosamines, may have genotoxic potential. These screening assays, while giving some idea of the relative toxicity of different DBPs and priority for further testing, do not give any information on human intake amounts that can be used to derive values in drinking water that would be of no concern to human health.

In order to assess the toxicity of these DBPs, the limited available, reliable toxicity data have been collated. The DBPs have been grouped chemically and where possible toxicity data for a single surrogate chemical member of each group has been identified and used to derive a tentative Tolerable Daily Intake (TDI) from which a health-based GV can be derived for use in the risk prioritisation. The surrogate chemical chosen was either the member of the group for which there was most toxicological information and yielding the most precautionary TDI or, in some cases, the only chemical in the group for which there were any data. DBP categories, surrogate DBPs and GVs are summarised in Table 1.

Notes:
[a]The GVs have been derived specifically for this prioritisation process and are neither intended nor suitable for human health risk assessment.
[b]For nitrosamines, DWI have created a multi-tiered approach to the regulation of NDMA in drinking water, identifying trigger levels at >1 ng/l, >10 ng/l and >200 ng/l.[19] Owing to their genotoxic potential, the most precautionary trigger concentration has been used as the GV.
[c]The GV for the surrogate iodinated THM, iodoform, is in the mg/l range (2.3 mg/l) and is based on a 2-year carcinogenicity study in rats. The recent concern over the toxicity of iodinated DBPs[4] is based on non-regulatory *in vitro* tests, which can be a useful tool. However, they are still not yet a substitute for testing done *in vivo* in whole animals.

Table 1 *DBP categories and health-based guidance values*

DBP category	Surrogate DBP	Reference value	Guidance value (μg/l)[a]
Halofuranones	3-Chloro-4-(dichloromethyl)-5-hydroxy-2(5H)-furanone (MX) Chemical with lowest reference value in regulatory study.	LOEL 0.4 mg/kg bw[14]. Lowest value based on carcinogenic endpoint in animal study	2.4
Haloacetonitriles	Dichloroacetonitrile Chemical with lowest WHO GV	TDI 2.7 μg/kg bw[15]	20[15]
Haloacetamides	Bromoacetamide Only chemical with any toxicological data	LD50 61 mg/kg bw[16]	12.2
Halonitromethanes	Dibromonitromethane Chemical with lowest LD50 Trichloronitromethane (Chloropicrin)	LD50 86 mg/kg bw[17] NOAEL 6-week study 25 mg/kg bw/day[18]	17.2 (150)
Nitrosamines	N-nitrosodimethylamine (NDMA) Chemical with assessment by authoritative bodies	DWI Guidance on NDMA[19]	0.001 (DWI Tier 4)[b]
Haloketones	1-Chloro-2-propanone Chemical with best regulatory data	TDI 0.0071 mg/kg bw/day[20]	43
Non-halogenated aldehydes and ketones	Methylethyl ketone (2-butanone) Lack of toxicological data, so TDI based on 2-butanol, metabolite of 2-butanone	TDI 0.6 mg/kg bw/day[20]	3600
Haloaldehydes	Chloral hydrate (trichloroacetaldehyde Chemical with assessment by authoritative body	TDI 0.0045 mg/kg bw/day	100 (WHO health-based but not formal guideline)
Iodinated THMS	Iodoform (triiodomethane) Chemical with best toxicological data *in vivo*	NOAEL 39 mg/kg bw/day (NTP, 1978)[21]	2300[c]
Iodinated acids	Iodoacetic acid Only chemical with any toxicological data	LD50 83 mg/kg bw[22]	16.6
Cyanogen halides	Cyanogen chloride Chemical with assessment by authoritative body	TDI 0.045 mg/kg bw/day as cyanide (corresponding to 0.11 mg/kg bw/day as cyanogen chloride)	600[15]; health-based on cyanide)
Bromate	-	-	10 (UK standard)
Regulated THMs	-	-	100 (UK standard)
Haloacetic acids (HAA9)	-	-	80[23])
Chlorate	-	-	700[15]
Chlorite	-	-	700[15]
Nitrite	-	-	500 (UK standard)

2.3 Prioritising DBPs

Prioritisation of DBPs at a generic level needs to take into account:

- Occurrence (indicative concentrations)
- Toxicology (Guidance Values)

The results of this derivation are summarised in Table 2. Example concentrations from WRc experience and from published sources (as outlined in Section 2.1) are used to derive an 'indicative' concentration solely for risk assessment purposes, considered to be typical for the category of DBPs. The ratio of this indicative concentration to the toxicologically-derived GV is calculated for each DBP category. The higher this ratio, the greater is the risk associated with the DBP type, i.e. typical concentrations are higher relative to the toxicological considerations. This ratio is used to derive a risk ranking for each DBP category, based on:

- Ratio <0.1, Risk = 1 (low)
- Ratio 0.1 to 1, Risk = 2 (medium)
- Ratio > 1, Risk = 3 (high)

Results of risk ranking of DBPs using this scoring system are shown in Table 2.

Table 2 *Prioritisation of DBPs*

DBP category	Guidance value (µg/l)	Example concentrations (µg/l)	Indicative concentrations for risk assessment (µg/l)	Indicative concentration-to-toxicity ratio	Risk ranking
Bromate	10	1-5	1-5	0.1-0.5	2
Regulated THMs	100	25-50	25-50	0.25-0.5	2
Halofuranones	2.4	0.1-0.3[2]	0.2	0.08	1
Haloacetic acids	80[a]	8-75[11] 11-134[12] 12-40[13]	10-80	0.12-1	2
Halonitromethanes	17.2	0.3-5[3] <1[6] Mean = 0.4[7]	1	0.06	1
Haloacetonitriles	20	0.2-0.9[3] <1[6] Mean = 3.2[7]	1	0.05	1
Haloketones	43	<1 µg/l[3]	1	0.02	1
Haloaldehydes (chloral hydrate)	100	<10 µg/l[3]	5-10	0.05-0.1	1
Haloacetamides	12.2	1-6[3] Mean = 1.5[6]	1-6	0.08-0.5	2
Non-halogenated aldehydes and ketones	3600	0.1[3]	0.1	0.00003	1
Iodinated THMs	2300	0.2-15[3] 1-2% of regulated THM value[11] <10[4]	0.5-1	0.0002-0.0004	1
Iodinated acids	16.6	<0.1[4]	0.1	0.006	1
Nitrosamines	0.001	0.001-0.025[b 10]	0.001	1	2
Cyanogen halides	600	Mean = 5.8[7]	5	0.008	1

Notes: [a]Based on the suggested value for HAA9 in the report reviewing standards for the EU Drinking Water Directive[23].

[b]The higher concentrations resulted from contamination of the coagulant used at treatment works.

Table 3 shows these DBPs prioritised in order of risk, together with considerations for significance under UK water quality and treatment conditions. The use of chlorine as the primary disinfectant is significant for all of these, and so is not included as a risk factor in the table. Generally, the presence of higher bromide levels in the raw water would also increase the risk for a range of DBPs, because of the greater toxicological concern over the brominated compounds.

Table 3 *DBP risk and UK importance*

DBP category	Risk	Risk factors of UK relevance
Regulated THMs	2	High DOC before chlorination; pH >7
Bromate	2	Bromide in raw water; Ozonation; Chlorination before high dose UV
Nitrosamines	2	DON (algae or sewage influence) in raw water. Chloramination. Use of polyamines in treatment.
Haloacetic acids	2	High DOC remaining before chlorination (likely to be dealt with through meeting the THM standard unless chlorination pH < 7).
Haloacetamides	2	DON (algae or sewage influence) in raw water. Chloramination.
Halofuranones	1	High DOC before chlorination; pH >7
Halonitromethanes	1	DON (algae or sewage influence) in raw water. Ozonation. Future use of high dose UV (for pesticides).
Haloacetonitriles	1	DON (algae or sewage influence) in raw water. Chloramination.
Haloketones	1	Ozonation.
Haloaldehydes	1	Ozonation.
Non-halogenated aldehydes and ketones	1	Ozonation.
Iodinated THMs	1	Chloramination; Iodide in raw water.
Iodinated acids	1	Chloramination; Iodide in raw water.
Cyanogen halides	1	Chloramination; DON; Ozonation.

3 MITIGATION STRATEGIES

A number of mitigation strategies were also investigated as a means to reduce the formation of DBPs.

3.1 Optimisation of coagulation

Other than minimisation of chlorine doses and contact times, probably the greatest mitigation effect from water treatment arises from operating coagulation to maximise, within practical limits, the removal of DOC from the water. This also provides benefits of removing DON to reduce the formation of nitrogenous DBPs. Measures to optimise coagulation should take into account:

- Identification of appropriate coagulants, doses and pH conditions to maximise DOC removal for the range of raw water quality conditions expected, including periods of high algal levels.
- Maintenance of these dose and pH conditions during periods of changing water quality, with either automatic dose control or manual intervention depending on the variability in raw water quality.
- Checks and monitoring to ensure chlorine demand does not increase unexpectedly.

3.2 Activated carbon

Activated carbon is not normally used primarily for removal of DOC, but can be effective in this respect, typically removing 10-30% of the DOC. Where carbon is installed for other purposes such as pesticide removal, removal of DOC and minimisation of DBP formation can be an additional benefit

3.3 Reservoir Management for Algal Control

Reservoir design and management can reduce algal growth and associated water supply problems, including DBP formation from algal associated organic matter.

3.4 Distribution Network Management

Distribution network management to reduce DBP formation should aim to reduce residence time in the system and/or the oxidant concentration. The practical scope for this may be limited in many situations, but consideration can be given to:

- Managing booster chlorination to minimise residuals.
- Control of service reservoir levels to minimise retention time, subject to maintaining adequate storage of water to meet demand.
- Management of loops and dead-ends to minimise retention time.

3.5 Catchment Management

DBP formation can be eliminated or reduced by a corresponding elimination or reduction of precursors in source waters. Catchment management to reduce colour/organics in source waters is a complementary approach to conventional water treatment processes to minimise DBP formation.

4 DBP RISK ASSESSMENT TOOL

The DBP Risk Assessment Tool (Figure 1) produced to complement this project, is an MS Excel-based spreadsheet that enables the potential for the formation of DBPs at water treatment works to be quickly assessed.

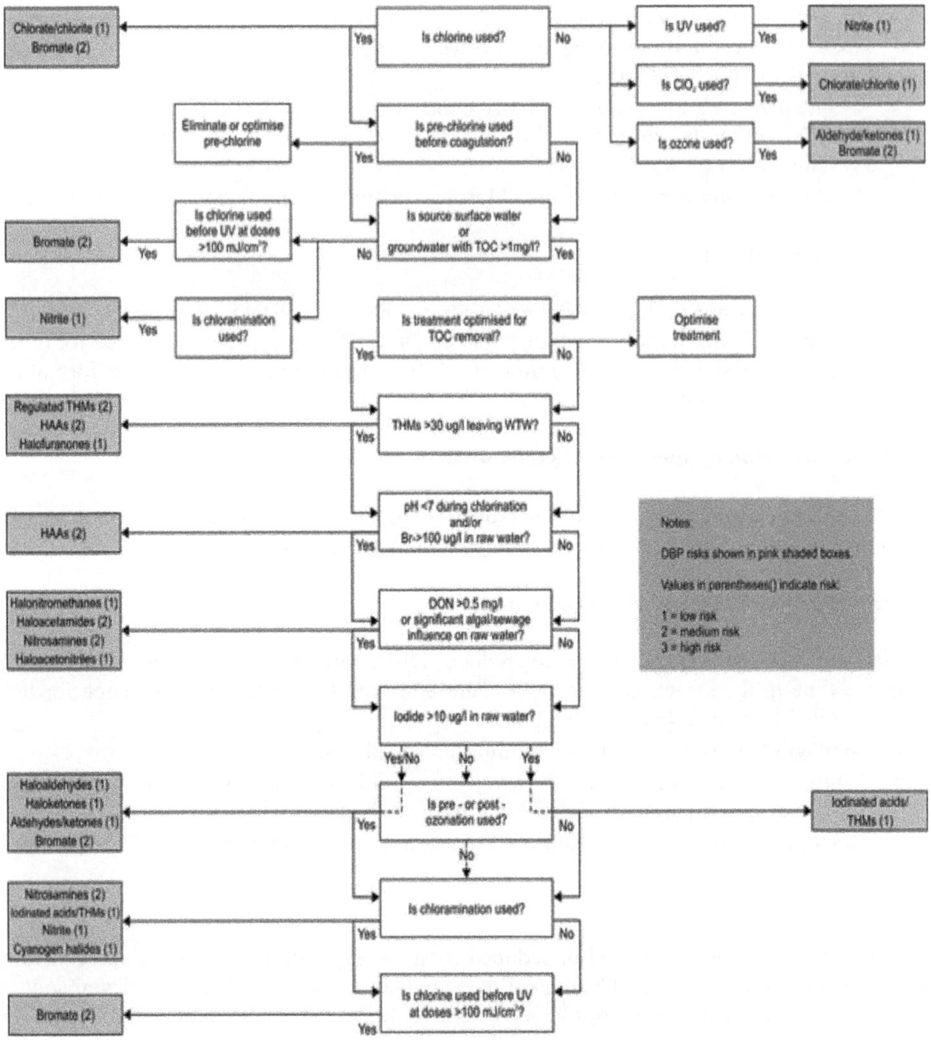

Figure 1 *DBP Risk Assessment tool: Process flowsheet*

The user enters information for a specific water treatment works, including details of raw water quality, treatment chemicals and processes. Based on the information entered, the potential for DBP formation is assessed and a total score is assigned which quantifies the level of risk and allows comparison with other works.

A benchmark total score for a significantly challenged surface water treatment works would be in the order of 25, increasing to 50 or 100 if pre-chlorination before coagulation was practised and/or treatment required optimisation for TOC removal. A similar benchmark total score for a significantly challenged groundwater treatment works would be in the order of 6.0.

Acknowledgement

This work was funded by UK Water Industry Research (UKWIR), Report No. 14/TX/05/2 'Disinfection By-product Risk Assessment: DBPs of Concern' and the accompanying Risk Assessment Tool are available from UKWIR (http://ukwir.co.uk/ukwirlibrary/97123).

References

1 DWI *DWI PR14 Guidance – Disinfection By-products.* 2012, Drinking Water Inspectorate, Defra, London.
2 DWI *Drinking Water 2011: A Report by the Chief Inspector of Drinking Water. 2012,* Drinking Water Inspectorate, Defra, London.
3 H.S. Weinberg, S.W. Krasner, S.D. Richardson, A.D. and Thruston, *Report EPA/600/R-02/068, National Exposure Research Laboratory, U.S. Environmental Protection Agency, Athens, GA,* 2002.
4 S.D. Richardson, F. Fasano, J.J. Ellington, F.G. Crumley, K.M. Buetner, J.J. Evans, B.C. Blount, L.K. Silva, T.J. Waite, G.W. Luther, A.B. McKague, R.J. Mitner, E.D. Wagner. and M.J. Plewa, *Environmental Science Technology*, 2008, **42,** 8330-8338.
5 S.A. Parsons, E.H. Goslan, S.A. Rocks, P. Holmes, L.S. Levy and S. Krasner *Study into the formation of disinfection by-products of chloramination, potential health implications and techniques for minimisation. Final Report, Scottish Government.* 2009
6 M. Templeton, M., Nieuwenhuijsen, N Graham, T. Bond, L. Huang. and Z. Chen, *Project No. DWI 70/2/243, Imperial College London. Drinking Water Inspectorate, Defra,* London 2010.
7 M. Templeton, R. Kanda, N. Graham, N.H. Kamal T. Bond, *Project No. DWI 70/2/268). Imperial College London. Drinking Water Inspectorate, London,* 2012.
8 W.A. Mitch, S.W. Krasner, P. Westerhoff, and A. Dotson *Occurrence and formation of Nitrogenous Disinfection By-Products*, Water Research Foundation/USEPA. ISBN 978-1-60573-053-0. 2009
9 G.R. Dillon, S. Blake, P. Rumsby, L. Rockett, T Hall. P. Jackson, and A. Rawlinson. *WRc Report No. Defra 7348, Drinking Water Inspectorate, Defra, London*, 2008
10 G.R. Dillon, T Hall, L Rockett, J. Jönsson, and R. Kanda *Project No. DWI 70/2/239. WRc Report No. Defra 8196.05. Drinking Water Inspectorate, Defra, London,* 2012
11 S.A. Parsons, and E.H. Goslan, *Project No. WT123 Cranfield University. Drinking Water Inspectorate, Defra, London,* 2009.
12 E.H. Goslan, S.W. Krasner, M. Bower, S.A. Rocks, P. Holmes, L.S. Levy, and S.A. Parsons, *Water Research*, 2009, **43**. p 4698-4706.
13 N.J. Graham, C.D. Collins, M. Nieuwenhuijsen and M.R. Templeton *The Formation and Occurrence of Haloacetic Acids.* 2009, DWI, Defra, London. http://dwi.defra.gov.uk/research/completed-research/reports/DWI70_2_194.pdf
14 S.L. Vaittinen, H. Komulainen, V.M. Kosma, A. Julkunen, J. Mäki-Paakkanen, K. Jansson, T. Vartiainen and J. Tuomisto, *Food Chem. Toxicol.* 1995, **33**, p 1027-1037.
15 *Guidelines for Drinking-water Quality*, 4th Edition. 2011, World Health Organization.
16 ACROS 2-Bromoacetamide, 98%. Materials Safety Data Sheet. 2009.
17 ChemID Advanced database. Available from: http://chem.sis.nlm.nih.gov/chemidplus/ [Accessed 2014]
18 Chloropicrin In Drinking-water. Background document for development of WHO Guidelines for Drinking-water Quality, WHO/SDE/WSH|/03.04/52. World Health Organization. 2003

19 Guidance on the Water Supply (Water Quality) Regulations 2000 (2001 in Wales) specific to N-nitrosodimethylamine (NDMA) concentrations in drinking water. Drinking Water Inspectorate, Defra, London. 2008

20 UKWIR/WRc Toxicity Datasheets. UK Water Industry Research Ltd.

21 NTP Bioassay of Iodoform for Possible Carcinogenicity. TR-110. National Toxicology Program. 1978

22 R.J. Lewis *Sax's Dangerous Properties of Industrial Materials.* 9th Edition, Vol. 1-3., 1996.

23 C. Jorgensen, H.B. Boyd, J Fawell, and O. Hydes *Final Report on Establishment of a List of Chemical Parameters for the Revision of the Drinking Water Directive.* 2008, Ref: ENV.D.2./ETU/2007/0077r. DHI.

JUGGLING ICE-CUBES: THE REGULATION OF DISINFECTION BY-PRODUCTS

Matthew Bower

Drinking Water Quality Regulator for Scotland, 1D South, Victoria Quay, Edinburgh, EH6 6QQ, UK

1 DISINFECTION BY-PRODUCTS – A HISTORY OF REGULATION

Disinfection by-products in drinking water have been the focus of attention by regulators and health professionals for approximately 40 years. This is a comparatively short time when it is considered that chlorine has been used to disinfect water supplies since the last years of the nineteenth century. Following on from attempts by John Snow to use chlorine to stop a cholera outbreak in Soho, London, chlorine was used to disinfect whole water supplies in towns such as Maidstone in the UK and Hamburg in Germany. Chlorination really proved itself at the start of the 20th century when it was used to treat contaminated water behind an outbreak of typhoid in which 131 people died in the city of Lincoln, UK.[1]

It was not until 1974, when chloroform was discovered in Amsterdam's water supply by Rook that the idea that chlorine may have a less benevolent side was born.[2] Regulation soon followed – after a survey of water supplies, the USEPA introduced the Total THM Rule in 1979. This was an annual average of 100μg/l for total THMs. Europe was less prescriptive – the 1980 EEC Drinking Water Directive recommended DBP concentrations be as low as possible, but did not set an enforced standard – some European countries, however, chose to set their own standards.[3]

Throughout the 1980s and 1990s, the number of disinfection by-products identified in drinking water grew, along with medical evidence, albeit weakly positive, that they were detrimental to human health. In 1998 Richardson documented in excess of 500 DBPs potentially present in drinking water. When WHO published its Guidelines for Drinking Water Quality second edition in 1993, health based guidance was available for many disinfection by-products including bromate, chlorate, individual THMs and chlorinated acetic acids, even if a lack of reliable health based evidence meant that it was not possible to set a definitive guideline value, even on a provisional basis.[4]

By the early 1990s many countries had set regulatory standards for trihalomethanes. England and Wales introduced a rolling three-month standard of 100 μg/l for total THMs in their 1989 drinking water quality regulations.

In 1998 USEPA produced the Stage 1 disinfection by-product rule, which set standards for THMs and five of the Haloacetic acids. These were regulated as an annual system-wide running average. At the same time, the new European Drinking Water Directive set a mandatory absolute standard of 100μg/l for total THMs. Although this gave member states a period of time to implement that standard, many already had some sort of regulation in place, a number of which were a good deal tighter than the Directive requirement.

2 CURRENT DBP STANDARDS AROUND THE WORLD

Most nations with a developed regulatory system for drinking water have set maximum standards for total THMs, this comprising the sum of concentrations of the four component compounds. Even across Europe, the approach taken varies considerably. The UK and Sweden have 100μg/l as an absolute standard. The Netherlands takes a tighter value of 25 μg/l, but this is a 90th percentile, with 50μg/l as an absolute standard. Germany has a standard of 50μg/l, but monitoring may take place at the treatment works provided THMs ex-works do not exceed 10μg/l.

The picture elsewhere in the world is similarly varied. Many standards are around 100μg/l, but in some places, notably Australia and South Africa, may be higher. Some countries also set standards for individual THMs. New Zealand uses the health-based formula espoused by WHO, which sums the ratios of the concentration of each THM to its guideline value.

2.1 The US Disinfection By-product Rule

The long-standing Disinfection By-product rule in the USA sets an upper limit of 80μg/l for total THMs. Under the Stage 1 DBP rule, this was measured as a zone-wide running annual average. With the recent introduction of the Stage 2 DBP rule, the numerical standard is unchanged, but the method of measurement has shifted from one averaging values across the whole system to a locational running annual average. There is the expectation that sampling locations will be selected to present a "worst case" scenario, to include points in the distribution system where DBPs are likely to be at their highest. Such locations are required to be established via an Initial Distribution System Evaluation. This approach effectively tightens the standard and removes the flexibility that water suppliers had to set their own sampling locations.

2.2 Regulation of Other Disinfection By-products

The US DBP Rule is not limited to THMs, and includes haloacetic acids and bromate and chlorite where the disinfection method is likely to form these. Some countries have standards for other DBPs such as chlorophenols (Canada, New Zealand, Australia), NDMA (Canada, some US states) and haloacetonitrile (Canada, Japan).[5]

3 APPROACHES TO THE REGULATION OF DBPS

It has long been accepted by most regulators that there are too many DBPs to apply and enforce a numerical standard for every DBP individually. It has generally been assumed that by minimising concentrations of THMs, the concentrations of other DBPs was also being minimised. To some extent this has proven to be true, especially where the minimisation of THMs has involved enhancements in the removal of natural organic matter (NOM) which yields the precursors of so many DBPs.

The regulation of any parameter can drive inappropriate behaviours, and this is something that regulators need to be conscious of and strive to minimise. The regulation of DBPs provides several examples of this:

3.1 Compromised Disinfection

Firstly, and most obviously, there is a risk that in attempting to minimise by-products of the disinfection process, the amount of disinfectant added to the water will be reduced and the effectiveness of disinfection compromised. This is clearly unacceptable, and misguided, as the health risk from undisinfected water is many times more significant than that arising from any DBP. This pitfall is well documented, and most regulations such as the EU Drinking Water Directive and guidance from WHO make the paramount status of disinfection very clear.

3.2 Misplaced Focus on a single DBP or DBP group

Regulation of DBPs in most parts of the world has focussed largely on THMs, something which has allowed them to assume a somewhat artificial importance, unrelated to relative health risk. In some cases this has been at the expense of an awareness of the existence and modes of formation of other DBPs, especially among utility staff without a scientific background. This can also lead to superficial solutions which tackle a THM problem without adequately addressing other, non-regulated, DBPs. The solution designed to tackle one DBP could actually make the problem of another DBP worse, if an holistic view is not taken.

3.3 Selective Sampling

The concentration of most DBPs will vary through distribution systems, depending on age of the water and the disinfectant concentration. For example, THM concentrations tend to be greatest at the end of distribution system. Many regulatory sampling programmes require randomised sampling throughout the supply zone, however where this is not in place, or not tightly regulated, a less scrupulous utility might be tempted to skew sample locations to those areas where DBPs are likely to be lower. In the USA, the Stage 2 DBP rule has attempted to address this by requiring sampling at "worst case" locations.

4 THE FUTURE OF DBP REGULATION

The current approaches to regulating DBPs have undoubtedly achieved much. For example, in Scotland where raw waters contain large amounts of NOM, the percentage of samples containing THMs in excess of the 100mg/l standard has reduced from greater than 7% to significantly less than 1% in ten years. (See Figure 1). Although there is undoubtedly further improvement to be achieved, the regulation of DBPs has driven investment in new treatment processes, optimisation of existing ones, and an enhanced understanding of formation mechanisms.

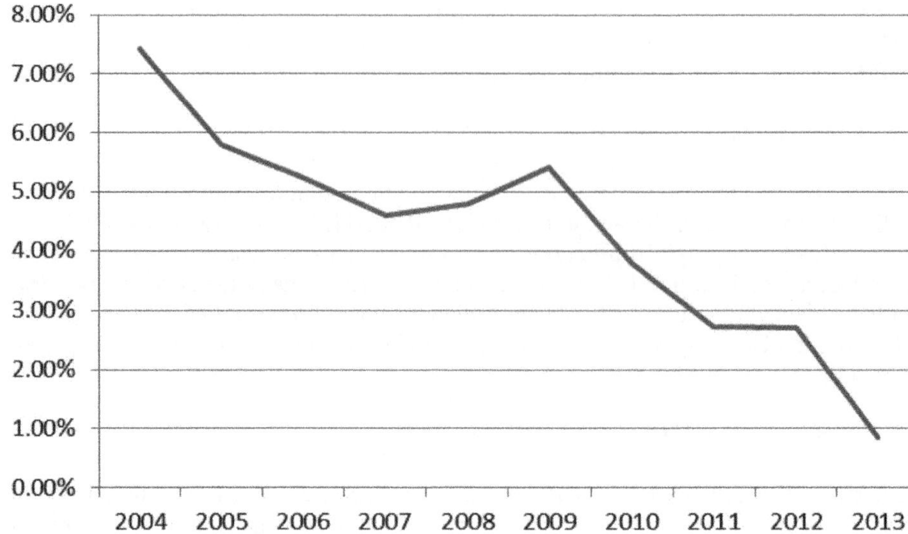

Figure 1 *Failure rate of regulatory compliance samples for THMs taken from consumers' taps in Scotland*[6]

With a growing awareness of the wide range of DBPs, their formation characteristics and health implications, it is evident that it is neither practical nor desirable to concentrate regulation on a single group of DBPs, and a more rounded approach is required. The current European Drinking Water Directive acknowledges this, with the requirement that "contamination from disinfection by-products is kept as low as possible without compromising the disinfection" in addition to setting the THM standard of 100µg/l. A number of regulatory approaches have been tried to demonstrating compliance with this requirement, including the setting of tighter national standards, reporting on compliance with lower *de facto* values, and requirements for strategies to optimise disinfection and minimise DBP formation.

It is likely that future regulation will increasingly utilise these and other tools with the aim of driving the correct behaviours and approaches to risk across whole supply systems, rather than chasing compliance with a single standard. This approach clearly aligns well with the Drinking Water Safety Plan approach to risk, advocated by WHO in their Bonn Charter.[7] As water utilities increasingly move towards this approach to managing risk, the control of the root causes of DBPs should form part of these plans.

4.1 Requirements for a Risk-based Approach

A risk-based approach to the control of DBPs is undoubtedly more difficult to monitor and regulate than a regime based on simple compliance with a standard. It will require regulators to take time to understand and evaluate the measures a utility has in place to control DBP formation and assess whether risks are being adequately managed. This will require expertise and an open mind – it may be helpful for regulators to discuss and define acceptable mechanisms and approaches with utilities in advance of any expected implementation period. Some form of monitoring will be required and it may be necessary to agree the scope and extent of any minimum sampling requirements. A regulator needs to demonstrate consistency, and it may be desirable to retain or develop numerically based metrics to enable simple comparisons to be made and monitor progress, however these should not be the primary focus of regulation.

There is the potential that a move away from relatively simple compliance-based regulation to a risk-based approach could be misunderstood as a relaxation by stakeholders in political and health sectors as well as by consumers. This must be avoided, and regulators have a key role in explaining benefits and providing reassurance in order to maintain confidence that risks are being adequately monitored and addressed.

Although the routine monitoring and analysis for the full range of DBPs is not feasible, it is vital that analytical laboratories maintain and continue to develop their ability to accurately identify and quantify DBPs and their precursor compounds in drinking water. Analytical expertise is vital in developing DBP minimisation techniques and monitoring effectiveness.

5 CONCLUSION

The concept of DBPs in drinking water is a potentially alarming one for the non-technically minded water consumer and the water industry around the world needs to demonstrate that the health risks are understood and well-managed. The comparative risk from undisinfected water needs to be emphasised, and must underpin any initiative for DBP control. A move to a more risk-based approach to regulation should enhance the minimisation of all DBPs, but may require a shift in attitude and competency by regulators and water utilities alike.

1. T. Pacey, *To Fetch a pail of Water*, Trevor Pacey / Tucann Design and Print, 1998
2. J. J. Rook, Formation of Haloforms During Chlorination of Natural Waters. *Water Treatment Examination,* **23**, 1974.
3. A. Nikolaou, *Haloforms and Related Compounds in Drinking Water*, Springer Science & Business Media, 2003.
4. *Guidelines for Drinking-water Quality*, Second Edition, World Health Organization, Geneva, 2003.
5. P. Jackson, T. Hall, W. Young, and P. Rumsby. *A Review of Different National Approaches to the Regulation of THMs in Drinking Water*, DWI research project DWI 70/2/216, DEFRA report 7831, 2008.
6. *Drinking Water Quality in Scotland 2013,* Annual report by the DWQR for Scotland, Scottish Government, 2014.
7. *Bonn Charter for Safe Drinking Water*, International Water Association, 2004.

Subject Index